SAP PRESS e-books

Print or e-book, Kindle or iPad, workplace or airplane: Choose where and how to read your SAP PRESS books! You can now get all our titles as e-books, too:

- By download and online access
- For all popular devices
- And, of course, DRM-free

Convinced? Then go to www.sap-press.com and get your e-book today.

ABAP® in the Cloud

 PRESS

Stefan Haas, Bince Mathew
ABAP RESTful Programming Model: ABAP Development for SAP S/4HANA
2020, 563 pages, hardcover and e-book
www.sap-press.com/4988

Haeuptle, Hoffmann, Jordão, Martin, Ravinarayan, Westerholz
Clean ABAP: A Style Guide for Developers
2021, 351 pages, hardcover and e-book
www.sap-press.com/5190

Paul Hardy
ABAP to the Future (3rd Edition)
2019, 864 pages, hardcover and e-book
www.sap-press.com/4751

Stefan Haas, Bince Mathew
ABAP Programming Model for SAP Fiori: ABAP Development for SAP S/4HANA
2019, 461 pages, hardcover and e-book
www.sap-press.com/4766

Colle, Dentzer, Hrastnik
Core Data Services for ABAP
2019, 490 pages, hardcover and e-book
www.sap-press.com/4822

Gairik Acharya, Aleksander Debelic, Shubhangi Deshmukh, Aayush Dhawan

ABAP® in the Cloud

Development and Operations with SAP® BTP, ABAP® Environment

Rheinwerk Publishing

Editor Meagan White
Acquisitions Editor Hareem Shafi
Copyeditor Yvette Chin
Cover Design Graham Geary
Photo Credit Shutterstock.com: 125708681/© Kendall Rittenour
Layout Design Vera Brauner
Production Graham Geary
Typesetting SatzPro, Krefeld (Germany)
Printed and bound in the United States of America, on paper from sustainable sources

ISBN 978-1-4932-2063-2
© 2021 by Rheinwerk Publishing, Inc., Boston (MA)
1st edition 2021

Library of Congress Cataloging-in-Publication Data
Names: Acharya, Gairik, author.
Title: ABAP in the Cloud : development and operations with SAP BTP
  ABAP environment / by Gairik Acharya, Aleksander Debelic,
  Shubhangi Deshmukh, Aayush Dhawan.
Description: 1st edition. | Bonn ; Boston : Rheinwerk Publishing, 2021. |
  Includes bibliographical references and index.
Identifiers: LCCN 2021000559 | ISBN 9781493220632 (hardcover) | ISBN
  9781493220649 (ebook)
Subjects: LCSH: Cloud computing. | ABAP/4 (Computer program language)
Classification: LCC QA76.585 .A2458 2021 | DDC 004.67/82--dc23
LC record available at https://lccn.loc.gov/2021000559

Contents at a Glance

Dear Reader,

The first night in a new apartment or house is always a little surreal. You have all the same basic rooms (or maybe a new room or two if you're lucky) and your stuff is still there, but nothing quite fits right yet. Perhaps you did a big decluttering before you moved and some things you're used to reaching for are missing—even though you don't really need them. You know that everything will find its proper place eventually, but it's going to take you a few weeks to settle in.

ABAP in the cloud (aka Steampunk, aka SAP BTP, ABAP environment, aka SAP Cloud Platform, ABAP environment) is a little bit like that. While SAP has done a bit of decluttering, most of your familiar ABAP features are still here, and the challenge is just to see how they fit into this new space. With our very own fab four as your guides, you should settle in very nicely.

What did you think about *ABAP in the Cloud: Development and Operations with SAP BTP, ABAP Environment*? Your comments and suggestions are the most useful tools to help us make our books the best they can be. Please feel free to contact me and share any praise or criticism you may have.

Thank you for purchasing a book from SAP PRESS!

Meagan White
Editor, SAP PRESS

mwhite@rheinwerk-publishing.com
www.sap-press.com
Rheinwerk Publishing · Boston, MA

Contents

3   Consuming External APIs

4   Operating Applications

Preface

ABAP is transforming in multiple ways—changing basic building blocks, defining new platforms, and introducing new capabilities at an unprecedented speed.

While several ABAP books exist in the market, a strong demand for information on SAP Business Technology Platform (SAP BTP), especially SAP BTP, ABAP environment, meant than an expanded book was an obvious need. While planning for this book, our primary focus was to carefully determine the chapters and sections that could provide real value to our readers. Our focus was not just to provide information about application development but also advice on innovating more quickly using this new platform. We aimed to encompass all aspects of development, including development patterns, side-by-side extensions, external application programming interface (API) calls, and application administration and monitoring. As we progressed through the chapters, we explored many new ideas and enriched the content in numerous ways. This book shares many real-world scenarios based on our own experiences in different projects.

We hope this book will be useful to the technical leaders, architects, and developers learning how to apply the many capabilities of SAP BTP, ABAP environment. Senior business executives will hopefully also benefit from the high-level concepts explained throughout the book.

Structure of the Book

This book starts with an introduction covering a brief history SAP BTP (until recently called SAP Cloud Platform) and basics about SAP BTP, Cloud Foundry and SAP BTP, ABAP environment specifically. Then, we'll walk you through different development approaches that can be leveraged in SAP BTP, ABAP environment, including the development and consumption of APIs. Finally, we'll cover all the important topics related to the operation of applications developed and deployed with SAP BTP, ABAP environment and conclude this book with a look at the future roadmap.

Although readers may already have some knowledge and experience with the various topics covered in these chapters, we strongly recommend going through the chapters in sequence to avoid missing important information that may be required to better understand more advanced topics.

Chapter 1: Getting Started

The first sections of this chapter will cover introductory topics important for a better understanding of the ecosystem in which SAP BTP, ABAP environment is positioned, including all the components and tools required to start the actual work, such as the following:

- **SAP BTP, Cloud Foundry environment**
  After an introduction to SAP BTP and the Cloud Foundry environment, we'll follow along with the step-by-step process for setting up an SAP BTP trial account. We'll also briefly cover SAP BTP enterprise environments.

- **SAP BTP, ABAP environment**
  We'll provide an overview of the architecture of the platform before following along with the step-by-step process for provisioning trial and enterprise environments.

- **ABAP Development Tools (ADT) for Eclipse**
  As the development environment in which the majority of development work is performed, we'll start with an introduction to Eclipse and ADT for Eclipse, followed by installation and initial configuration steps.

- **Identity and access management**
  After finalizing the preparatory steps, we'll turn to identity and access management (IAM). While earlier sections covered how to prepare the environment, we'll close this chapter with a discussion of setting up development users with the required authorizations.

Chapter 2: Developing Applications

Now that you have a good understanding of the basics of SAP BTP, ABAP environment and fulfilled all the prerequisites for getting the environment ready, we'll dive more deeply into the development process itself with a look at the following topics:

- **ABAP RESTful application programming model**
  As the latest generation of the ABAP programming model, we'll cover the architecture, scope, and usage of the ABAP RESTful application programming model. In this section, we also introduce you to the concept behind side-by-side extensions.

- **Setting up the development environment**
  In Chapter 1, we covered all the necessary preparatory steps, except for setting up the SAP Business Application Studio, which we'll cover in this chapter. You'll use the SAP Business Application Studio in the following sections to develop frontend components.

- **SAP Fiori application development in the SAP Business Application Studio**
  We'll go step by step and develop our first SAP Fiori application on SAP BTP, ABAP environment, including creating all the necessary prerequisite objects such as tables, core data services (CDS) views, and OData services.

- **Application development**
  We'll focus on developing applications using the ABAP RESTful application programming model, covering different scenarios (managed, managed with draft support, and unmanaged). Then, you'll learn how to expose your services as externally available Web APIs for backend service development. Another important topic is the reusability of your services and how they will be exposed, which you'll master while setting up API release states.

- **Deploying application in SAP BTP, ABAP environment**
  You'll learn how to make your SAP Fiori apps available for consumption, ready to be included in the SAP Fiori launchpad.

- **Custom entities in the ABAP RESTful application programming model**
  You'll learn how to create custom entities and consume them in your applications.

- **Advanced development techniques**
  We'll explore how you can use the XCO library to quickly generate the necessary repository objects. We'll also discuss how the RAP generator can put the XCO library into action to accelerate the development process.

Chapter 3: Consuming External APIs

In this chapter, you'll learn how to leverage business services running on different platforms, both on-premise and in the cloud, to build intelligent solutions. As a way for different applications to communicate with each other, APIs play essential role, and you'll learn how to consume external APIs through the following topics:

- **Connecting to on-premise SAP systems**
  To leverage services that are running on on-premise SAP systems, you'll learn how to build applications and extensions in SAP BTP, ABAP environment. We'll go through the end-to-end process, starting with connectivity configuration and consuming remote function call (RFC), Simple Object Access Protocol (SOAP), and OData services available in on-premise SAP systems. We'll also cover the inbound RFC concept, which enables the consumption of remote-enabled function modules developed in SAP BTP, ABAP environment from on-premise systems.

- **Connecting to cloud systems**
  You'll also learn how to leverage services exposed in cloud-based systems, for example, SAP S/4HANA Cloud. We'll start with the SAP API Business Hub, a central location to explore, discover, and consume APIs, and then go through an example of consuming an API exposed in SAP S/4HANA Cloud.

- **SAP BTP services**
  After our introduction to the SAP API Business Hub, we'll dig more deeply on how you can easily leverage the available APIs in your own code.

- **Services on non-SAP platforms**
  We'll conclude this chapter by crossing the boundaries of the SAP world and enhancing our solutions even more by leveraging services available from other providers. These services bring unique capabilities that can help you extend your business processes beyond the conventional.

Chapter 4: Operating Applications

While earlier chapters focused on application development, in this chapter, we'll cover the important area of operations across an end-to-end application lifecycle, covering development, deployment, and continuous improvements. We'll focus on the following topics:

- **Application lifecycle management**
  We'll start by showing you how to manage an end-to-end application lifecycle. We'll cover topics like using abapGit and Git-enabled Change and Transport System (gCTS); migrating ABAP code from on-premise systems to SAP BTP, ABAP environment; and transporting objects between SAP BTP, ABAP environment tenants.

- **Monitoring**
  A critical role in application operations, monitoring includes application debugging, followed by technical monitoring in SAP BTP, ABAP environment. You'll also learn how to use an SQL trace to analyze your SQL statements. Finally, we'll cover some general security aspects of applications by applying roles and authorizations to our solutions as well as accessing security reporting using authorization trace and IAM reporting.

- **Security management**
  We'll cover topics related to application-level security, and you'll learn how to prevent data or code from being stolen or hijacked. You'll also learn how to manage certificates for secure communications, clickjacking protection, and content security.

- **Application jobs**
  You'll learn how to define, schedule, and monitor jobs based on predefined or custom templates.

Chapter 5: Next Steps

SAP BTP, ABAP environment, is constantly evolving. In this chapter, we'll take a look at the future roadmap to better understand what new features and capabilities you can expect.

Acknowledgments

Writing this book would not be possible without the great help and support from SAP and IBM, who share with us their vast knowledge and experience; constantly supported our work on this book; and provided the infrastructure to explore, understand, and develop content.

We would especially like to thank:

- Andre Fischer and Karl Kessler, SAP BTP, ABAP environment product management, for supporting us with feedback and guidance at all stages of book development
- Devraj Bardhan, IBM's SAP Innovation Global Leader, for thought leadership, encouragement, and motivation
- Meagan White and Hareem Shafi, at SAP PRESS, for their guidance and help and for working closely with us in all phases of book development
- Our families for their patience and understanding

Chapter 1
Getting Started

In this chapter, we'll cover the prerequisite knowledge you'll need for a good understanding of the SAP Business Technology Platform (SAP BTP) in general and SAP BTP, ABAP environment specifically. You'll also learn how to set up the development environment, including how to create the development user required in the SAP BTP, ABAP environment itself.

First, let's quickly go through the history of the SAP Business Technology Platform (SAP BTP), previously called SAP Cloud Platform. At SAP TechEd Las Vegas 2012, SAP announced plans for what was called at that time "SAP HANA Cloud," a cloud platform based on SAP's in-memory technology. Soon thereafter, at SAPPHIRE NOW 2013, SAP announced SAP HANA Cloud Platform, the platform that will serve as a foundation for all of SAP's cloud-based solutions. Later, in 2017, this foundation was renamed SAP Cloud Platform, and then more recently to SAP Business Technology Platform, SAP's platform as a service (PaaS) as we know it today.

In the beginning, SAP BTP was only available for the Neo environment—SAP's proprietary environment, running exclusively in SAP data centers. Neo is a development environment for Java, HTML5, and SAP HANA extended application services (SAP HANA XS). Neo also offers the ability to use virtual machines (VMs), enabling scenarios that go beyond the native capabilities of the platform itself.

Cloud Foundry on SAP BTP was introduced at SAPPHIRE NOW 2017. Unlike Neo, the Cloud Foundry environment is open source and offers support for a wide variety of programming languages. Initially supported by SAP were Java and Node.js, and in 2018, ABAP was added. Besides these languages, Cloud Foundry supports other languages as well, following the *bring your own language (BYOL)* approach.

In this chapter, you'll learn about the positioning of SAP BTP, Cloud Foundry environment and SAP BTP, ABAP environment. Before diving into SAP BTP, ABAP environment, we'll take a step back and cover the basics in Section 1.1. You'll also learn about ABAP Development Tools (ADT) for Eclipse development environments in Section 1.3. We'll go step-by-step through the initial setup both on SAP BTP and on a desktop to prepare you for hands-on development. At the end of this chapter, in Section 1.4, we'll cover initial set up of identity and access management (IAM).

1.1   SAP BTP, Cloud Foundry Environment

Cloud Foundry is an open-source platform, managed by the Cloud Foundry Foundation (*https://www.cloudfoundry.org/foundation/*). The Cloud Foundry environment can be hosted on any infrastructure, on-premise or in the cloud. All the major cloud providers—IBM, Amazon Web Services (AWS), Google Cloud Platform, Microsoft Azure, etc.—offer commercial options for Cloud Foundry as an industry-standard platform as a service (PaaS). Internally, Cloud Foundry consists of virtual machines (VMs), on which applications run in a distributed manner, thus achieving scalability and high availability. In other words, multiple copies of applications will run on multiple VMs. With each application deployed, Cloud Foundry provides everything that the VMs need to compile and run an application locally.

In the following sections, we'll start with an overview of the SAP BTP Cloud Foundry environment, followed by step-by-step description on how to provision both trial and enterprise SAP BTP accounts.

1.1.1   Overview

SAP BTP, Cloud Foundry environment is available in many regions and runs as an infrastructure as a service (IaaS) managed by SAP partners. IaaS providers include Microsoft Azure, Amazon Web Services (AWS), Google Cloud Platform, and Alibaba Cloud. The underlying architecture is container based, which results in great flexibility in terms of the variety of languages supported.

The current SAP BTP regions are shown in Figure 1.1. The SAP data centers support Neo environments, while data centers running on hyperscaler infrastructures support Cloud Foundry environments.

Through SAP BTP, SAP can deliver services to help you increase development productivity. These services help developers store data easily; create great user experiences (UXs); integrate with services beyond the platform's boundaries; perform analytics; leverage transformational technologies such as the Internet of Things (IoT), blockchain, machine learning, and big data; and maintain security. These services can be grouped in the following way:

- **SAP Extension Suite**
  This group can be further subdivided into development efficiency, digital experience, and digital process automation. This group of services supports the development process from different perspectives and includes development acceleration, a harmonized digital experience, and process intelligence and automation.

- **SAP Integration Suite**
  This group of services enables the integration of cloud and on-premise applications by connecting processes, data, people, and devices.

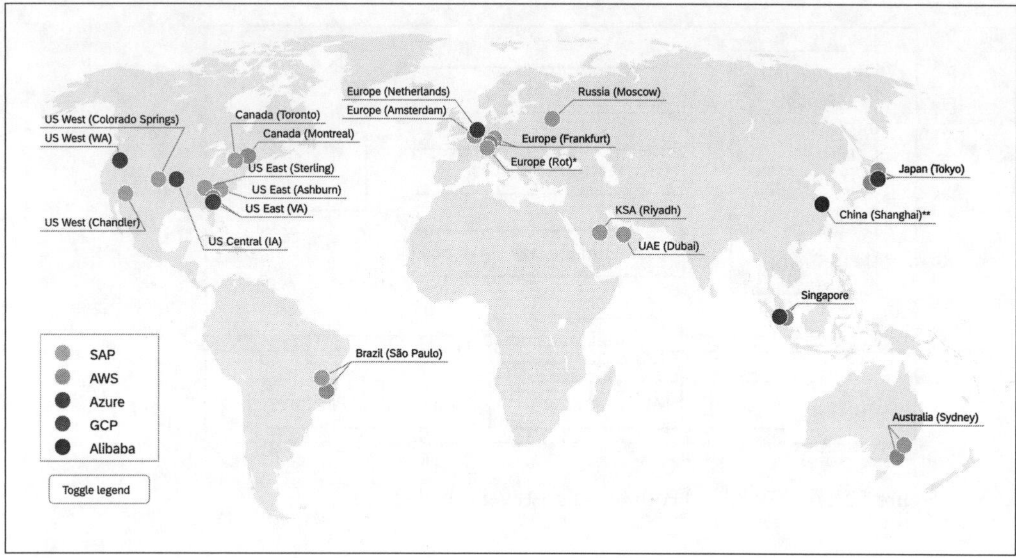

Figure 1.1 SAP BTP Regions

> **Note**
>
> The most current information about the available services on SAP BTP is available at *https://discovery-center.cloud.sap/serviceCatalog*.

Due to the Cloud Foundry architecture, applications and services operate in isolation. As a result, you can develop microservices in different languages and deploy them together with the necessary SAP services, all within the same Cloud Foundry environment.

Outside the Cloud Foundry environment on SAP BTP, SAP provides many tools for development (i.e., SAP Business Application Studio) and operations (i.e., the SAP BTP cockpit).

> **Note**
>
> For years, SAP Web IDE was the primary development environment but was replaced in early 2020 with SAP Business Application Studio. Although still available in the Neo environment, SAP Web IDE is no longer a recommended solution and will not be further maintained.

Figure 1.2 shows the high-level architecture of Cloud Foundry on SAP BTP.

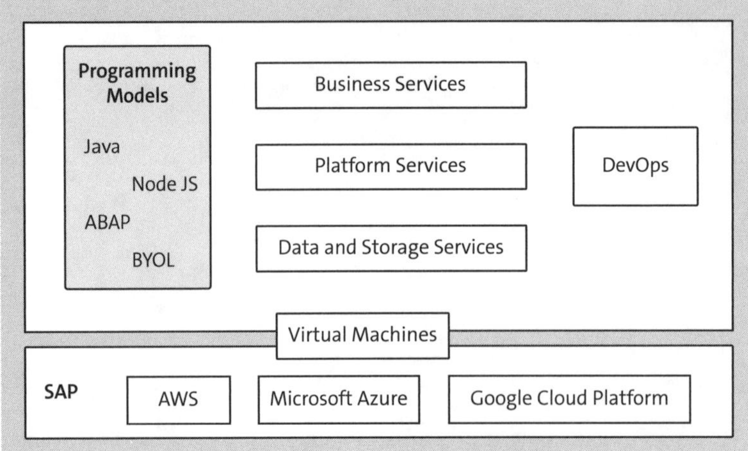

Figure 1.2 SAP BTP: Cloud Foundry High-Level Architecture

As shown in Figure 1.2, in the SAP BTP, Cloud Foundry environment, various services are provided. At its foundation, the infrastructure is hosted by leading cloud providers. The next layer is the data and storage level, with services such as SAP HANA, Object-Store, MongoDB, etc. On top of this layer, we have platform and business services, together with support for different programming models and DevOps capabilities, enabling easier and faster development of intelligent cloud applications.

For more information about Cloud Foundry on SAP BTP, go to *https://help.sap.com/ viewer/product/CP/Cloud* and look for the **Basic Platform Concepts** section.

As a part of the strategy to move to Cloud Foundry, SAP is migrating its software as a service (SaaS) products to Cloud Foundry as well. SAP's roadmap will move completely towards Cloud Foundry in the next few years, ultimately deprecating the Neo environment.

1.1.2   Provisioning a Trial SAP BTP, Cloud Foundry Tenant

To start setting up the free development environment, first you'll need access to a Cloud Foundry environment on SAP BTP that has an entitlement for SAP BTP, ABAP environment. If you don't have access to such an environment, you can sign up for a trial account on SAP BTP.

In this section, we'll walk you through the end-to-end process of provisioning a trial account for the Cloud Foundry environment. The process on a productive environment is similar, with some differences in configuring entitlements, in provisioning the Cloud Foundry subaccount, etc.

The first step in provisioning a trial Cloud Foundry account is to get a trial account on SAP BTP by navigating to *https://cloudplatform.sap.com/*. You'll be presented with two options, as shown in Figure 1.3: **Start free trial** or **Log in to trial**.

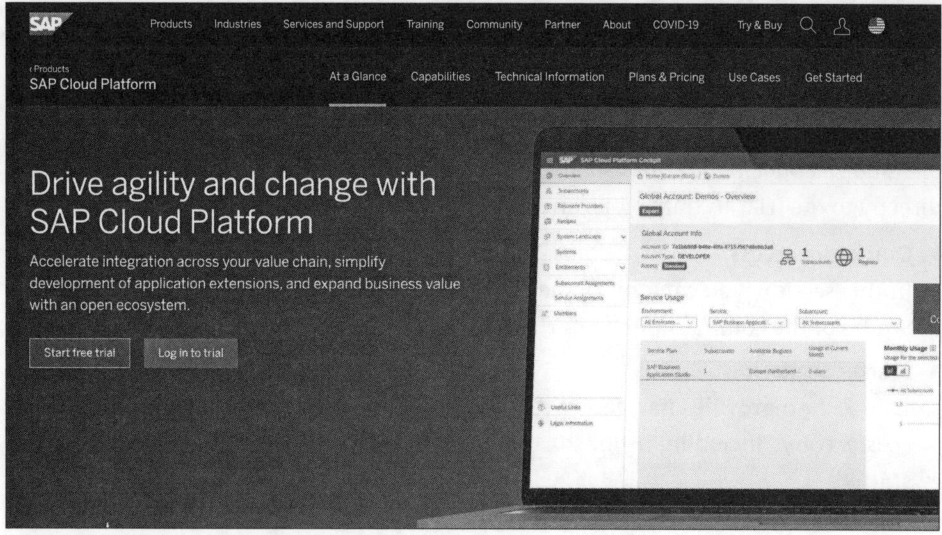

Figure 1.3 SAP BTP: Home Page

If you already have an S-user, you can select **Log in to trial**, which will display a **Legal Disclaimers** screen. After you confirm the **Privacy Statement** and **Terms and Conditions** for SAP BTP, your trial account will be ready. If you don't already have an S-user, you'll initiate the registration process by clicking **Start free trial**. On the next screen, select the **Register for Free** option, which will navigate you to the form shown in Figure 1.4.

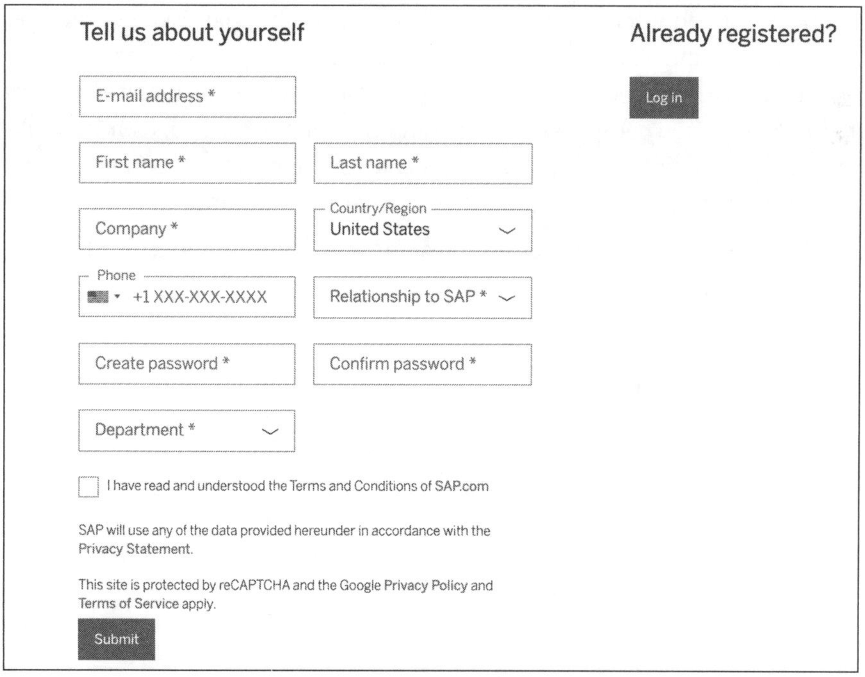

Figure 1.4 SAP BTP: Trial Account Registration Form

Once you fill out and submit the form, an email will be sent to the address you specify to complete the registration process. At the end of this step, you'll receive a P-user (a public user).

Now, you can log on to the cockpit for your trial account by navigating to *https://cockpit.hanatrial.ondemand.com*. On the screen shown in Figure 1.5, besides navigating directly to the trial account, you can quickly navigate to the following sections:

- **Quick tool access**
  In this section, you can quickly navigate to tools like the SAP Business Application Studio or the command-line interface (CLI).

- **Starter Scenarios**
  This section provides navigation to commonly used tutorials from *https://developers.sap.com/*, including a link to the Tutorial Navigator, to help you quickly get started.

- **Learn the Basics**
  These guided tours walk you step by step through the SAP BTP cockpit, showing you how to deploy a sample app and how to create a service instance.

- **More Information**
  This section includes URLs to general topics like documentation, community, etc.

By clicking on **Enter Your Trial Account**, you'll be redirected to your Cloud Foundry trial environment. The first thing you may notice is the calendar icon, with a highlighted number on top that indicates how many days you have left until your trial account expires. Initially, the counter is set to 30 days. A Cloud Foundry trial account is limited to 365 days. Once the initial 30 days expire, you'll have the opportunity to extend your trial, without data loss, for another 30 days. This trial account can be renewed repeatedly until the 365-day trial period expires, and once this limit is reached, your trial subaccount will be deleted, including all your data and your applications. After deletion, you can create a new trial account or purchase an enterprise account.

By default, the following entities have been created for your trial account:

- One global account
- One subaccount
- One Cloud Foundry space

> **Note**
> The global account/subaccount (organization)/space hierarchy in fact represents the domain model that is a key component of every Cloud Foundry platform.

Figure 1.5 SAP BTP: Trial Account Home Page

On the **Global Account** page, shown in Figure 1.6, you can perform the following tasks:

- **Manage subaccounts**
  You can create/edit/delete subaccounts.

- **Manage directories**
  This beta feature enables the grouping of subaccounts for management and moni-toring purposes.

- **Manage resource providers**
  You can manage external resources used within the environment (for example, for the SAP Serverless Runtime service).

- **Access boosters**
  Boosters (previously called *recipes*) are interactive guided tools that support predefined scenarios to enable specific environments.

- **Maintain entitlements**
  You can manage available service entitlements distributed across subaccounts.

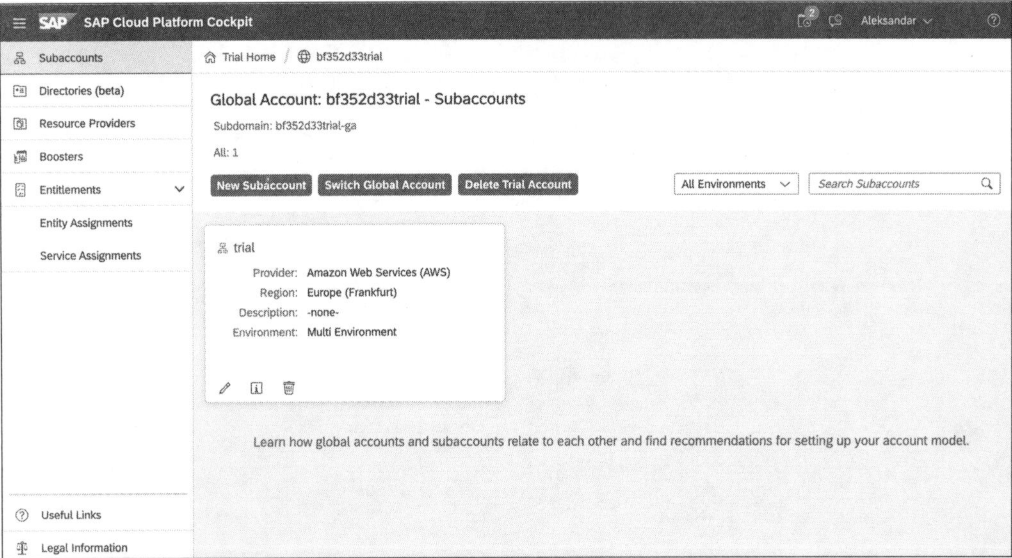

Figure 1.6 SAP BTP: Trial Global Account Home Page

If you enter the **Service Assignments** section under **Entitlements**, you can, for example, manage ABAP service entitlements. In an SAP BTP trial account, you can access a shared service plan, as shown in Figure 1.7.

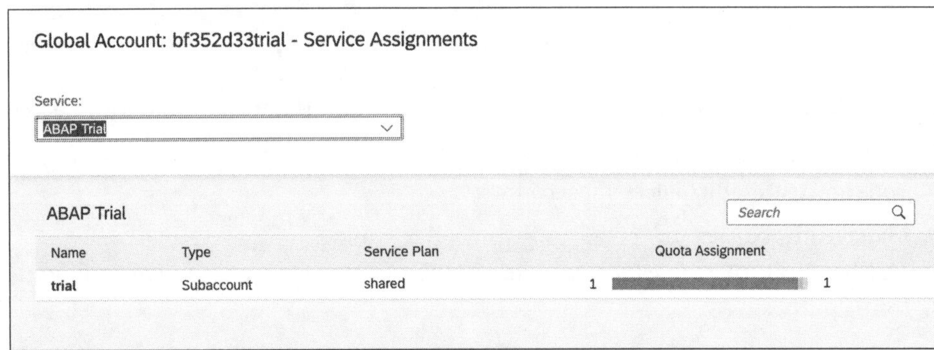

Figure 1.7 ABAP Service Entitlement on an SAP BTP Trial Account

In an SAP BTP trial account, all the necessary entitlements are already maintained and are assigned to the default subaccount. The SAP BTP trial account offers a variety of services to meet the requirements of almost all possible scenarios. This service inventory includes ABAP, SAP API Management, SAP Workflow Management components, SAP HTML5 Application Repository service for SAP BTP, the SAP Integration Suite, Hyperledger Fabric on SAP BTP, Multichain on SAP BTP, SAP HANA Cloud, SAP Intelligent Robotic Process Automation, SAP Cloud Transport Management, and many more. This list continues to change and expand. Also included with the trial account is 4GB of the SAP Application Runtime service, which is sufficient for writing simple applications.

An SAP BTP Cloud Foundry subaccount is provisioned on a data center provided by AWS or Microsoft Azure, with choice of three data centers (regions): two for AWS (Europe – Frankfurt and US East) and one for Microsoft Azure (Singapore). We'll use an AWS data center since SAP BTP, ABAP environment is currently available only in data centers provided by AWS. (Other cloud providers are included on the roadmap.) However, besides the main subaccount, you can create an additional subaccount, for which you'll have the choice again between AWS and Microsoft Azure.

Unlike the SAP BTP trial global account, a productive global account offers only service entitlements that you've purchased. You'll have to provision each subaccount manually, but you can choose different cloud providers. (Restrictions apply for certain regions.) However, one limitation is still in place—you must align your choice of subaccount provider with services you want to use. Of course, in a productive global account, you can provision subaccounts from different cloud providers and in different data centers/locations.

Once you enter your subaccount, you'll be presented with a dashboard like the one shown in Figure 1.8.

As shown in Figure 1.8, in a trial Cloud Foundry subaccount, Cloud Foundry has been activated, and one space has been created for you. In an enterprise Cloud Foundry subaccount, you'll need to complete these tasks manually.

This step concludes the process of setting up an SAP BTP trial account with a Cloud Foundry subaccount.

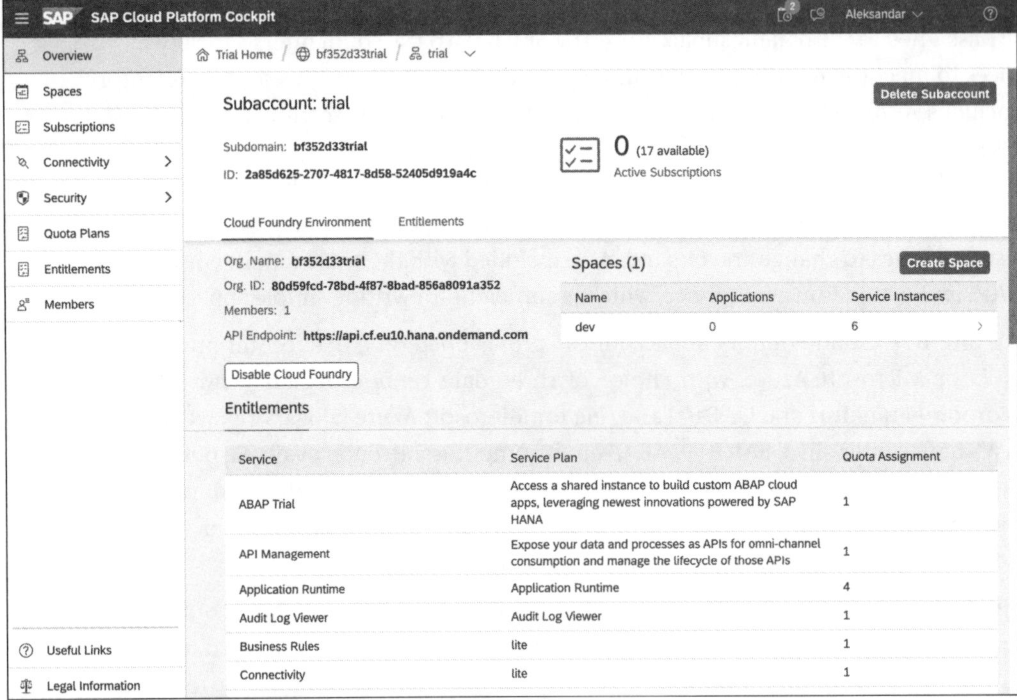

Figure 1.8 Cloud Foundry Subaccount Dashboard

1.1.3   Managing an Enterprise SAP BTP, Cloud Foundry Environment

When you purchase resources on SAP BTP, you'll receive *entitlements* for these services, registered under your SAP BTP global account. To list the entitlements available within a global account, you can access two views:

- **Subaccount Assignments**
  A list of entitlements for each subaccount/subaccounts, providing an easy way to manage entitlements for one or more subaccounts.

- **Service Assignments**
  A read-only list that shows how much of each service entitlement is currently consumed and how much is left unassigned. In this view, you can filter by service to determine how much entitlement is currently assigned to specific subaccounts.

An example of an entitlement list displayed in the **Service Assignments** view is shown in Figure 1.9.

As shown in Figure 1.9, notice how entitlements are represented in this system, which was provisioned before the new *standard* service plan was introduced in May 2020, which enabled flexible sizing of the ABAP environment. Before the new service plan standard, only the 16_abap_64_db service plan was available, which provided 16 GB of

ABAP persistency and 64 GB of SAP HANA persistency. With the standard plan, you can purchase the ABAP persistency (abap_compute_unit) and SAP HANA persistency (hana_compute_unit) you require, in units of 16 GB. For example, as shown in Figure 1.10, we're using 2 units of old 16_abap_64_db, which has been automatically converted to 2 abap_compute_unit and 8 hana_compute_unit.

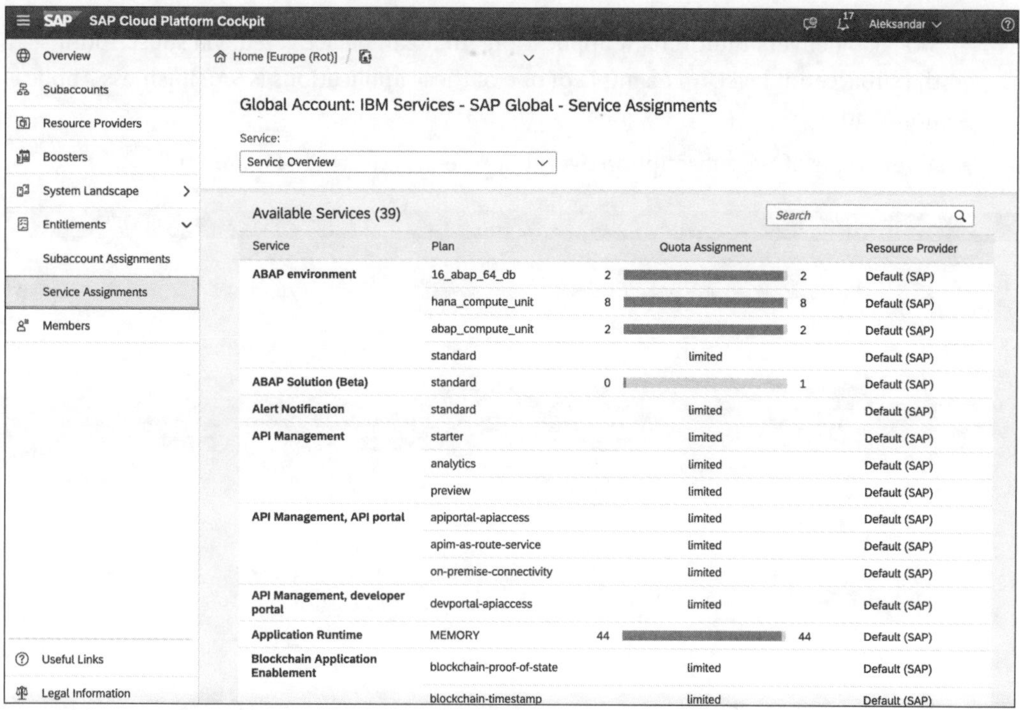

Figure 1.9 Entitlements: Service Assignments View

As shown in Figure 1.10, we can see our newly purchased SAP BTP, ABAP environment entitlement.

Service	Plan	Assign Quota ⓘ	Subaccount Assignment	Remaining Global Quota
ABAP environment	standard		2 shared units ⓘ	2 shared units ⓘ
	abap_compute_unit	☑	2 units	0 units
	hana_compute_unit	☑	8 units	0 units

Figure 1.10 Entitlements: SAP BTP, ABAP Environment Service

The hierarchy in a productive SAP BTP, Cloud Foundry environment is same described earlier in Section 1.1 with our trial account: global account → subaccount → space. The biggest difference with a trial environment is that many tasks must be performed manually, in the following sequence:

1. Create a subaccount.

2. Enable Cloud Foundry within the subaccount.

3. Assign service entitlements.

4. Create a Cloud Foundry space.

5. Create service instances within the Cloud Foundry space.

SAP also delivers multitenant applications that can be activated, via subscriptions, at the subaccount level. An example of one of these applications is SAP Business Application Studio.

An example of the subscription overview screen is shown in Figure 1.11.

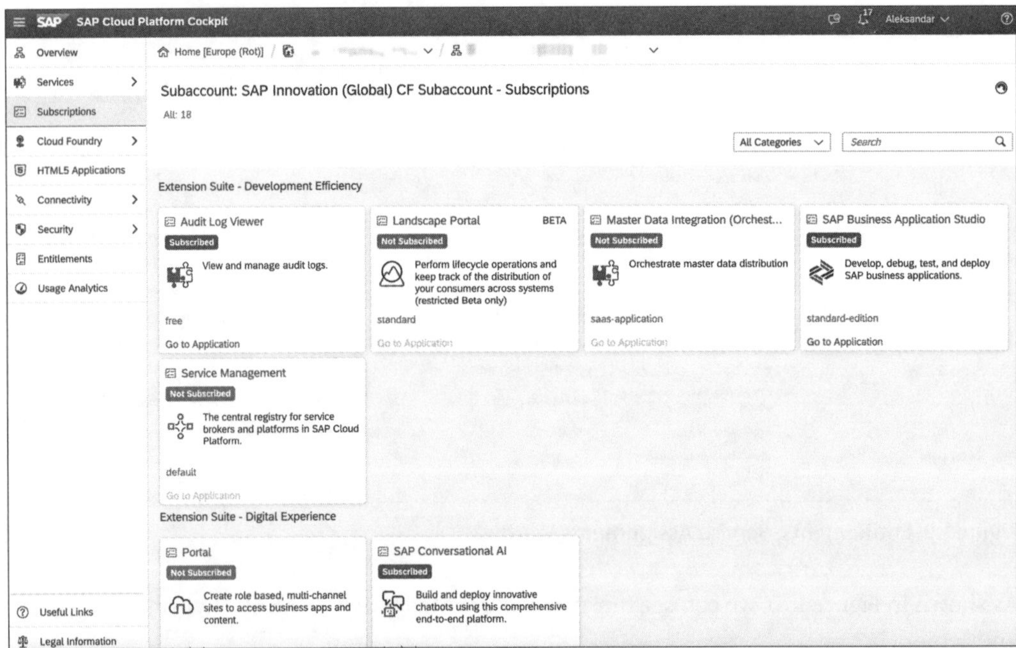

Figure 1.11 Subscriptions Overview

In the following section, we'll focus on provisioning and configuring services and subscriptions required for development in SAP BTP, ABAP environment.

1.2   SAP BTP, ABAP Environment

ABAP has been the essential building block powering all SAP solutions for decades. Then, at SAP TechED 2017, SAP announced what the ABAP developer community was waiting for—the release of SAP BTP, ABAP environment. ABAP was the missing link since SAP BTP already could offer development and runtime environments for Java and Node.js. Now, ABAP developers can leverage their existing knowledge and experience to continue building applications and side-by-side extensions on SAP BTP.

In the following sections, we'll start with an overview of SAP BTP, ABAP environment to help you understand the architecture, use cases, and service plans. Then, we'll go through the step-by-step workflows for provisioning both trial and enterprise environments.

1.2.1   Overview

SAP BTP, ABAP environment is positioned as a PaaS, running exclusively on Cloud Foundry. Figure 1.12 shows the SAP BTP, ABAP environment reference architecture.

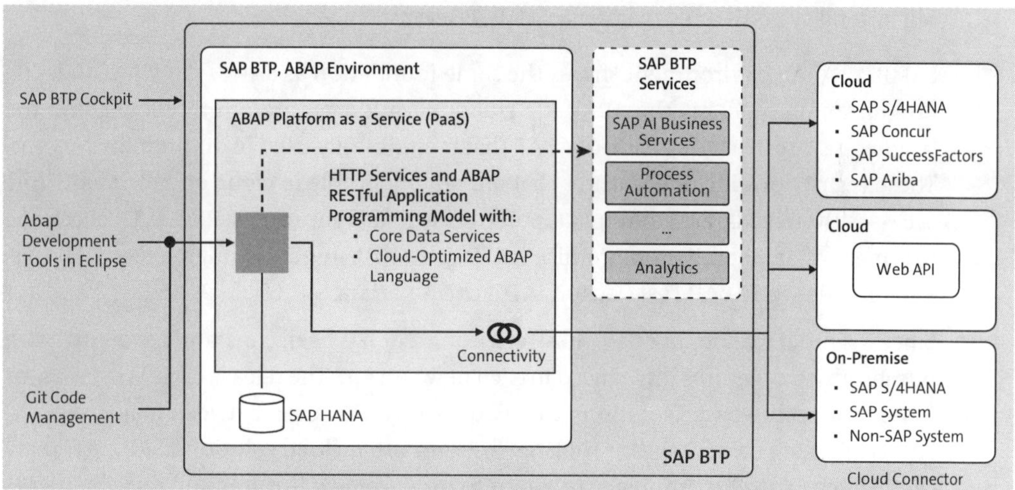

Figure 1.12 SAP BTP: ABAP Environment Reference Architecture

In the center of this architecture is SAP BTP, ABAP environment. Its positioning within SAP BTP, Cloud Foundry enables close integration with other services and capabilities available in SAP BTP.

To manage ABAP environment service instances and for development, you'll use the following tools:

- For creating and managing the ABAP environment, you'll use the SAP BTP cockpit.

- For development, you'll use ABAP Development Tools (ADT) for Eclipse for your development tasks.

- For code exchange and versioning, you'll use GitHub via the abapGit plugin available for ADT or the Git-enabled Change and Transport System (gCTS). Both abapGit and gCTS usage will be described in more detail in Chapter 4.

SAP BTP, ABAP environment runs on the SAP HANA database, with ABAP-managed access to released database objects. (Native SAP HANA access is not allowed.) Connectivity options cover both cloud-based and on-premise scenarios but also include the capability to expose web application programming interfaces (APIs). These topics will be covered in more detail in Chapter 2 and Chapter 3.

Some use cases for SAP BTP, ABAP environment can be summarized into following options:

- **Side-by-side extensions**
  For building new extensions and migrating existing extensions decoupled from the system core, using APIs for communication, this option supports the *keep the core clean* paradigm.

- **New application development**
  For developing applications from scratch, this option supports integration with backend systems and leveraging SAP BTP services.

SAP BTP, ABAP environment shares the same foundation as SAP S/4HANA Cloud. The programming model is the ABAP RESTful application programming model, and the language is a restricted version of ABAP that is commonly used in on-premise systems. These restrictions include features that are not applicable in cloud environments (file access, system calls, etc.) and unsupported features (dynpros, lists, etc.). To check the readiness of any specific on-premise ABAP code that might be migrated to SAP BTP, SAP provides the ABAP Test Cockpit (ATC) check variant SAP_CP_READINESS.

Since its initial release, SAP BTP, ABAP environment has been constantly evolving, with numerous new features introduced in each new version. The release cycle is same as for SAP's other cloud-based solutions, that is, quarterly. However, between major releases, SAP is releasing fix packages as well. Again, as for other cloud solutions, SAP customers can use the Customer Influence program to propose new features and improvements and vote for features submitted by other customers and partners.

> **Note**
>
> More information about SAP BTP, ABAP environment, is available at *https://help. sap.com/doc/abapdocu_cp_index_htm/CLOUD/en-US/index.htm*.

> **Note**
>
> "Steampunk" was the internal project name for SAP BTP, ABAP environment. However, in 2019, a year after the platform GA release, it was released to public and is commonly used since. SAP BTP, ABAP environment is also sometimes called ABAP PaaS.

SAP BTP, ABAP environment belongs to a group of services under the SAP Extension Suite that increase development efficiency. SAP BTP, ABAP environment is available under three service plans:

- **SAP BTP enterprise agreement**

  This service plan is a consumption-based license model. Under the SAP BTP enterprise agreement license model, the following metrics apply:

  - Hours of persistent memory in 16 GB blocks

  - Hours of runtime memory in 16 GB blocks

  With this service plan, your organization pays for each hour consumption of the provisioned resources.

- **Subscription**

  This service plan allows access to a cloud service up to a licensed maximum. With the subscription model, again, you're presented with the same units, so you can define the persistent and runtime memory to which you'll subscribe in 16 GB blocks. A subscription fee is charged, regardless of actual usage.

- **Trial**

  This service plan is used for platform exploration and evaluation and was released at TechEd Las Vegas 2019. With this trial offering, you'll need to follow certain "rules of the game" such as the following:

  - Shared offering: Users of a trial instance in fact share the same instance. As a result, everyone can see and change everyone else's objects and code.

  - Educational purpose only: Only for self-education, not for other purposes.

  - Limited lifetime: The same as the lifetime of Cloud Foundry trial account, that is, 365 days.

  In a trial account, the following features are limited:

  - Limited access to applications on the administrator's SAP Fiori launchpad

    - No apps for identity and access management (IAM)

    - No apps for communication management

    - No app for software component lifecycle

  - No support for custom identity providers

  As a result, you cannot use a destination service or connectivity to on-premise systems. Moreover, you cannot assign business roles or use custom authorization fields and objects.

1.2.2   Provisioning a Trial SAP BTP, ABAP Environment

In Section 1.1.2, we provisioned a trial Cloud Foundry environment. Now, we'll create a trial ABAP service instance. Although you can use the **Prepare an Account for ABAP Trial** booster, we'll walk you through the step-by-step process so you can get better understanding of how provisioning works:

1. First, navigate to the **dev** space, as shown in Figure 1.13.

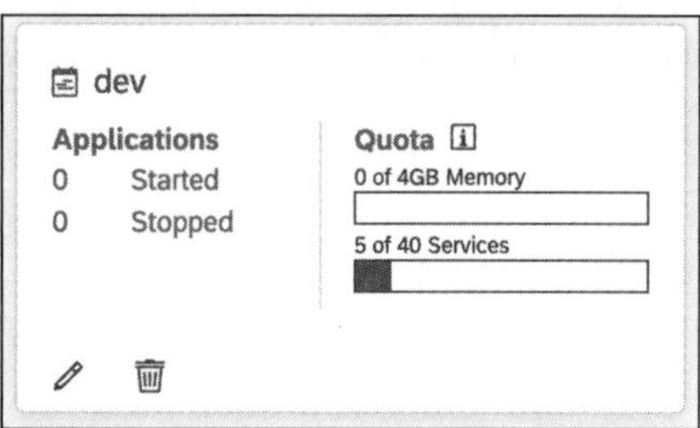

Figure 1.13 Trial Cloud Foundry Space

2. In the menu on the left, select the **Service Marketplace**. In the main section of the screen, you'll see a list of available services in your subaccount, as shown in Figure 1.14.

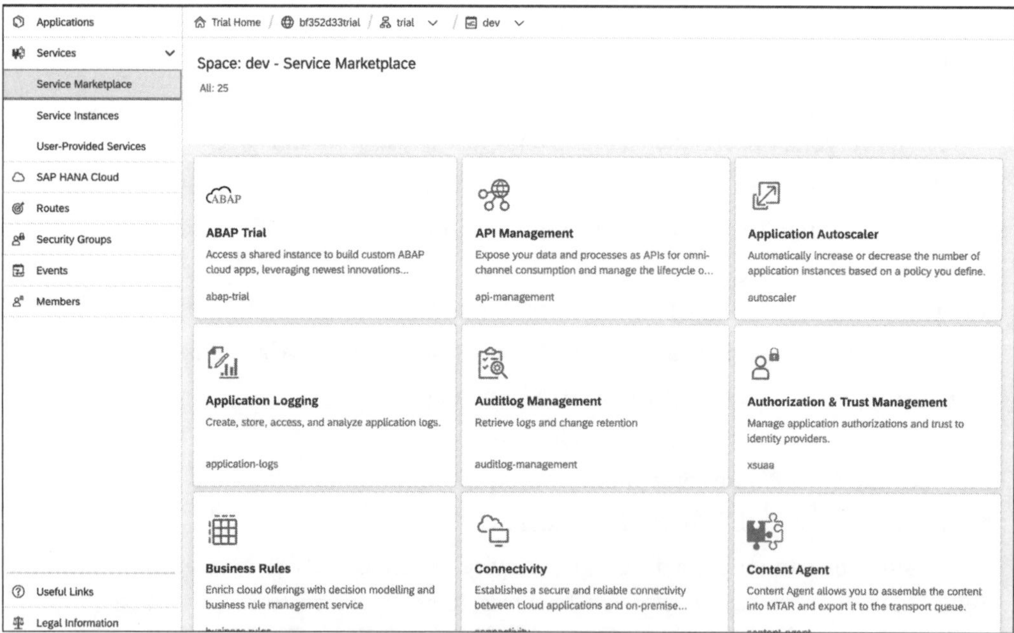

Figure 1.14 Service Marketplace in the Trial Cloud Foundry Subaccount

3. Now, select the **ABAP Trial** service. You'll then see the **ABAP Trial** service overview page on the right side of the screen, as shown in Figure 1.15.

   In the main working area, besides a short description, you can click the URLs to navigate to documentation and support. Also, information about the current service plan (**shared**) is displayed.

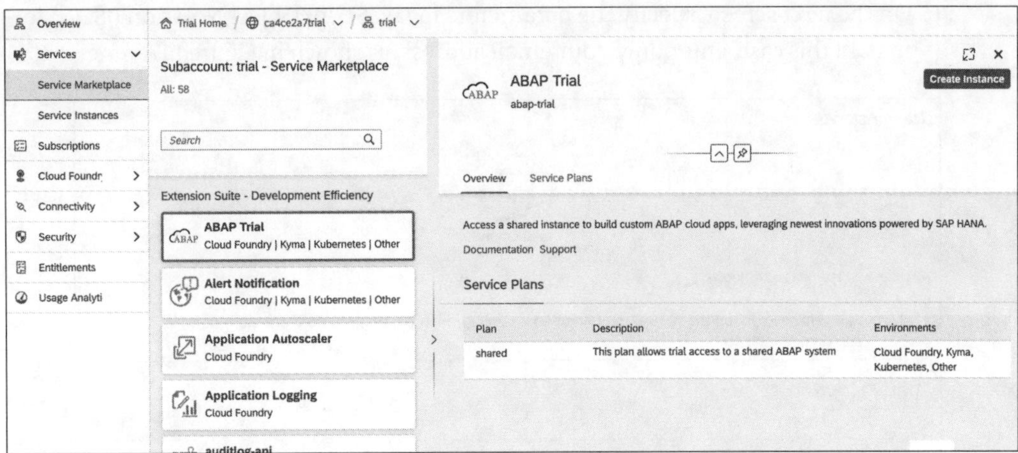

Figure 1.15 ABAP Trial Service Overview Screen

4. Select the **Create Instance** option, and a popup window will appear that will navigate you through the series of screens where you'll enter the necessary information and create the trial ABAP service instance.

5. On the first screen, choose the service plan (the only option available in our example is **shared**) and enter a name for the instance, as shown in Figure 1.16. Then, click **Next**.

Figure 1.16 Creating an Instance: Choosing the Service Plan and Entering an Instance Name

6. On the next screen, specify the parameters in JavaScript Object Notation (JSON) format. In this case, enter only your email address, as shown in Figure 1.17.

Figure 1.17 Create Instance: Specifying Parameters

7. You're now on the final screen, where you'll be presented with an overview of all the parameters relevant for service instance creation, as shown in Figure 1.18. To initiate the process of instance creation click on **Create Instance**.

8. With this step, your trial ABAP service instance is created and will be included in the list of service instances. You can see the newly created SAP BTP, ABAP environment service instance listed in Figure 1.19.

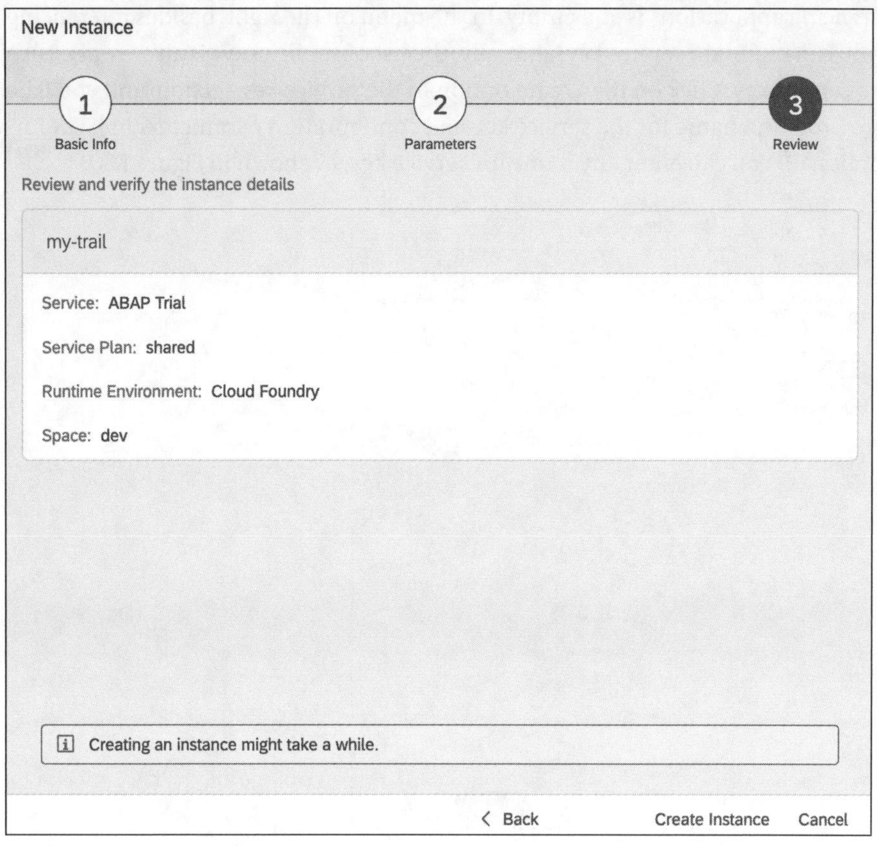

Figure 1.18 Creating an Instance: Specifying an Instance Name

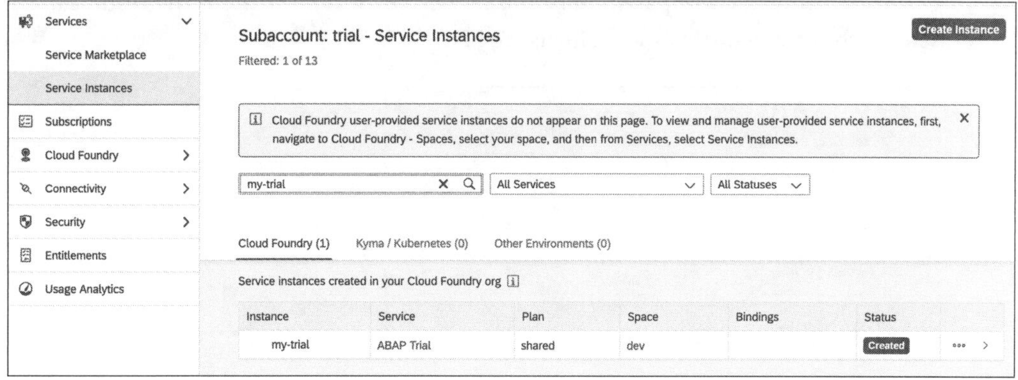

Figure 1.19 ABAP Trial Service Instance

Let's navigate to the service instance by clicking on the URL to the right, marked with the greater than sign (>). Since this instance is a newly created service instance, the list of referencing applications is still empty. In the menu on the right, besides referencing apps, you can maintain service keys. Initially, the list of service keys is also empty. Let's create a service key. Click on the **Create** option in the service key section, and you'll be asked to provide a name for the service key and configuration parameters in JSON format (optional). You can enter any name for service key, as shown in Figure 1.20.

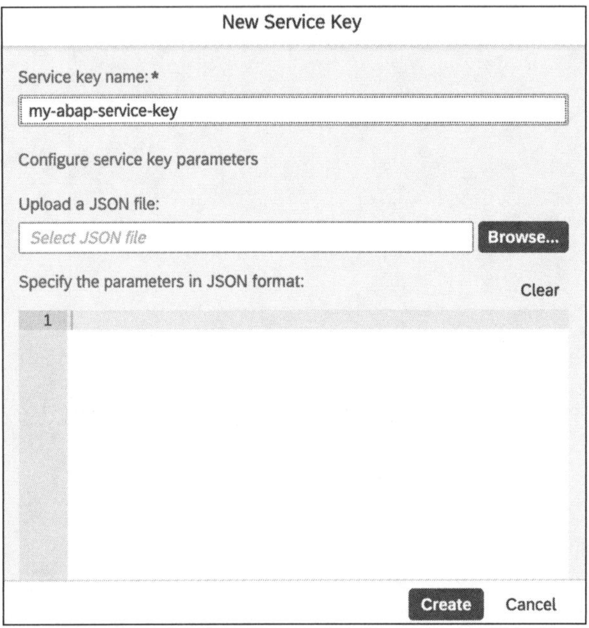

Figure 1.20 Create Service Key Popup Window

Once you confirm these parameters, the service key will be created. The result can be seen in the dashboard, as shown in Figure 1.21.

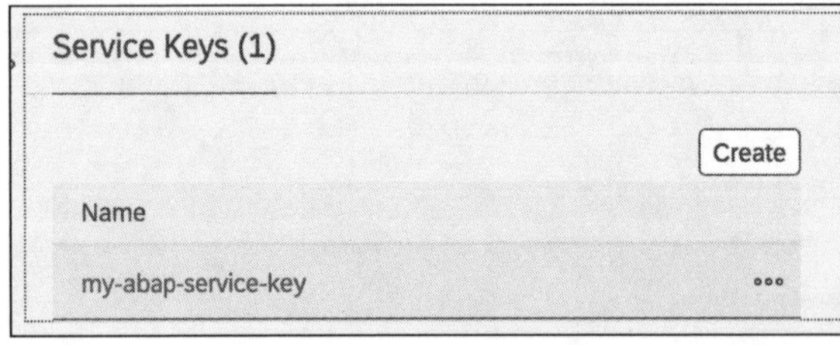

Figure 1.21 Service Key Overview

At this point, you can copy the service key content to the clipboard or download it to a file. The service key contains important information to be consumed by applications. In our case, one possible usage is in ADT for Eclipse, which will be explained in Section 1.3.

This step concludes our section on the creation of a trial service instance.

1.2.3   Provisioning an Enterprise SAP BTP, ABAP Environment

A prerequisite to provisioning an enterprise-level SAP BTP, ABAP environment is to configure entitlements. Some example configured entitlements for SAP BTP, ABAP environment are shown in Figure 1.22.

Service	Plan	Quota Assignment		Resource Provider
ABAP environment	16_abap_64_db	2	2	Default (SAP)
	hana_compute_unit	8	8	Default (SAP)
	abap_compute_unit	2	2	Default (SAP)

Figure 1.22 ABAP on SAP BTP Entitlements

The steps for provisioning a service instance in an enterprise environment are similar to the steps for a trial environment. An additional step is the subscription activation for **Web access for ABAP.** With this subscription, you'll have web access to your ABAP service instances, including administrative user interfaces (UIs), as shown in Figure 1.23.

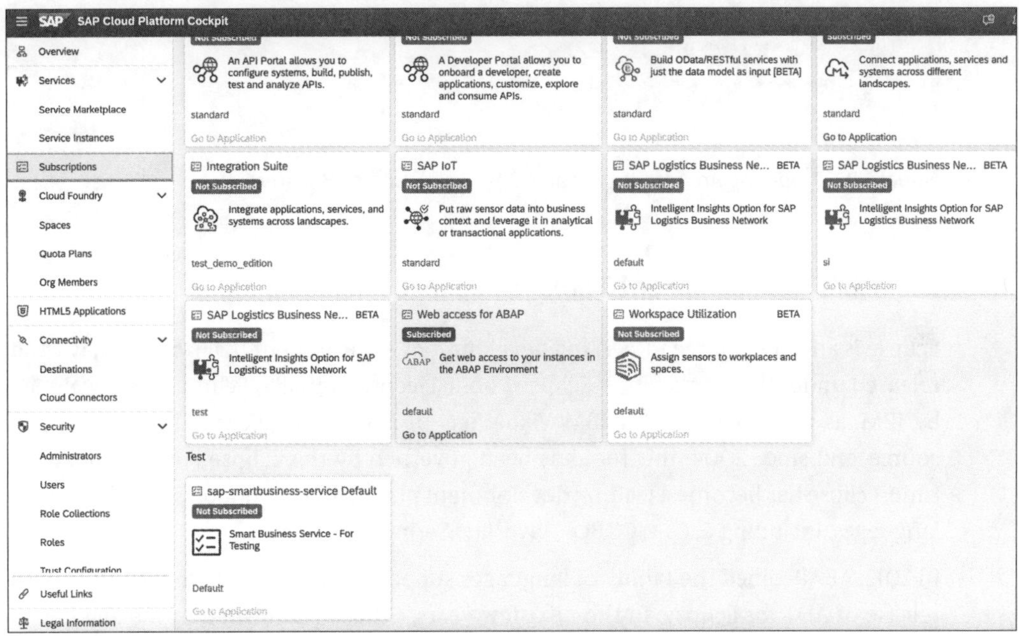

Figure 1.23 Web Access for ABAP Subscription

As with a trial environment, you can use a booster, **Prepare an account for ABAP Development,** which will automate the provisioning and configuration of all required components. The components covered by this booster are shown in Figure 1.24.

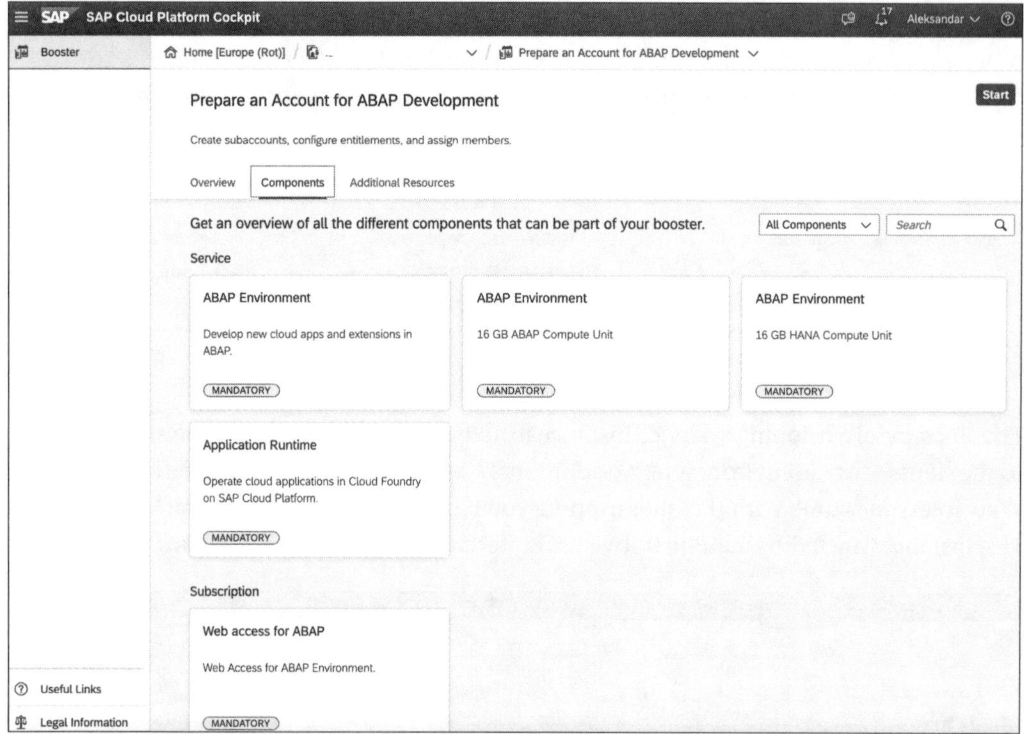

Figure 1.24 Preparing an Account for an ABAP Development Booster

1.3   ABAP Development Tools for Eclipse

Eclipse is an open-source integrated development environment (IDE), built in Java and released under the terms of the Eclipse Public License. Initially, Eclipse was developed by IBM, as a replacement for IBM VisualAge. In 2001 IBM, released Eclipse to open source, and since 2004, this tool has been governed by the Eclipse Foundation. Over time, Eclipse has become a leading development platform that supports a wide range of languages, including C, C++, COBOL, Java, JavaScript, Perl, Ruby, Python, R, etc.

In 2012, ABAP joined the family of languages supported in Eclipse, following the initial release of ADT for Eclipse. In the next few years, Eclipse languished in the shadow of SAP's proprietary IDE, ABAP Workbench. However, with the introduction of ABAP core data services (CDS) views, Eclipse has become the mandatory development environment. Now, SAP BTP, ABAP environment is the only option for the development of backend components (classes, tables, CDS views, services, etc.).

> **Note**
>
> To learn more about the current ADT version and supported Eclipse platform, navigate to *https://tools.hana.ondemand.com/#abap*.

Apart from running on a Java Runtime Environment (JRE) or one of supported operating systems (Microsoft Windows and Apple macOS), SAP has recently introduced another option: *SapMachine*. SapMachine is an OpenJDK release that SAP has made available for customers and partners who want to use OpenJDK to run their applications.

> **Note**
>
> More information about SapMachine can be found at *https://sap.github.io/SapMachine/*.

In the following sections, you'll learn how to install ADT for Eclipse and how to customize views. As a result, you'll prepare and configure the development environment according to your preferences.

1.3.1   Installation and Initial Setup

As a prerequisite for ADT, first, you'll need to install Eclipse. We recommend using the latest supported version. Make sure you always use the supported version of Eclipse to avoid any issues during ADT installation and in later use. To install ADT, in Eclipse, follow the menu path **Help • Installing New Software...**, as shown in Figure 1.25.

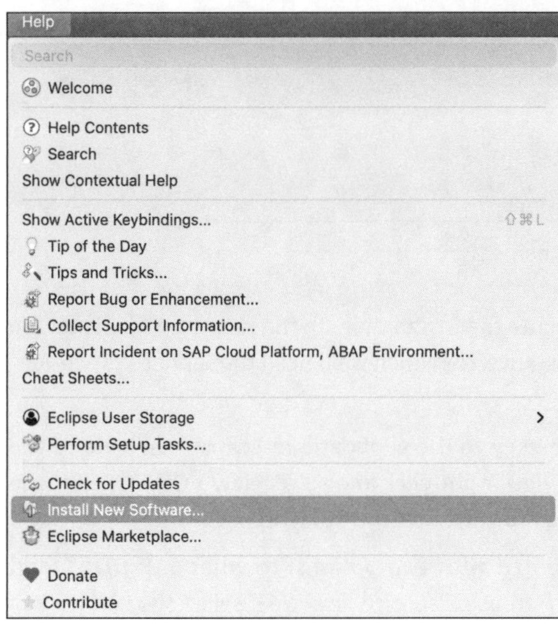

Figure 1.25 Installing New Software

In the **Work with** field, enter the URL "https://tools.hana.ondemand.com/latest." As a result, you'll see a list similar to the one shown in Figure 1.26.

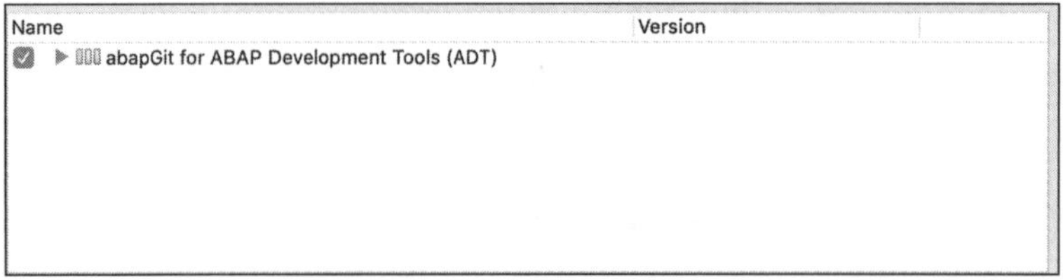

Figure 1.26 Installable Components List

For development in SAP BTP, ABAP environment, it is sufficient to select only **ABAP Development Tools**. However, you can choose to install other products as well; they are all optional in this case.

After confirming the features to be installed and reviewing the relevant license agreements, click **Finish** to complete the installation. You should always use the latest supported Eclipse and ADT versions to leverage the latest features.

Now that you've installed ADT for Eclipse, you're ready connect to your SAP BTP, ABAP environment instance. First, let's install the abapGit plugin, which will be needed for connecting to GitHub.

Again, choose the **Install New Software** option and enter the following URL into the **Work with** field: "https://eclipse.abapgit.org/updatesite/." You'll get an output message, as shown in Figure 1.27.

Figure 1.27 abapGit: Installable Components List

After selecting abapGit, you can execute the installation. In the next step, we'll add the trial SAP BTP, ABAP environment instance, for which we'll need the service key we generated in Section 1.2.2.

First, copy the content of the service key to the clipboard. In Eclipse, navigate to the **Project Explorer** view. While in this view, right-click and select **New • Other**. Select the project type, in this case, **ABAP Cloud Project**, as shown in Figure 1.28.

On the next screen, you'll be presented with two options: to enter SAP BTP, Cloud Foundry environment logon data or to use a service key. We'll select the service key option and click on **Next**. On the next screen, paste the service key from the clipboard, as shown in Figure 1.29.

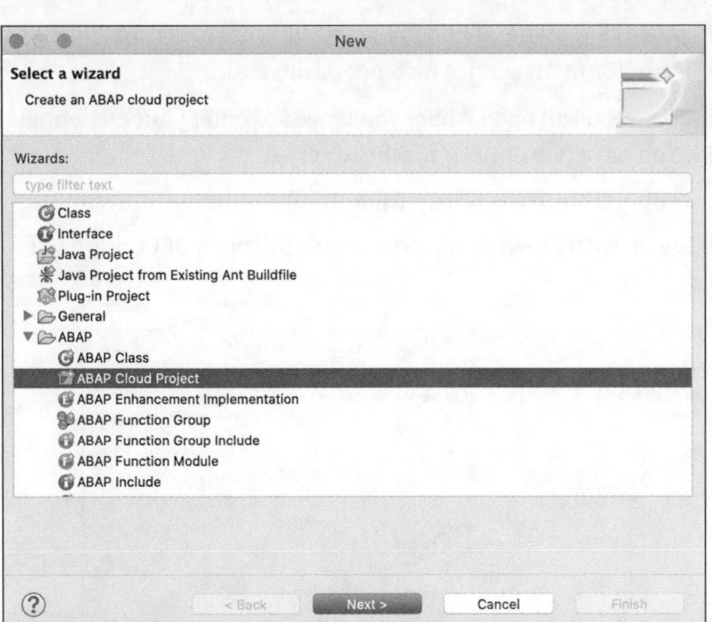

Figure 1.28 Creating an ABAP Cloud Project

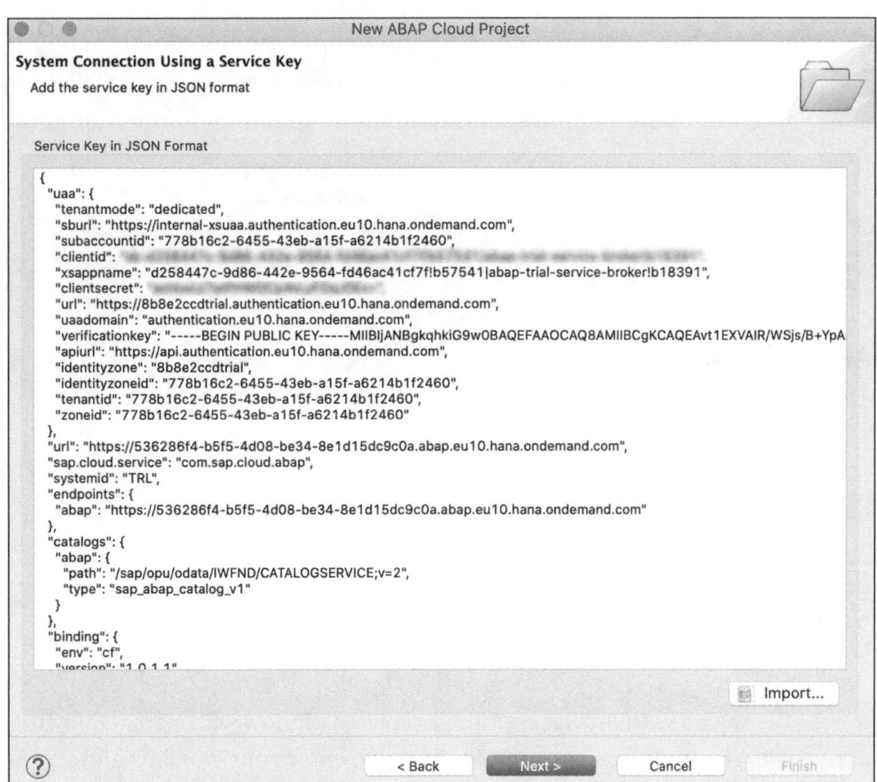

Figure 1.29 Creating an ABAP Cloud Project: Service Key

You can also import a service key as a file, if you previously downloaded the service key from the SAP BTP, ABAP environment service instance dashboard.

Click on **Next** again, and a screen will open where you'll need to enter your credentials, as shown in Figure 1.30. You have two options for this step:

- Entering logon details directly into the popup window.
- Selecting either the **Log on with Browser** option or the **Copy Logon URL** option to log on in a web browser.

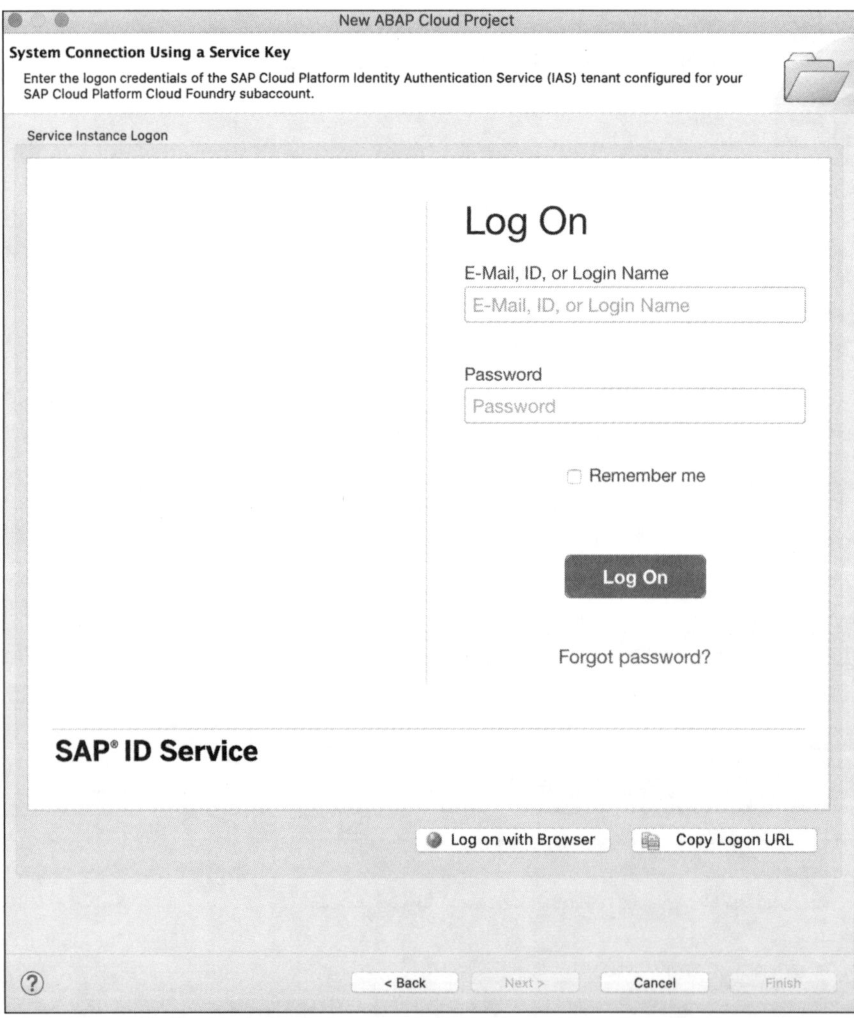

Figure 1.30 Service Instance Logon Screen

If you choose the **Log on with Browser** option, a browser window will open, requesting you enter your user name and password. In this case, enter your email ID and corresponding password (the password for your S-user or P-user, which was used for creating the trial SAP BTP account).

After successful authentication, you'll see the overview screen shown in Figure 1.31, which displays details about the service instance to which we've connected.

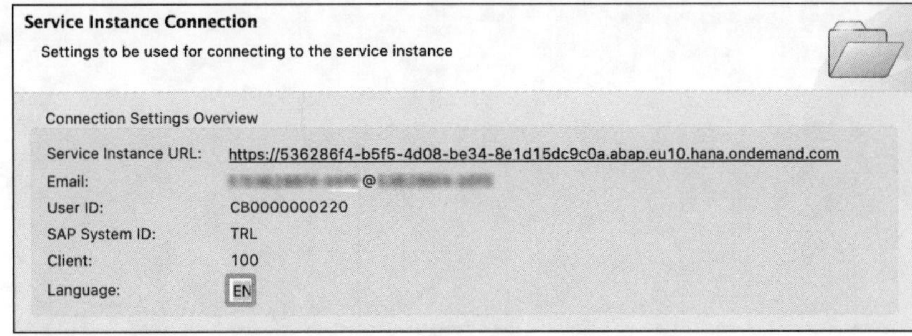

Figure 1.31 Service Instance Details

Now, click on **Finish**, and your trial SAP BTP, ABAP environment instance should become visible in the **Project Explorer** view, as shown in Figure 1.32. Eclipse will recognize that, for this project type, the **ABAP** perspective is the default perspective and will offer you the option of opening this perspective.

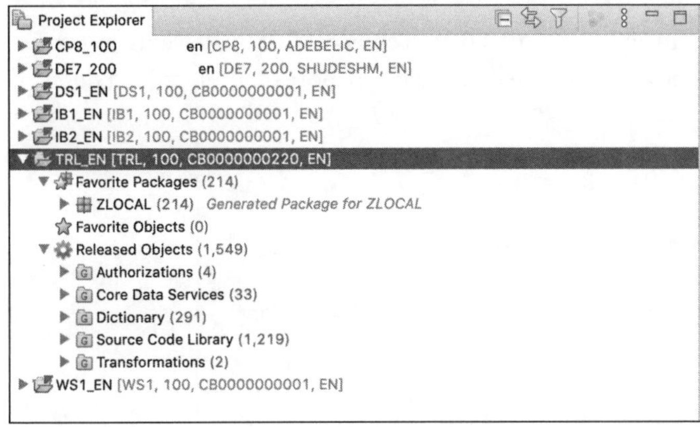

Figure 1.32 Trial SAP BTP, ABAP Environment Service Instance in the Project Explorer View

In the following chapters, you'll learn how to navigate through the different sections of a project, but for now, we'll just mention two important sections:

- **Favorite Packages**
  In this section, you can define which packages should be available for quick access.

- **Released Objects**
  This section is a list of released, standard SAP objects, which you can use and reference in your code.

Now, you'll need to create a package in which to store your development objects. Simply right-click on the **Favorite Packages** section in the SAP BTP, ABAP environment instance and select **New • ABAP Package**.

On the next screen, enter information about the package and click **Next**, as shown in Figure 1.33.

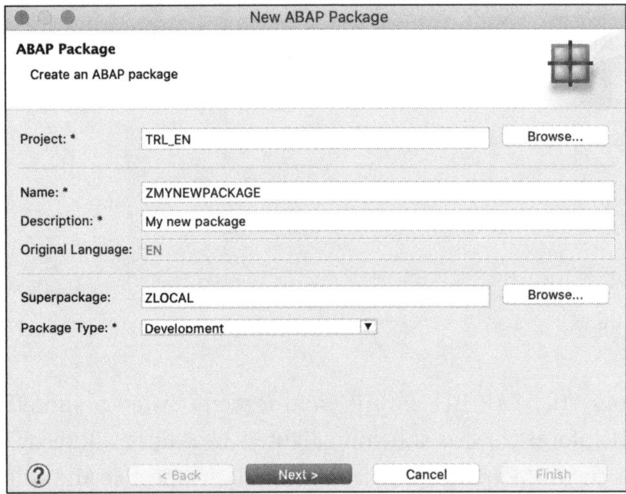

Figure 1.33 Creating a New Package

On the next screen, shown in Figure 1.34, you have a choice: choosing an existing transport request or creating a new one. To create a new transport request, enter a description and click on **Finish**.

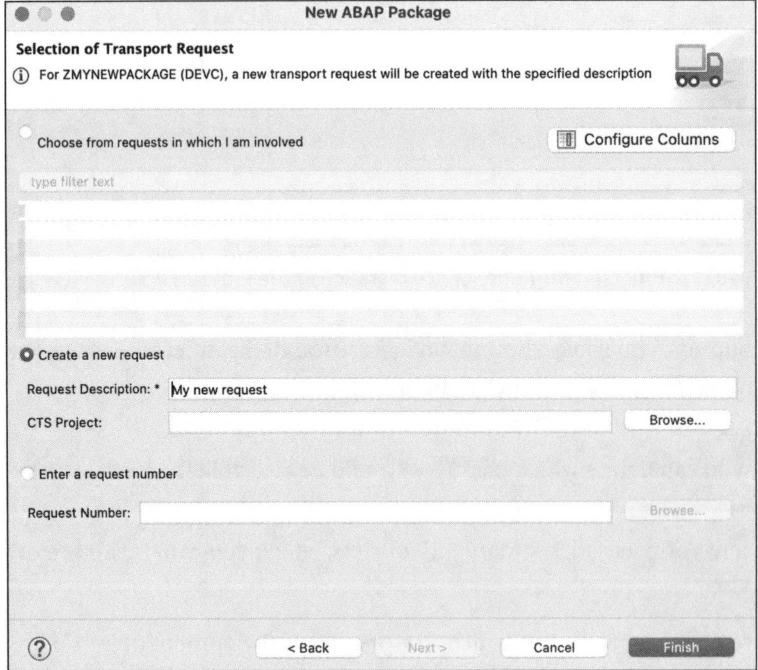

Figure 1.34 Creating a Transport Request

With this step, from a development tools and cloud environment perspective, you're ready to start developing on your SAP BTP, ABAP environment trial instance.

For an enterprise instance, the process of adding an ABAP cloud project is the same. However, unlike in a trial instance, in an enterprise instance, you can maintain additional users and roles for your development team.

1.3.2   Views

In ADT, different views are available so you can get more information about different components of the environment, such as objects, runtime, etc. A list of all available ABAP views is shown in Figure 1.35.

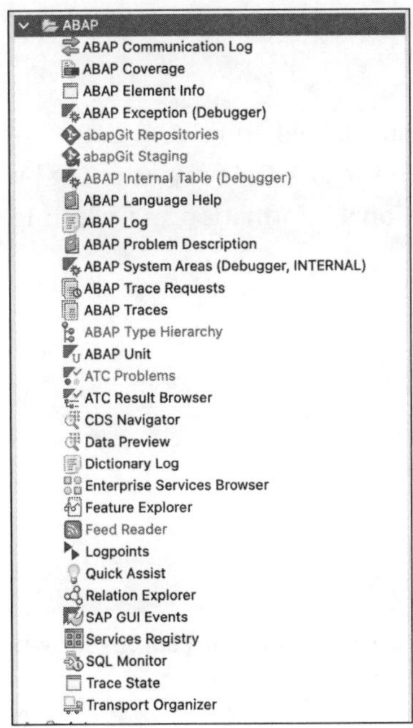

Figure 1.35 ABAP Views in ADT

Selected views will be displayed in lower-right section of the **ABAP** perspective, as shown in Figure 1.36.

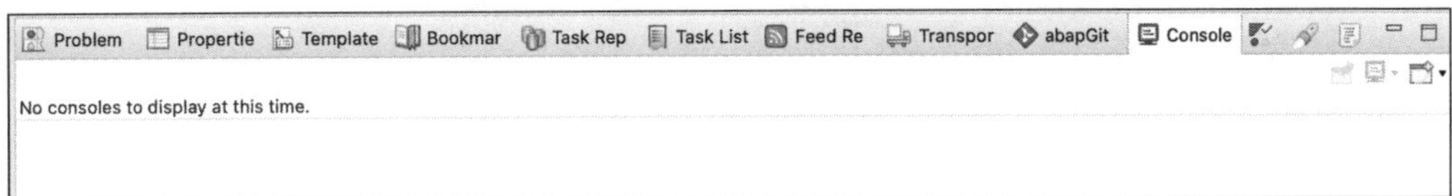

Figure 1.36 ABAP Perspective: Views

Some views can be further customized to provide better insight into the precise information you need.

One example is the **Feed Reader** view, shown in Figure 1.37. By default, in this view, you can only see ABAP dumps (runtime errors).

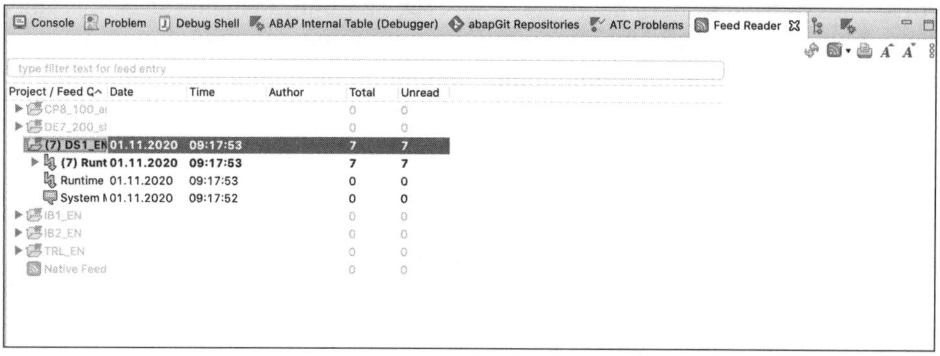

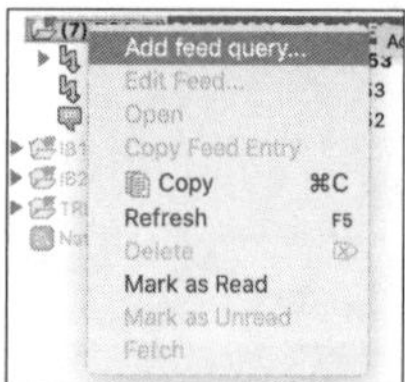

Figure 1.37 Feed Reader

However, during testing and debugging, perhaps you noticed some SAP Gateway-related issues. For debugging, you'll need to look into the SAP Gateway error log by right-clicking on the system for which you need additional information and selecting the **Add feed query...** option, as shown in Figure 1.38.

Figure 1.38 Adding a Feed Query

Now, select the query you want to add, in this case, the **SAP Gateway Error Log**, as shown in Figure 1.39.

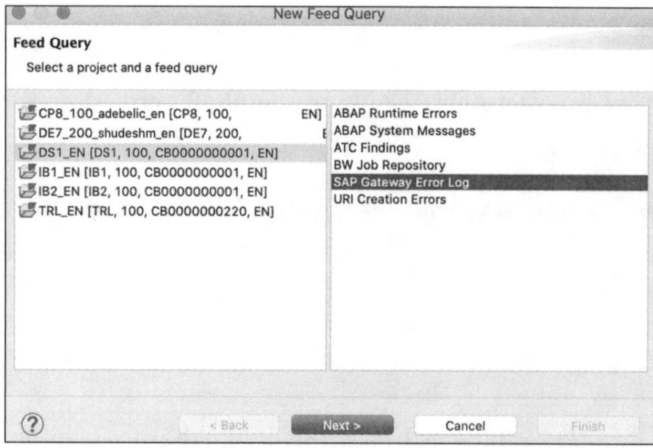

Figure 1.39 Selecting an Additional Query in the Feed Reader

In the popup window shown in Figure 1.40, you can customize some details of the feed, for example, its title and its refresh interval.

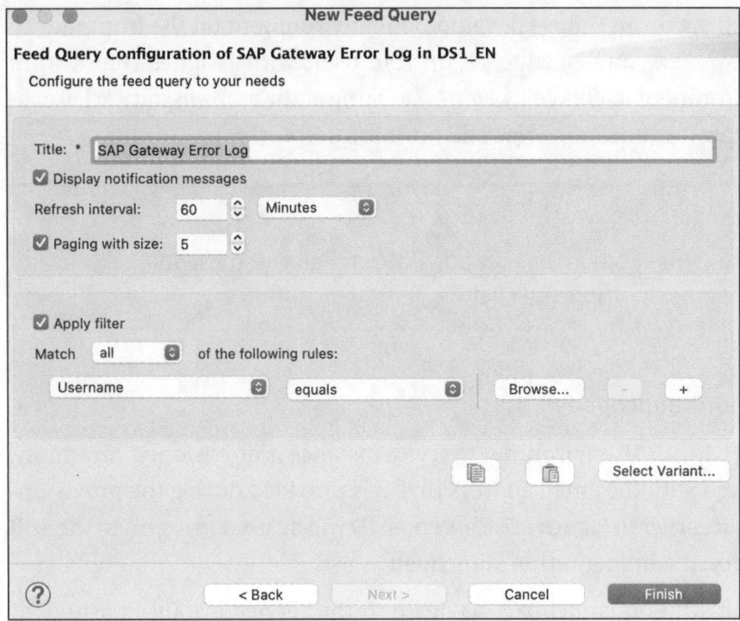

Figure 1.40 New Feed Query Options

Finally, your new query is available in the **Feed Reader**. With this information, you can more easily analyze SAP Gateway errors, as shown in Figure 1.41.

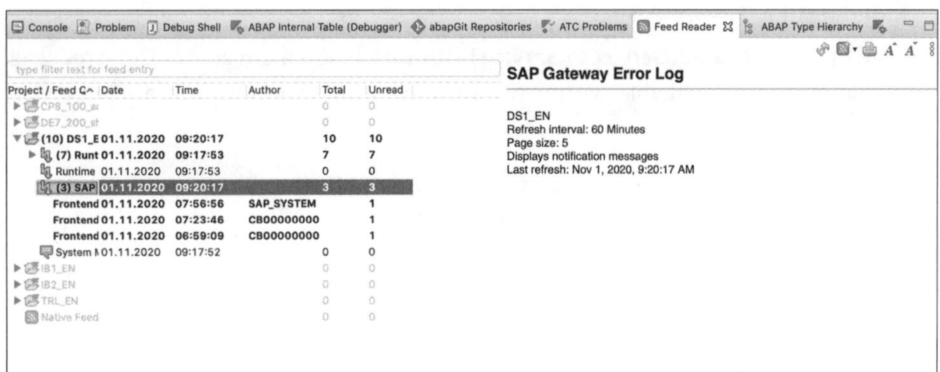

Figure 1.41 Feed Reader with SAP Gateway Error Logs

In the following section, we'll cover how to add more development users and assign the corresponding roles.

1.4   Identity and Access Management

In previous sections, we covered the provisioning of an SAP BTP, ABAP environment and the steps for setting up an Eclipse development environment on the frontend. To start developing, however, you'll need to set up your users with assigned roles within SAP BTP, ABAP environment itself. We'll begin by looking at the administration launchpad, where you'll maintain business roles and business users.

> **Note**
>
> The steps described in this section are only applicable for an enterprise ABAP environment; these tasks are not available in a trial account.

1.4.1   Administration Launchpad

Once your new SAP BTP, ABAP environment service instance is provisioned, an administrator user is created with the email address that was provided during the provisioning process, as shown earlier in Figure 1.17. This email ID will be used to log on to the SAP BTP, ABAP environment administration launchpad.

To access the administration launchpad, navigate to the service instance dashboard and then follow these steps:

1. Navigate to the Cloud Foundry subaccount in which an SAP BTP, ABAP environment service instance has been provisioned.

2. In the menu on the left, navigate to **Cloud Foundry** · **Spaces** and access the corresponding Cloud Foundry space.

3. Now, on the left, select **Services** · **Service Instances**. In the main area of the screen, all service instances available within the selected Cloud Foundry space will be displayed.

Once presented with the service instances list, you can further filter the list and locate the ABAP service instance, as shown in Figure 1.42.

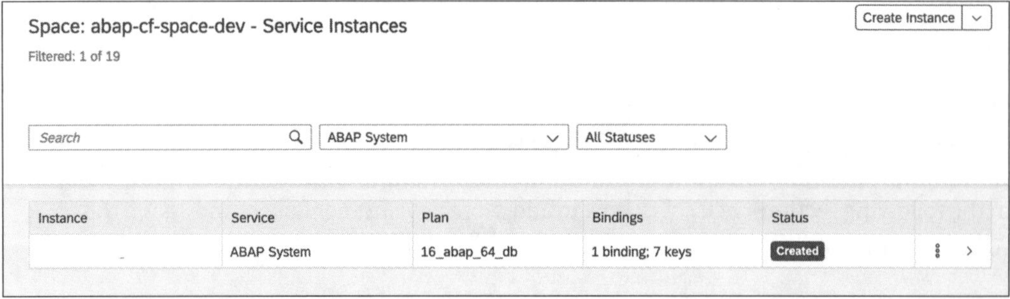

Figure 1.42 Service Instances List

To access the administration launchpad, click on button to the right of the service instance name and select the **View Dashboard** option.

> **Note**
>
> The administration launchpad can also be accessed from ADT by clicking on a project's details.

You'll now be redirected to the logon screen for the ABAP system service instance. For easy future access, we recommend bookmarking this URL.

To access Cloud Foundry resources, your email ID is used. However, you must have an S-user connected to this email ID. For logging on, you'll use the admin email ID and corresponding S-user password. This step will open the administration launchpad, as shown in Figure 1.43.

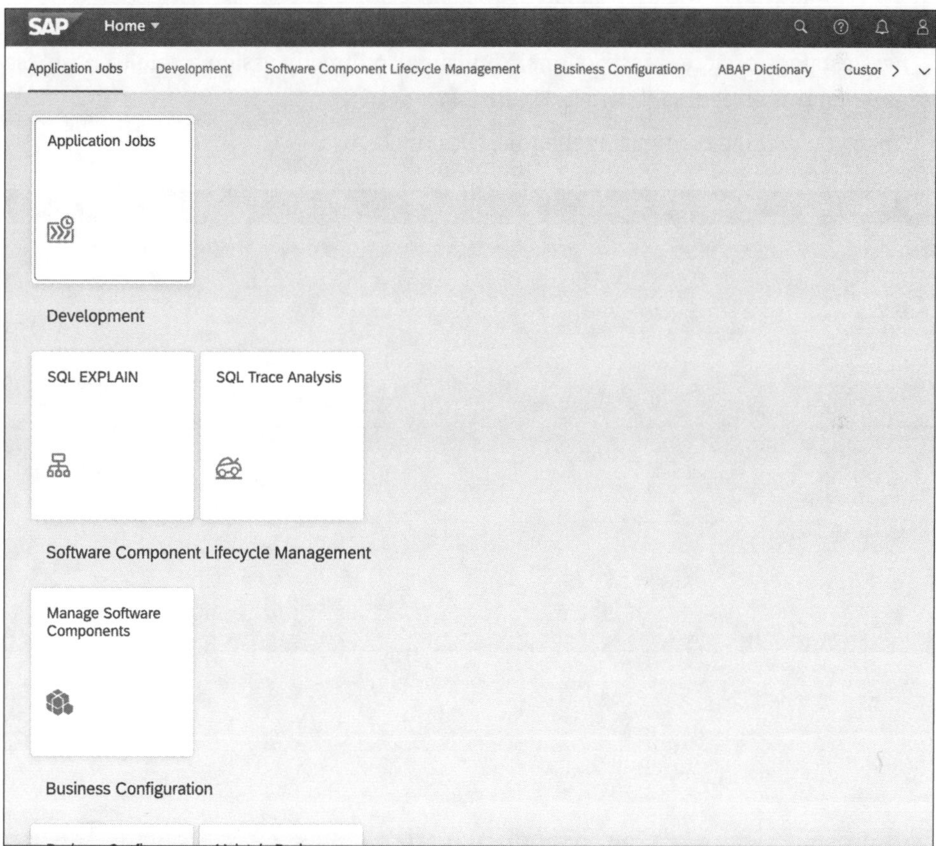

Figure 1.43 SAP BTP, ABAP Environment Admin Dashboard

1.4.2   Maintaining Business Roles

On the new SAP BTP, ABAP environment system, the first thing you must do is configure business roles. Initially, only one role is available: SAP_BR_ADMINISTRATOR. To avoid giving administrator privileges to other users, you'll create corresponding roles.

The easiest approach to creating business roles is to create a role for development users from a predefined template. The template for a development role is SAP_BR_DEVELOPER. Follow these steps to customize this role:

1. Click the **Maintain Business Role** tile.

2. Select the **Create from Template** option.

3. A list of available templates will be presented. Select **SAP_BR_DEVELOPER** from this list.

4. The system will propose a role ID and a description. You can confirm these options or change both fields.

5. On the next screen, you can explore and adjust role attributes, such as assigned business catalogs or access restrictions. You can also directly assign a role to business users on this screen, as shown in Figure 1.44.

6. Once you're done configuring the role, click on **Save**.

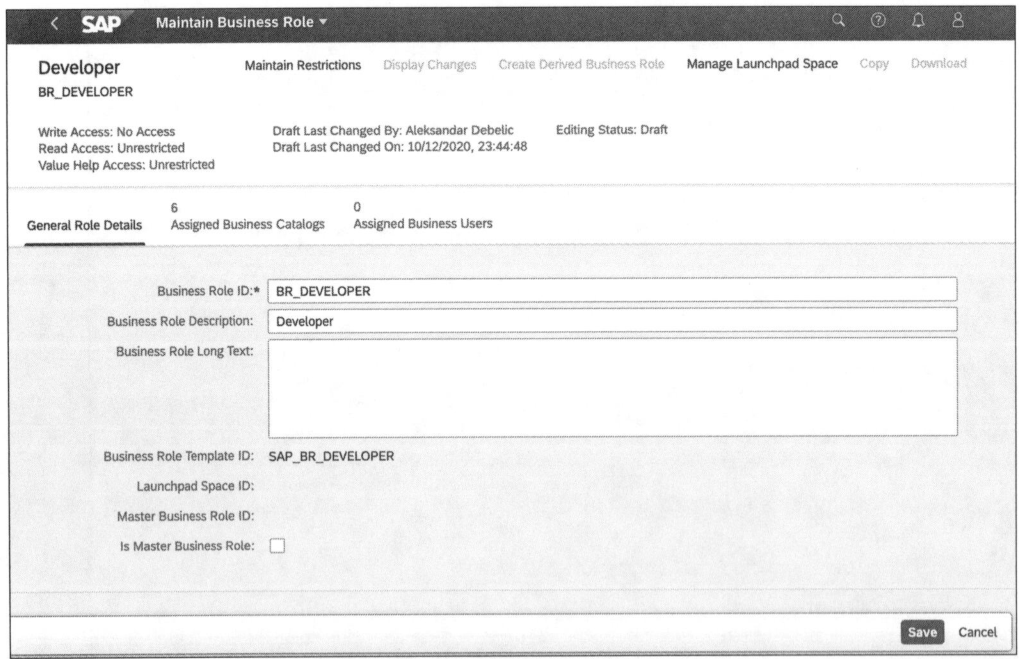

Figure 1.44 Maintaining Business Roles

The role is now ready for use and can be assigned to business users.

For advanced users, you can create roles manually. From the initial screen of the SAP Fiori app Maintain Business Roles, choose the **New** option and then follow these steps:

1. Enter a business role ID and a description. Then, click **Create** to proceed to the next screen.

2. You'll be navigated to the main business role maintenance screen, but no business catalogs have been assigned yet.

3. Navigate to **Assign Business Catalogs** tab and click on **Add**. You'll be presented with a list of available business catalogs, as shown in Figure 1.45.

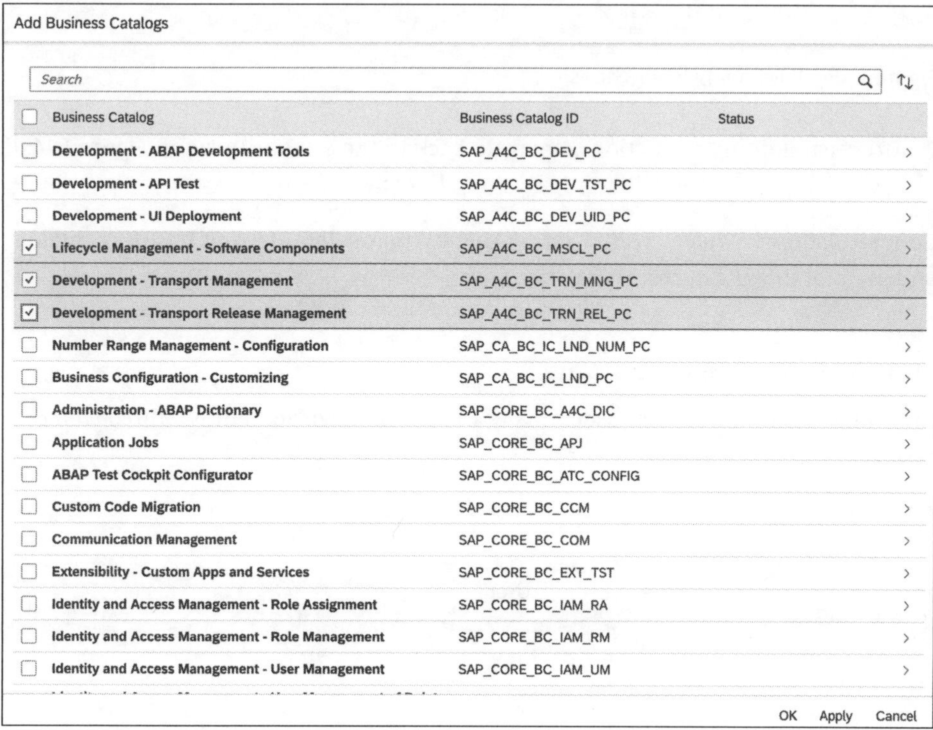

Figure 1.45 Add Business Catalogs Screen

4. Select the desired business catalogs and confirm by clicking **OK** or **Apply**.

5. Now, the only remaining task is to maintain restrictions (click on the **Maintain Restrictions** button).

6. By default, no restrictions exist on **Read** and **Value Help**, but for **Write**, notice the screen says **No Access**. To grant write access to the role as well, you'll need to change write access to **Unrestricted**, as shown in Figure 1.46.

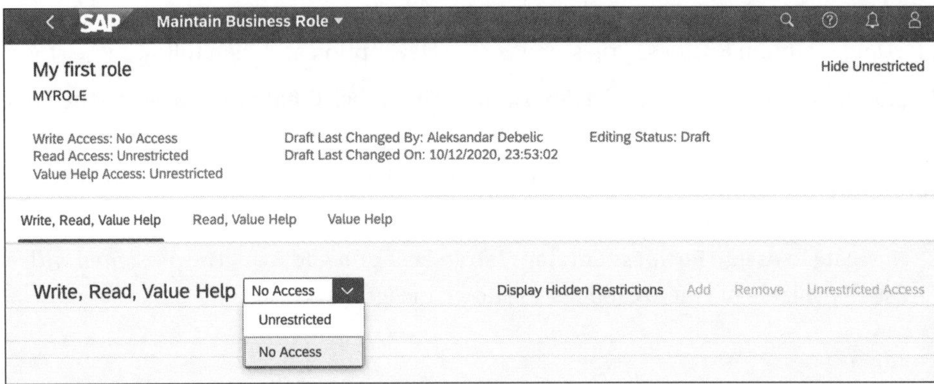

Figure 1.46 Maintaining Restrictions

7. After maintaining restrictions, navigate back to the screen shown in Figure 1.44. If you're happy with the role configuration, click **Save**.

Congratulations! You've created a new role by manually selecting business catalogs and maintaining the necessary restrictions.

1.4.3   Maintaining Business Users

In general, creating business users in SAP BTP, ABAP environment is a three-step process, as shown in Figure 1.47.

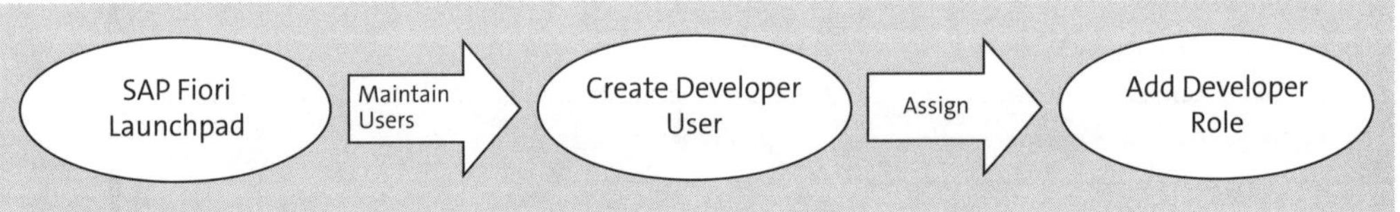

Figure 1.47 Business User Creation Process

To create a new employee, you'll use the SAP Fiori app Maintain Employees. On the initial screen, you'll see a list of existing employees, as shown in Figure 1.48.

By clicking on the **New** button, you'll be navigated to a screen where you'll enter details about new employee, as shown in Figure 1.49.

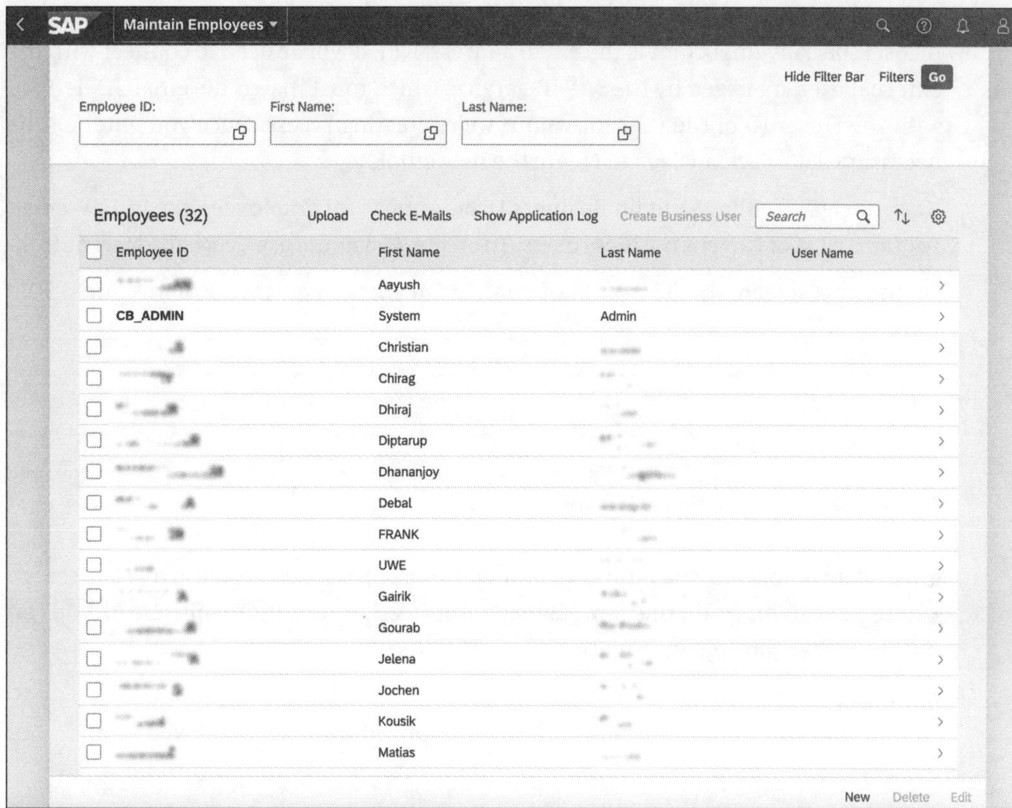

Figure 1.48 Maintain Employees SAP Fiori App

Figure 1.49 Employee Detail Entry Screen

Mandatory information includes maintaining the **Employee ID**, **Last Name** and **E-mail** fields. Especially important is the email address, which will be used to connect with the external user, managed by the SAP ID service, which must match the email ID defined with the S-user ID of the user for whom we're creating access. Once you enter all the necessary data, click on **Save** to create the new employee.

Now, navigate to the Maintain Business Users app. As for employees, the initial screen contains a list of current business users. To create a business user, click the **New** button.

On the next screen, you'll see a list of unassigned employees, as shown in Figure 1.50.

Create New Business User			
Search			
First name	Last name	Full Name	Person ID
Some	User		171

Figure 1.50 List of Unassigned Employees

Click on an employee, and after confirming the popup window, you'll arrive at a screen where you can maintain the new business user (see Figure 1.51). On this screen, you can maintain the following information:

- **Personal Data**
  This read-only section includes information pulled from the employee data.

- **User Data**
  In this section, you can modify a user name (the default is proposed by the system), specify the validity of a business user, and set/reset locks.

- **Regional settings**
  In this section, you can maintain details like decimal format, date format, time zone, time format, and language.

Now, let's focus on the business roles section. To use the business user for development, the user must have the corresponding role/roles. Click **Add Business Roles** to select the role/roles to be added. Once you've made your selections, click on **Save**. Now, the business user is ready for use.

To access the SAP BTP, ABAP environment from ADT in Eclipse, as described in Section 1.3, use the credentials that contain from email address and password for the corresponding S-user.

Finally, we're ready to start developing!

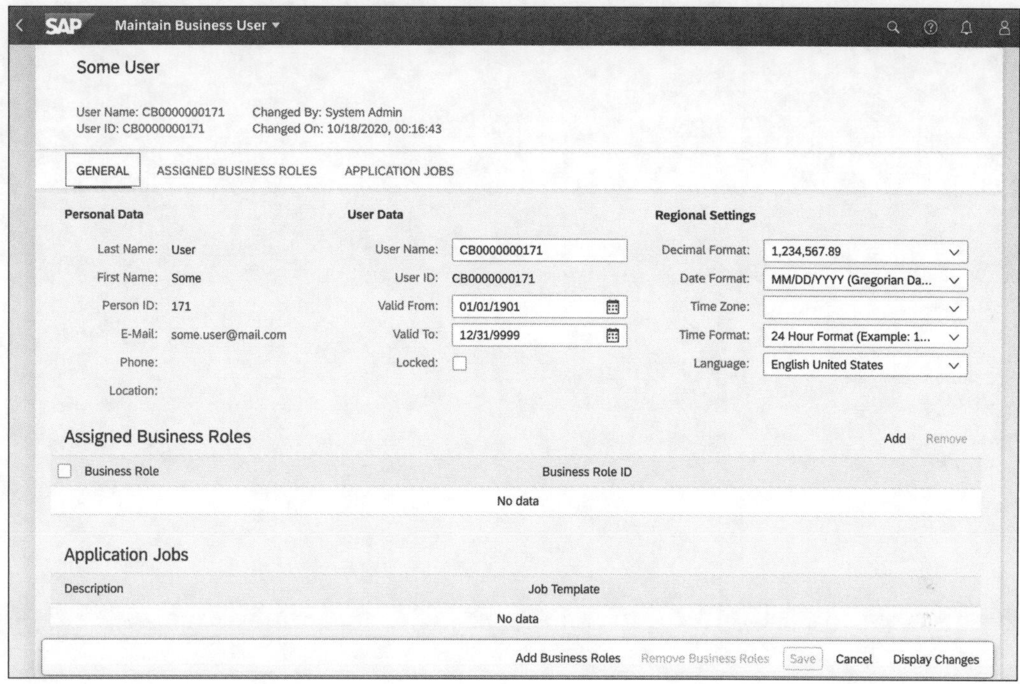

Figure 1.51 Maintain Business User App

1.5   Summary

In this chapter, we introduced you to SAP BTP, including a look at how it evolved from Neo to Cloud Foundry. You learned how SAP BTP, ABAP environment is positioned in the context of SAP BTP and how to provision trial and enterprise instances. We also walked you through the process of setting up a development environment, both on the SAP BTP side and on a desktop, including setting up access and roles for your development team.

In the next chapter, you'll learn all about the ABAP RESTful application programming model, and we'll cover managed, unmanaged, and draft development scenarios. You'll also learn about the development of frontend components and their deployment in Cloud Foundry and in the ABAP stack.

Chapter 2
Developing Applications

In this chapter, you'll learn the concepts behind the ABAP RESTful application programming model, which governs ABAP applications in the cloud. We'll walk you through detailed steps for building applications leveraging the ABAP RESTful application programming model, covering different end-to-end scenarios. We'll leverage ABAP Development Tools (ADT) for Eclipse and SAP Business Application Studio, and you'll learn how to deploy your developed apps to the ABAP stack. We'll also explore side-by-side extension concepts and use cases.

In Chapter 1, we covered the prerequisite setup required to proceed with hands-on development. At this point, you should have a good understanding of how to set up the SAP Business Technology Platform (SAP BTP), ABAP environment as well as how to configure the desktop development environment in ADT for Eclipse.

In this chapter, you'll learn about several approaches to application development in SAP BTP, ABAP environment, starting in Section 2.1 with an overview of the ABAP RESTful application programming model and different application development scenarios supported in SAP BTP, ABAP environment. You'll also learn about side-by-side extensions and how to identify use cases for this approach. We'll set up the development environment in Section 2.2, before moving on to hands-on discussions of developing applications for SAP Fiori (Section 2.3), in managed scenarios (Section 2.4 and Section 2.5) and in unmanaged scenarios (Section 2.6). To complete our discussion of application development, we'll briefly look at backend services in Section 2.7.

The next sections will show you how to set up API release states in Section 2.8, and finally, we'll show you how to deploy applications to SAP BTP, ABAP environment in Section 2.9.

In Section 2.10, you'll learn how to use custom entities in the ABAP RESTful application programming model. At the end of this chapter, in Section 2.11, we'll look at some advanced development techniques, such as the XCO library and its object consumption for application building and the RAP generator, an open-source tool to help you accelerate your work in SAP BTP, ABAP environment.

2.1   ABAP RESTful Application Programming Model

With over 100,000 businesses running its solutions, SAP is over 40 years old. Naturally, you may be concerned about what a new programming model will mean for your existing repositories or your existing code. How will this new programming paradigm support your existing systems? Will this new model support additional requirements or enhancements or increase in scope in the form of applications or users?

Let's explore how these questions are addressed by an understanding of the objectives of this new programming paradigm and model, such as the following objectives:

- Enabling existing code usage and efficient development are the main objectives of the ABAP RESTful application programming model. SAP BTP, ABAP environment provides comprehensive application development capabilities to extend, integrate, and build innovative applications in less time and without the effort of maintaining infrastructure.

- Providing 24/7 availability of cloud-based applications, with the ability to extend business applications without invoking complex design-time environments.

- Enabling a high degree of scalability and elasticity to meet, for example, heavy increases in users and requests.

- Retaining lifecycle and supportability features from previous models, while also supporting new innovations.

- Offering an efficient and standardized way for building all types of SAP HANA-optimized applications, which are role based for different personas, while supporting different devices on the user experience (UX) level, enabling extensibility and scalability, etc.

In the following sections, you'll learn in detail the concepts behind the ABAP RESTful application programming model, its architecture, its scope, and its usage in various types of applications. Let's start with a look at how ABAP has evolved.

2.1.1   Evolution of the ABAP RESTful Application Programming Model

The evolution of the ABAP RESTful application programming model can be divided into three stages, as shown in Figure 2.1. Let's briefly look at each stage next:

1. **Classic ABAP programming**
   The majority of SAP clients still use SAP ERP 6.0, usually with one of the Enhancement Packages (EHPs), as their core ERP solution. This solution uses SAP NetWeaver 7.50 or earlier as its technology platform and can run on any database. This solution will continue to be supported until 2027.

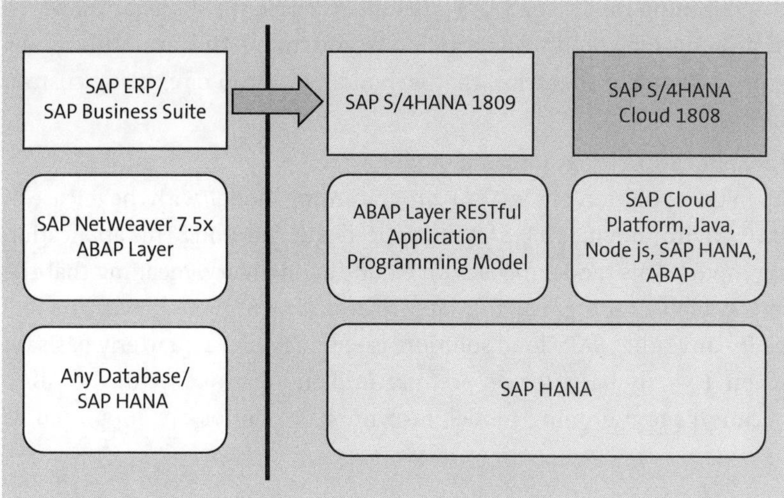

Figure 2.1 Evolution of the ABAP Programming Model

With SAP NetWeaver 7.40, SAP introduced its in-memory database SAP HANA and started optimizing the ABAP platform to leverage the features and capabilities introduced by this revolutionary database. In parallel, support for any database continued. New approaches, such as support for RESTful communication and role-based user interfaces (UIs) with SAP Fiori, were the drivers for change in the programming model after decades of stability.

2. **ABAP programming model for SAP Fiori**

Moving to SAP S/4HANA required a change in the ABAP programming model. The new programming model for SAP Fiori was introduced with SAP NetWeaver 7.5 and included some major changes, such as the following:

- Transaction SEGW/OData referenced data sources

- Core data services (CDS)

- The CDS-based Business Object Processing Framework (BOPF)

With the ABAP programming model for SAP Fiori, you could build SAP HANA-optimized, web-based applications. Even today, this approach remains the best practice for ABAP development in SAP S/4HANA.

In SAP NetWeaver 7.51 (that is, SAP S/4HANA 1610), the draft capability was introduced as a further improvement. This capability enables the development of SAP Fiori applications that use stateless communication by setting exclusive locks and storing intermediate data (something similar to parking documents in old SAP ERP transactions).

The ABAP programming model for SAP Fiori supports greenfield scenarios, which means application design and implementation from scratch. However, this model did not support brownfield scenarios, that is, building on top of existing custom transactional logic.

3. **ABAP RESTful application programming model**
   As a next step in the evolution of the ABAP programming model, with the release of SAP BTP, ABAP environment 1808, SAP introduced the ABAP RESTful application programming model. This model focuses on cloud-first delivery, meaning that the model became available on-premise in a later release (SAP S/4HANA 1909). Its feature scope, as for any other SAP cloud solution, is enhanced on a quarterly basis on SAP BTP and on a yearly basis for on-premise implementations. With the ABAP RESTful application programming model, brownfield scenarios are supported as well.

The major features of the new ABAP RESTful application programming model include the following:

- Business services
- Behavior definition and implementation

This model consists of three main layers:

- The data modeling and behavior layer, for data modeling, behavior definitions, and behavior implementations
- The business service provisioning layer, for business object projection, service definitions, and service bindings
- The service consumption layer, for consuming OData

The ABAP RESTful application programming model is the next generation of the ABAP programming model and not only targets cloud-based ABAP implementations but on-premise implementations as well. The ABAP RESTful application programming model provides the ability to connect to other cloud systems and enables the efficient and standardized development of web-based applications and Web application programming interfaces (APIs).

The ABAP RESTful application programming model will be the foundation for ABAP development for the next 20 years. It meets modern-day user expectations by following a REST-based approach, wherein the application state is persisted in SAP HANA at the database level, instead of as session data on the application server level. This model enables collaboration and preserves work continuity by enabling the seamless use of many kinds of devices.

The ABAP RESTful application programming model follows the *code-to-data paradigm* by default, as part of its engine and its core, to reap all the benefits of SAP HANA. An easy-to-learn flow of role-based user interactions enables your employees to get the work done, by using template-based UIs that reflect user roles more precisely.

The ABAP RESTful application programming model provides views and perspectives on data specific to a role, while hiding unnecessary details. You can get instant and unlimited access to data, with support for SAP HANA features and performance benefits. This programming model supports cloud-based applications, and new features and fixes can be continuously delivered to your customers, without the need for downtime or outage. SAP BTP, ABAP environment lets you more easily extend core business processes and create new applications, with SAP Fiori user experience principles, while leveraging other solutions and capabilities available on SAP BTP, such as machine learning, the Internet of Things (IoT), etc.

Typically, your applications will be SAP Fiori apps that consume OData services built using the ABAP RESTful application programming model. The key components include CDS views, behavior definitions, and behavior implementations that talk to the SAP HANA database to fetch the data.

The ABAP RESTful application programming model is based on a modern development toolset, which is fully Eclipse based and supports syntax checks, syntax highlighting, pretty printing, navigation, search and quick fixes, and other productivity features. Code quality is ensured by providing static code checks, via the ABAP Test Cockpit (ATC), with remote and local scenarios and unit testing including an isolation framework as well as debugging, profiling, static and dynamic logging, and runtime monitoring and analysis capabilities.

With the ABAP RESTful application programming model, you can push down data-intensive logic to SAP HANA and run user interaction logic inside the browser. The ABAP RESTful application programming model thus can be divided into three parts: SAP Fiori applications; the SAP HANA database; and the application layer, which consists of CDS views, behavior definitions, behavior implementation classes, and OData, as shown in Figure 2.2.

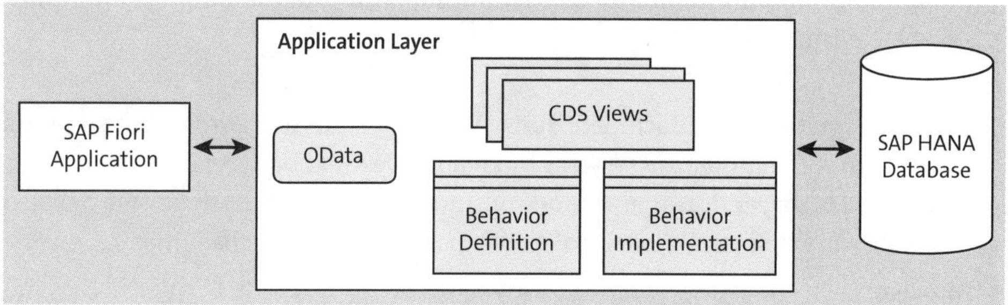

Figure 2.2 ABAP RESTful Application Programming Model Flow

You'll use *code transformation tools* to transform your existing code for use in SAP BTP, ABAP environment, which involves the following tasks:

- **Decouple**

  Existing custom code from your backend systems can be read and updated/ enhanced via remote interfaces and adapt to released APIs. With SAP BTP, ABAP environment, custom applications can be developed on a separate platform as a service (PaaS). Decoupling helps increase value by building and innovating applications beyond traditional SAP capabilities via cloud services. Depending on your business requirements, your cloud environment can be scaled up or down without impacting your SAP systems. This approach helps you deliver innovative applications in the form of services, with minimal impact on your core SAP system, thus reducing the risks that came come with upgrading.

- **Adopt**

  While using the new programming paradigm, note that the ABAP RESTful application programming model emphasizes the use of core data services (CDS) and code pushdown concepts. CDS views are created on database tables or on other views. CDS views play an important role in virtual data models (VDMs), which we'll describe in detail later in Section 2.1.3 and Section 2.4.1. To send data from the cloud, you can use CDS views in the form of an OData service or an API. Previously, performance was negatively impacted since data was fetched from databases to the application layer to perform calculations, which resulted in terrible performance with huge volumes of data. Now, since calculations can be performed at the database level, this means smaller chunks of data are transferred to the application server— the essence of the code pushdown concept. By implementing this concept, the overall performance of your operations should be enhanced many times over.

- **Eliminate**

  To eliminate depreciated ABAP functionalities, before moving your custom code from existing functionalities on on-premise systems, you'll need to check for code readiness, your on-premise code's suitability for the cloud, which includes checking the code's scope for cloud compatibility and checking for already released objects in the cloud. To check for code readiness, you can use the ABAP Test Cockpit (ATC) tool provided by SAP. Functionalities that are not supported for the cloud must be eliminated to ensure compatibility.

- **Use**

  To use integration and attachment services, SAP BTP, ABAP environment supports integration with on-premise systems by creating destination services, SAP Fiori services, and OData services that can be consumed in your applications and can also integrate with predictive models to enable machine learning capabilities.

- **Manage**

  To manage development and deployment process, you can use Git and adopt continuous integration practices. Applications will be created in the cloud and can be deployed in SAP BTP either in the Neo or in Cloud Foundry environment. You'll learn how to deploy applications in detail in Section 2.9. Code can be pushed to or pulled from a Git repository, which we'll explain in detail in Chapter 4.

- **Utilize**
Commercialization infrastructure for partners provided by SAP BTP will be utilized. Building new cloud-based applications and extending existing on-premise and cloud-based applications are both possible with the services and capabilities provided. Multiple environments (such as Cloud Foundry, ABAP, and Kyma) and multiple regions provide flexibility and the ability to utilize services best suited to meet your business requirements, regardless of environment.

Benefits

The ABAP RESTful application programming model has the following four main benefits:

- Enables you to build transactional, responsive SAP Fiori applications that adapt to desktops, mobile devices, and tablets
- Leverages the capabilities of the in-memory computing engine to follow the code-to-data paradigm
- Enables robust programming since only the allowed resources are consumed
- Includes cloud-ready extensions

2.1.2   Key Elements of the ABAP RESTful Application Programming Model

Three main elements exist in the ABAP RESTful application programming model, as shown in Figure 2.3.

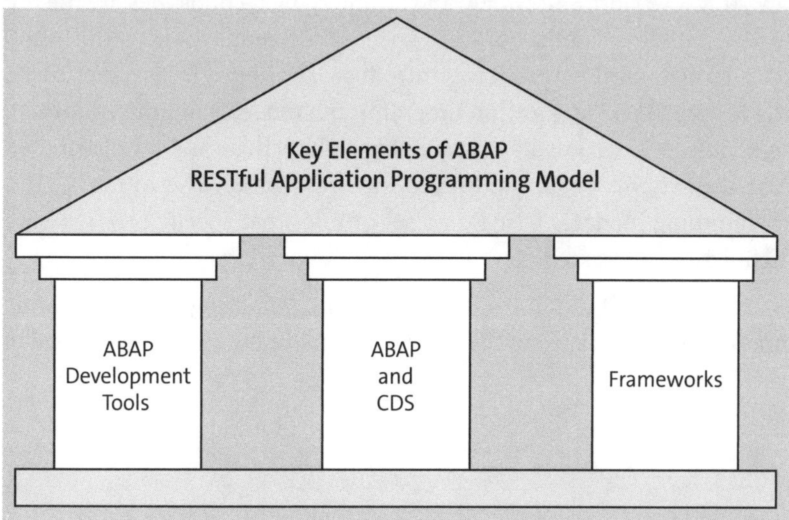

Figure 2.3 Key Elements of ABAP RESTful Application Programming Model

Let's briefly look at each of these factors next:

- **ABAP Development Tools (ADT) in Eclipse for the whole development process**
  The purpose of ADT is to enable seamless workflows—with a single click, often developers can advance to the next step. SAP GUI does not exist in the cloud, which solidifies Eclipse as the main integrated development environment (IDE) for ABAP. Developers no longer have to read lots of documentation and can start development more quickly. With this approach, all implementation tasks can be integrated into a single development environment, which helps optimize the development flow.

- **Language support for ABAP and CDS**
  Direct support for business object development (data modeling and transactional enablement) in CDS and in the ABAP language is aligned to support development following the ABAP RESTful application programming model. Separate frameworks like BOPF and framework-specific tools are no longer needed. No code is generated, and therefore, code redundancy and code inconsistency are avoided. Standard implementation tasks via APIs support useful IDE features like types, signatures, static code checks, autocompletion, and element information. You can avoid runtime errors using static code checks enabled by typed APIs.

- **Powerful frameworks for runtime components**
  You can control and standardize technical implementation tasks with dedicated code exits for application-specific business logic, which are provided on protocol-agnostic layers.

2.1.3   Architecture of the ABAP RESTful Application Programming Model

The user experience is a cornerstone in today's digital environments and evolving landscape. Supporting SAP Fiori, SAP HANA, and all emerging technologies are the top requirements when building applications. To meet these requirements, the ABAP RESTful application programming model was introduced. Designed for browser-based applications, the ABAP RESTful application programming model is optimized for SAP HANA. This methodology is designed for developing architectures and for distributed systems that use the Internet to communicate with each other. The ABAP RESTful application programming model is based on a set of rules that specify how network resources are defined and addressed.

As shown in Figure 2.4, the ABAP RESTful application programming model is mainly divided into three components. Each of the following components will be described in detail in later sections:

- Data modeling and behavior
- Business service provisioning
- Service consumption

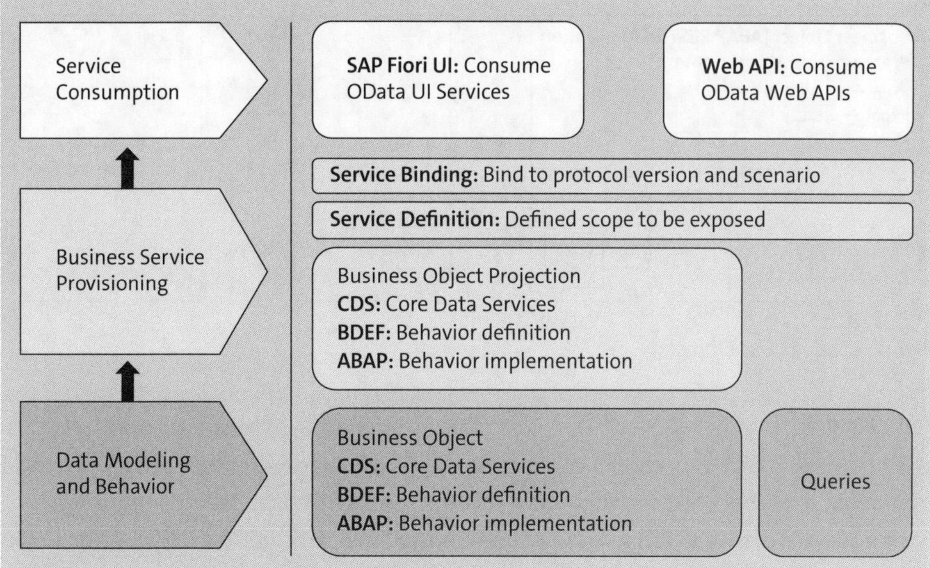

Figure 2.4 ABAP RESTful Application Programming Model

Data Modeling and Behavior

The ABAP RESTful application programming model uses core data services (CDS) to define data models. CDS provides a framework for defining and consuming semantic data models. Depending on the use case, data models support transactional or query access to a database. Thus, data models are used as business objects and in analytical queries and reports.

Business objects and queries together create a complete data model and application. This reuse layer consists of data modeling, database tables, views, and available application logic.

A *behavior* describes what can be done with the data model, for example, if the data can be updated. In transactional scenarios, a business object behavior defines which operations and what characteristics belong to a business object. For read-only scenarios, the behavior of the data model is defined by its query capabilities.

Business Objects

Business objects are composed of a data model, a behavior, and a behavior (runtime) implementation, as shown in Figure 2.5. The data model of a business object consists of a root node and one or more child nodes.

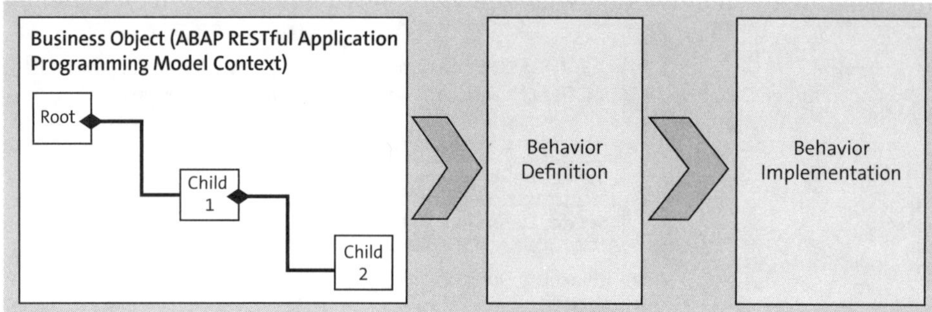

Figure 2.5 Business Object

Data Model

The business object data model consists of a hierarchical tree of nodes that are linked by associations that define a whole-part relationship, which is called a *composition tree*. Each node of a composition tree is an element modeled with a CDS entity and arranged along a composition path. As shown in Figure 2.6, a sequence of compositions connecting entities with each other builds up the composition tree of an individual business object. Since a business object is a tree data structure, one of the entities will represent the root.

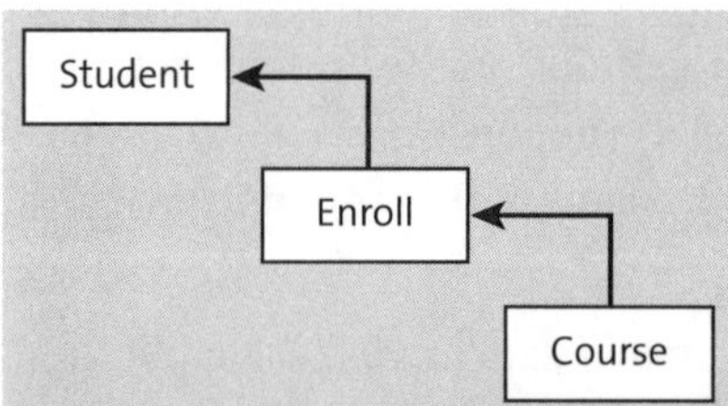

Figure 2.6 Virtual Data Model (VDM) Structure

CDS views are associated with each other to create virtual data models (VDMs), where one CDS view is the root and the others are children. A VDM can take data directly from an SAP HANA database table and expose this data with a service definition.

With this VDM, you can build both transactional and analytical applications and also create APIs that can be called and consumed from anywhere. With VDMs, you'll avoid redundant ABAP CDS views and can structure data into logical units from a business perspective to build a consistent model.

A VDM consists of the following four layers:

- **Database tables**
  These persistent data objects are used to store data in the form of rows and columns. Database tables are created through ADT. A database table field or a group of fields can be defined as the primary key to distinguish records. This primary key can be

then used as a foreign key in other tables to establish relations with other tables via cardinality.

- **Basic interface views**
  Views of this type are defined to consume data by directly accessing database tables. Basic interface views are the lowest level of a VDM. Basic interface views represent projections on top of SAP database tables, mirroring data from a single database table and transferring this data into a CDS annotation. No data redundancy can occur within a basic interface view. This type of view exposes all relevant business data, enhancing this data with annotations, but are not meant for conversions or calculations of any sort.

- **Composite interface views**
  This type of view is built on top of basic views or other composite views. Unlike basic interface views, composite views can have data redundancy. These views may project, calculate, join, filter, and aggregate data based on basic views or other composite views.

- **Projection view**
  This type of view is located on the top layer of a VDM and is used by UI tools, for example, SAP Business Explorer, SAP Lumira, SAP BusinessObjects Web Intelligence, SAP Analysis for Microsoft Office, etc.. The result is a report that can be accessed by business users. These views can consume other basic views or composite views to read data and create a final data set to be fed into UI tools for reports based on business requirements. Measures that must be calculated after aggregation will be defined in this view.

Behavior Definition

A behavior definition defines which operations are offered by the business object. Three types of operations can be specified:

- Standard operations include CRUD (create, read, update, and delete) operations and predefined behaviors.

- Specific operations like actions or functions that may have parameters and different semantics.

- Exclusive locks for concurrency handling, which is quite important. ETags are supported for exclusive locks.

OData is a stateless communication protocol, which means the session details are not required to be maintained on the server, but applications are stateful, which means that session details need to be maintained. Thus, to support stateful applications on a stateless communication channel, the state, as a draft, is defined as its own resource. As a result, the backend server can stay stateless between two requests. In this way, the draft concept can be implemented in the ABAP RESTful application programming model.

Behavior Implementation

How behavior should be processed during runtime is defined in a behavior implementation. One behavior definition can have multiple behavior implementations.

The following three scenarios are possible:

- Scenario 1: You already have complete application code available.
- Scenario 2: You are starting from scratch.
- Scenario 3: You have some application code available.

All these scenarios have one thing in common—the runtime business object model. Behavior implementation involves three phases:

- Interaction phase
- Transaction buffer
- Save sequence

The *interaction phase* involves three steps: modify, read, and lock. Let's say you want an end user to be able to edit or change some data or perhaps add data by clicking an **Edit** button. A transaction opens and interacts with backend changes that are stored in the *transaction buffer*, which represents the state of the instance data. Business objects always need transactional buffers to keep runtime changes in it. On clicking the **Save** button, the data from the transaction buffer is written to the database. Now, the interaction phase ends, and the *save sequence* starts. The save sequence phase is what makes the new or changed data persistent.

Business Service Provisioning

In the ABAP RESTful application programming model, business service provisioning is used to project a data model to a certain UI (which may be role-specific/task-dependent or may project to any user interface) or to a web API. Regardless of the provisioning target, the business service is based on the same data and functionality contained in a service definition. The business service is an artifact through which the data model and its related behavior are exposed or projected.

In the service binding, a service must include a protocol and protocol variant, for example, OData v2 or v4, as well as a specific use case, which will define the capabilities the OData service must fulfill. A service binding is protocol-specific. This separation allows data models and service definitions to be integrated into various communication protocols without the hassle of reimplementation. As shown in Figure 2.7, a business service consists two main components: the data model and the business service itself.

A data model consists of the business objects built on database tables. CDS views and queries represent the transactional behavior of the data. Also included in a data model are search helps and text tables to be maintained for different languages.

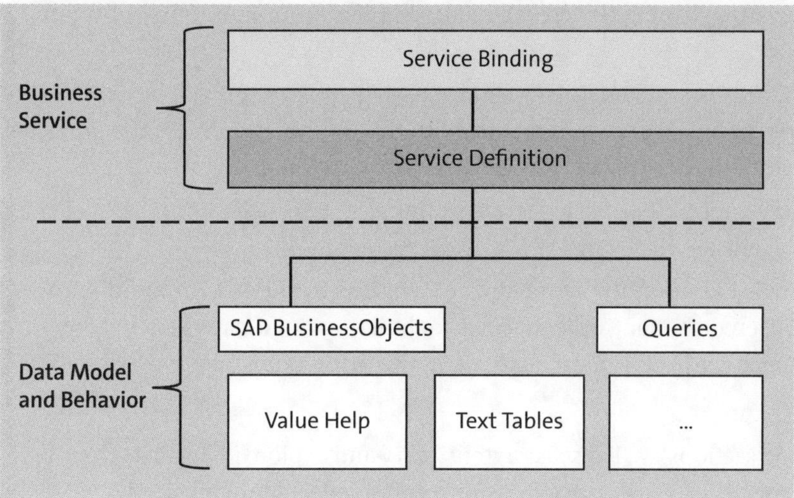

Figure 2.7 Business Service

A service is defined for the CDS view and will be bound through the service binding functionality to be consumed by UI applications.

2.1.4   Application Scenarios

As shown in Figure 2.8, you can build an application in two ways.

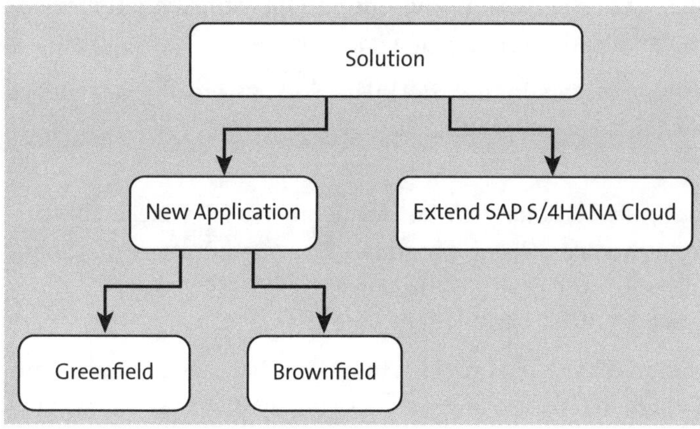

Figure 2.8 Application Scenarios

Let's briefly look at each option next:

1. When building a new application, two different approaches are available: the green-field approach and the brownfield approach. With a greenfield approach, an application is built from scratch and is completely new application. With a brownfield

approach, existing logic is reused and made compatible for SAP S/4HANA technologies.

2. In SAP S/4HANA or SAP S/4HANA Cloud, existing applications can be extended to meet additional business requirements. With this option, two different approaches are available, both adhering to the "keep the core clean" paradigm:

   - In-app/key-user extensibility for extending SAP S/4HANA using SAP-provided options available within an SAP S/4HANA system itself.

   - Side-by-side extensibility for building extensions in SAP BTP. In addition to SAP BTP, ABAP environment, other options for building cloud-native extensions exist as well such as Java, Node.js, and even the bring your own language (BYOL) approach.

In the following sections, we'll discuss extensibility and explore in detail its three basic types: classic extensions, key-user extensions, and side-by-side extensions.

Side-by-Side Extension Concept

In 2018, SAP introduced the concept of the clean digital core while developing extensions for SAP S/4HANA or for other SAP applications. Keeping the core clean means developing extensions in such a way that safeguards your application for future SAP S/4HANA upgrades. In theory, the concept is simple, but implementation can be difficult. Traditionally, SAP has provided several options for extending their business applications. These extension tools and techniques can be used by developers for many years. However, SAP in practice has not controlled what is developed via these traditional extension options, resulting in complex and long running upgrade processes.

With the help of extension concepts in SAP S/4HANA, SAP has broadly categorized extensions into the following three categories:

- **Classic extensions**
  Classic extensions are your traditional Z* ABAP developments (e.g., executable programs, classes, function modules, dialog programs, core modification, and many others). These extensions call SAP objects without using any APIs.

- **Key-user extensions**
  Key-user extension options have been provided by SAP so that they use stable integration points. Thus, these extensions are not impacted by upgrades. In layman's terms, these extensions are apps and options through which you can enhance other business applications. Examples of an in-app extensions include the Custom Fields and Logic app, the Custom Business Objects app, and the Custom CDS Views app. Additionally, UI adaptation options are also available as a part of in-app extensions to modify the appearance of the screen (e.g., hiding fields, changing field descriptions, grouping fields, etc.). More information is available at *http://s-prs.co/v523601*.

- **Side-by-side extensions**
  Side-by-side extensions are simply applications built outside of your digital core.

Instead of building applications in SAP S/4HANA, you'll build the application in SAP BTP using publicly released SAP S/4HANA APIs. Thus, your extensions will be loosely coupled to your core, thus making your core stable; modifications will be compatible and will follow zero downtime principles.

Figure 2.9 shows the three different types of extensions and where each is positioned.

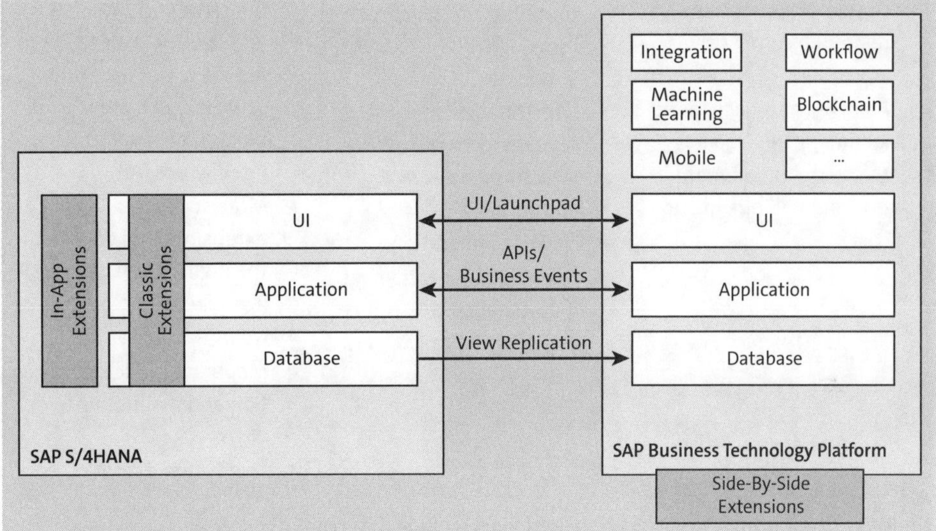

Figure 2.9 Extensibility

With in-app/key-user extensibility, you can enhance existing applications using tools delivered within the SAP S/4HANA system itself, leveraging released extension options, with little or no code development. By leveraging key-user extensibility options, you're still adhering to the "keep the core clean" paradigm. In-app/key-user extensibility options can be grouped in the following categories:

- User interface adaptations
- Custom fields
- Analytics/forms
- Business logic
- Custom business objects
- Custom user interfaces

Side-by-side extensibility refers to extensions built outside the core system, in SAP BTP. A side-by-side extension can be designed for a single instance or for multitenant scenarios and can follow the software as a service (SaaS) approach. You'll develop side-by-side extensions when business requirements cannot be met by in-app/key-user extensibility capabilities. Table 2.1 shows the pros of the different kinds of extensibility.

Classic Extensibility	Key-User Extensibility	Side-by-Side Extensibility
<ul><li>Classic way to develop WRICEF applications</li><li>Provides development flexibility to enable non-standard functionalities to support processes related to competitive advantage</li><li>A proven approach, popular among developers</li><li>Moderate initial migration and implementation cost</li></ul>	<ul><li>New way to develop WRICEF applications</li><li>Common tools for developers and super users</li><li>Minimize business disruptions during upgrades by using whitelisted APIs only</li><li>SAP S/4HANA lifecycle management provides a structured low-risk approach</li><li>No initial migration</li><li>Low implementation, maintenance, and future migration costs</li></ul>	<ul><li>New way to develop WRICEF applications</li><li>Addresses complex requirements using multiple runtime environments (e.g., ABAP, Java, and Node.js) and allows for the use of innovation services, such as IoT, blockchain, and artificial intelligence (AI)</li><li>More scalable, allowing extensions and growth of the digital core while scaling new technologies</li><li>No initial migration</li><li>Moderate implementation and maintenance costs</li><li>Low future migration cost</li></ul>

Table 2.1 Comparing Classic, In-App, and Side-by-Side Extensibility

SAP provides the following two programming models for building side-by-side extensions:

- SAP Cloud Application Programming Model, using either Java or Node.js
- ABAP RESTful application programming model, using ABAP

Depending on your existing investment, developer backgrounds, business decisions, etc., you can choose one of these programming models to build side-by-side extensions. Our focus in this book is the ABAP RESTful application programming model, and in the next section, we'll help you identify side-by-side extension use cases.

Use Case Identification

When working in an SAP S/4HANA implementation project following clean digital core principles, your main task is to identify in-app and side-by-side extension use cases, as listed in Table 2.2.

Use Cases/Requirements	Key-User Extensibility	Side-by-Side Extensibility
Custom code can be used by external user groups (not SAP users)?		Y
Custom functionality can be consumed by mobile devices?		Y
Scope is covered under in-app extensions?	Y	
APIs are available for consuming standard functionalities?	Y	Y
Custom code can be executed before or after standard business process?		Y
Custom code can be executed in between a standard business process?	Y	
Custom functionality can be enhanced by using innovative cloud services (e.g., machine learning, blockchain, IoT, etc.)?		Y

Table 2.2 Key-User versus Side-by-Side Extension Use Cases/Requirements

Since key-user extensions are not a focus of this book, let's explore some available side-by-side extension patterns, such as the following:

- **Convenience applications**
  For providing better user experiences by using existing functionalities. These applications may not provide many new features but instead reuse existing features to improve the user experience.

- **Proxy applications**
  Primarily Internet-facing applications (web or mobile) where your primary users are not SAP users. Examples include online stores, publicly available websites, etc.

- **Substitute applications**
  Replaces specific process steps with the new custom applications to address customer-specific requirements.

- **Preprocessing applications**
  For validating and manipulating data before triggering the main business application.

- **Postprocessing applications**
  Event-driven applications or workflow applications. For example, saving a sales order should trigger certain actions or a certain workflow.

- **Analytical applications**
  For connecting to various data sources to consume historical data as well to represent this data in an analytical way.

- **Standalone applications**
  An application that generates its own data and saves this data in SAP BTP itself. This application can connect to SAP S/4HANA for certain validations, actions, or field values.

Now that you understand the many possibilities available when building side-by-side applications, let's conclude this section by exploring three real business requirements that we implemented as side-by-side extensions:

- **Identifying inactive vendors (analytical application)**
  A business process requires the identification of inactive vendors in SAP if suppliers have no transactions in the past 18 months—no new or open purchase orders (PO), no open invoices, no credit memos, etc. Business users can review this report to decide whether to block/delete a vendor and whether a vendor should be considered inactive permanently.

- **Employee move request (preprocessing or standalone application)**
  This development is intended to identify the requirement of creating maintenance orders using an app when an employee moves from one location to another. Business users can execute this relocation via the app.

  The user's current location details included in the request will be retrieved from SAP SuccessFactors. This data is stored in specific fields while maintenance orders are created in SAP S/4HANA for further processing.

- **Physical inventory approval (postprocessing application)**
  Physical inventory is a business process in which physical stock is matched with book (system) stock. Often, carrying out physical inventory at least once in a year is a legal requirement. Physical inventory can be carried out both for a company's own stock (unrestricted, quality, or blocked stock) and for special stocks (customer consignment stock, vendor consignment stock, returnable packaging, etc.). This inventory is carried out separately for both type of stocks.

  Currently, no standard SAP functionality triggers an approval process when posting inventory differences. The process can be initiated in SAP S/4HANA, but the difference posting cannot take place until proper approvals are secured. In this scenario, a side-by-side extension can take care of the entire approval workflow and, upon completion, can call an API to post the differences in SAP S/4HANA.

2.1.5   SAP Cloud Application Programming Model versus the ABAP RESTful Application Programming Model

As mentioned earlier, for building side-by-side extensions, you can choose one of two approaches:

- Cloud-native extensions using SAP Cloud Application Programming Model
- Extensions on SAP BTP, ABAP environment using the ABAP RESTful application programming model

Both approaches have their pros and cons, and no approach is "one size fits all."

One major advantage of using the ABAP RESTful application programming model, from an investment standpoint, is the ABAP language. Additional training requirements for development teams on the customer side may be minimal if your team has been working with ABAP for years, perhaps even decades. In contrast, SAP Cloud Application Programming Model involves a steep learning curve. Also, as mentioned earlier, the ABAP RESTful application programming model is not only applicable for SAP BTP, ABAP environment—you can also use it for SAP S/4HANA development. Another additional value is the ability to consume on-premise remote function call (RFC) function modules easily by providing the option to generate a service consumption model for RFC function modules.

Since SAP BTP, ABAP environment evolves with each new quarterly release, new features and capabilities are introduced to bridge gaps, which are becoming more rare.

SAP Cloud Application Programming Model is the logical choice for development teams that are experienced with development in Java or JavaScript. Another advantage is the ability to use open-source languages and runtimes, including the capability to leverage libraries from third-party vendors and the open-source community in general.

Many more pros and cons could be discussed, which would be beyond the scope of this book. However, keep in mind that no unique solution can solve all problems and should be evaluated on a case-by-case basis.

Thus concludes our introduction to ABAP RESTful application programming model. Throughout the remainder of this chapter, we'll dive deeply into several development scenarios in SAP BTP, ABAP environment.

2.2   Working with the Development Environment

In this section, we'll start working with our development environment, picking up from our discussion in Chapter 1. Then, in the following sections, we'll proceed with developing your first SAP Fiori application. Following along with application development, we'll cover managed and unmanaged scenarios, including managed transactional apps with draft capability support.

This section explains in detail the prerequisites and the development environment for application development in SAP BTP, ABAP environment.

2.2.1   Setting Up an ABAP Cloud Project

In the first chapter, we described in detail how to set up a development environment. After completing those steps, open the **Project Explorer** view and create a new **ABAP Cloud Project** (SAP BTP, ABAP environment project), as shown in Figure 2.10.

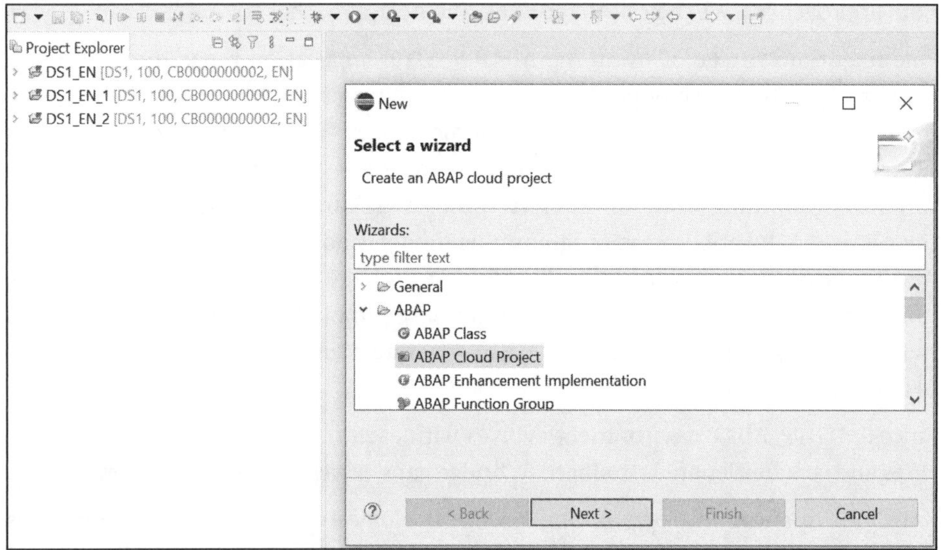

Figure 2.10 New ABAP Cloud Project

Create a custom package and capture these all objects in one transport request, as shown in Figure 2.11.

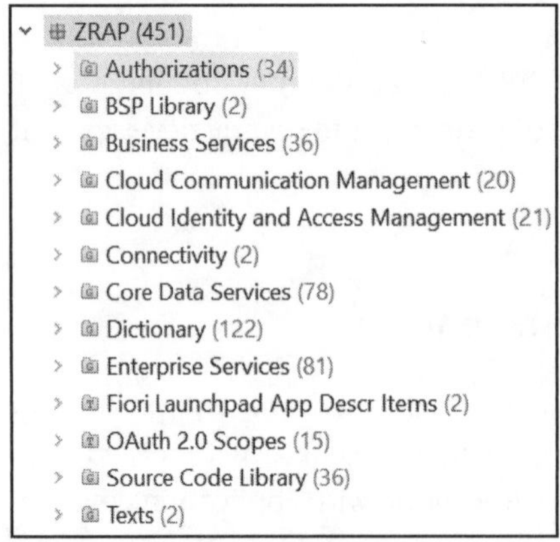

Figure 2.11 Custom Package

2.2.2   Setting Up SAP Business Application Studio

Consumer applications can be built through SAP Business Application Studio, the successor to SAP Web IDE. SAP Business Application Studio was released by SAP in February 2020 and is used to build applications like SAP Fiori apps, applications for SAP BTP mobile services, SAP S/4HANA extensions, and modeling workflows. For each of these application types, SAP Business Application Studio offers different development spaces that act like isolated virtual machines (VMs) in the cloud. SAP Business Application Studio can serve as an end-to-end development environment that leverages open-source and leading industry standards, enhancing productivity with direct command execution integrated with the command-line interface (CLI). Applications can be executed locally or can be deployed in the cloud with zero installation and maintenance. Thus, SAP Business Application Studio reduces the total cost of development and accelerates application development.

SAP Business Application Studio is available on the Cloud Foundry environment and can be used with Microsoft Azure and Amazon Web Services (AWS). In contrast, SAP Web IDE is available only on the SAP BTP Neo environment.

The "old" SAP Web IDE and SAP Business Application Studio differ in two broad ways:

- SAP Web IDE offers a generic workspace with tools while SAP Business Application Studio dev spaces act like isolated VMs in the cloud. Each dev space type contains tailored tools and preinstalled runtimes that simplify setup activities and save time.

- SAP Web IDE provides a browser-based experience, and the CLI is not integrated. In contrast, with SAP Business Application Studio, the development experience is improved with leading IDEs. SAP Business Application Studio provides superior debugging capabilities and productivity tools.

The licensing model for SAP Business Application Studio pay-per user. This solution is also available in an SAP BTP trial account.

In an SAP BTP, trial account, SAP Business Application Studio is immediately available as soon as the account is provisioned. The solution can be executed directly from the initial page of trial account, as shown in Figure 2.12.

Since SAP Business Application Studio is a multitenant application, a subscription must be activated. You can activate a subscription in a specific subaccount by navigating to **Subscriptions**, in the **Extension Suite – Development Efficiency** section.

After the subscription is activated, you'll need set up some roles. With SAP Business Application Studio, three roles are delivered:

- `Business_Application_Studio_Administrator` for managing user data

- `Business_Application_Studio_Developer` for loading and developing applications

- `Business_Application_Studio_Extensions_Deployer` for deploying simple extensions

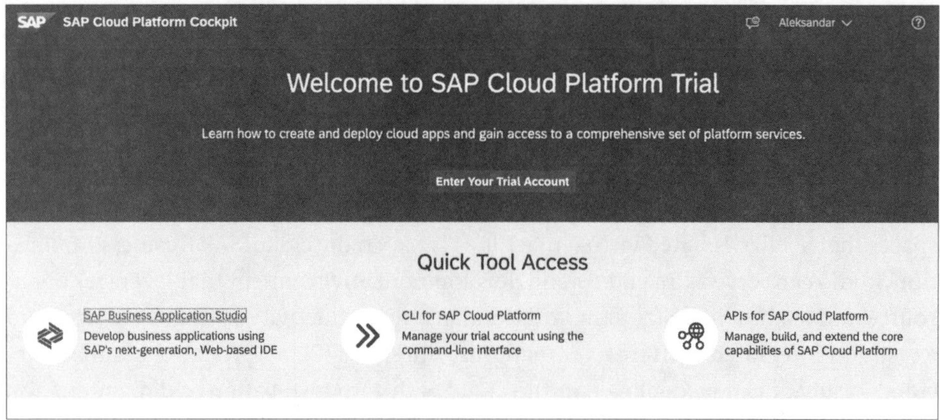

Figure 2.12 SAP Business Application Studio Set Up in an SAP BTP Trial Account

Before assigning roles to users, you must create the corresponding role collections. A *role collection* can contain multiple roles, but in the case of SAP Business Application Studio, the best approach is to keep separate role collections with corresponding roles. Role collection maintenance is performed on the subaccount level, under the **Security** menu item. The final result should look like the screen shown in Figure 2.13.

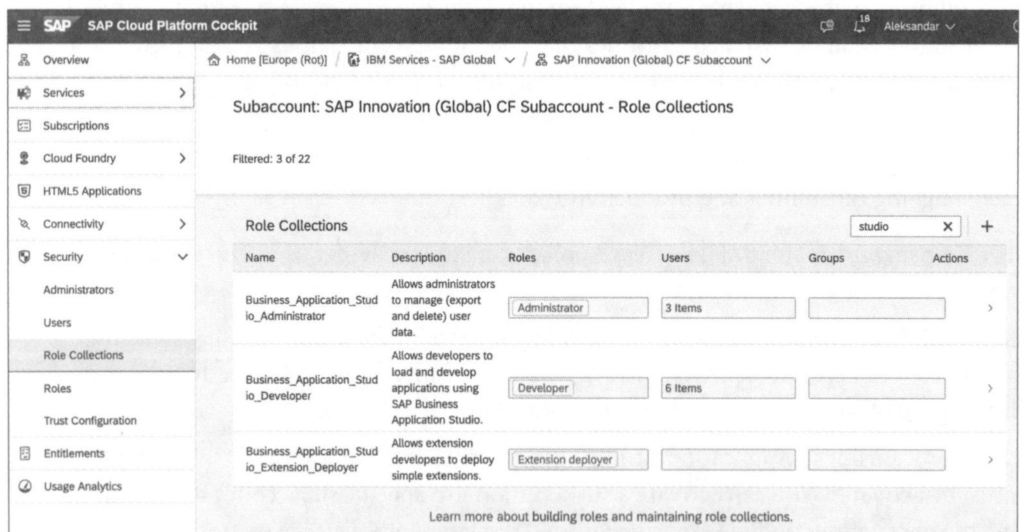

Figure 2.13 Role Collections

After role collections are defined, you'll assign them to users under the **Trust Configuration** menu item.

Now, we're ready to start configuring SAP Business Application Studio to make it ready for development. The following procedure is the same for both trial and productive environments.

To access SAP Business Application Studio in a trial account, as mentioned earlier, you can navigate directly from the home page. Another option is to navigate to the **Subscriptions** section under the subaccount, and then, under SAP Business Application Studio, click on **Go to Application**, which works for both trial and productive accounts.

When starting SAP Business Application Studio for the first time, no dev spaces will be assigned, as shown in Figure 2.14.

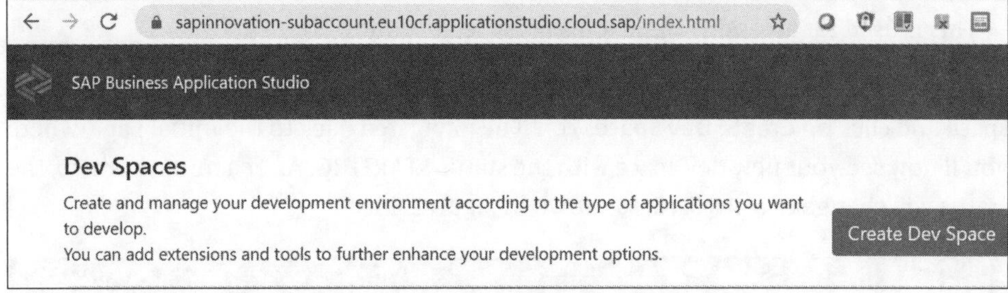

Figure 2.14 SAP Business Application Studio: Initial Page

Click the **Create Dev Space** button, and you'll be navigated to a page offering a variety of options, as shown in Figure 2.15. On this page, you'll define the kind of application you want to develop and specify additional extensions you want to include.

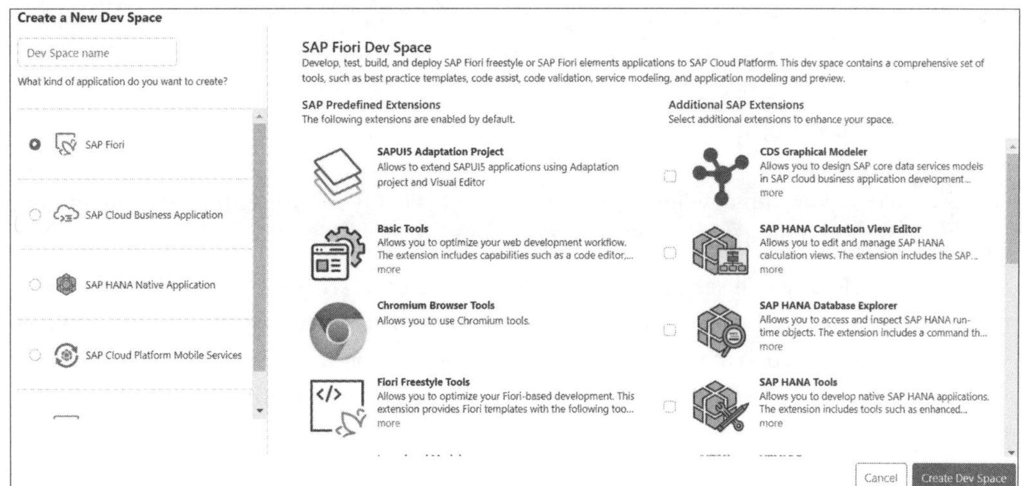

Figure 2.15 Creating a New Dev Space: Initial Page

At this point, you can select one of following options:

- **SAP Fiori**
  This is used for development of freestyle SAP Fiori applications or SAP Fiori elements applications.

- **SAP Cloud Business Application**
  This is used for building business services and business applications that extend SAP S/4HANA, using SAP Cloud Application Programming Model.

- **SAP Cloud Platform Mobile Services**
  Now SAP Mobile Services, this is used for developing iOS and Android apps in the cloud.

- **Basic**
  This is used for accessing basic tools provided by SAP.

Now, you'll select additional extensions as needed. Then, enter a name for your dev space and click on **Create Dev Space**. You'll be navigated back to the initial page, where you'll now see your new dev space with the status **STARTING**. After a minute or two, the status will change to **RUNNING**, as shown in Figure 2.16.

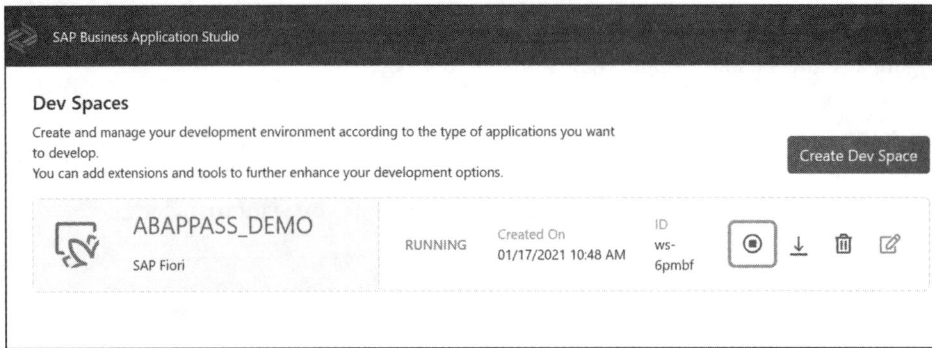

Figure 2.16 Dev Space

With that step, our development environment is ready for use. Enter the development environment by simply clicking on the dev space name.

On the right, you'll see tools that allow you to stop/start the dev space, download its content, delete the dev space, or make changes. (The change option is only available when the dev space has been stopped.) Also, notice that the dev space stops itself after a certain amount of inactivity. If this occurs, simply restart the dev space and start using it again.

2.3   SAP Fiori Application Development in SAP Business Application Studio

With the ABAP RESTful application programming model, all types of SAP Fiori applications can be developed. In this section, we'll describe in detail how to build applications.

The application we will build in this example is solution for managing student enrollment for training courses. As you'll see in the following sections, for any application, several objects must be created and several tasks be completed in the ADT, including the following:

1. Create database tables.

2. Create custom classes to insert data into tables.

3. Create a CDS view on tables.

4. Define an OData service through a service definition.

5. Bind the service.

6. Create the UI.

2.3.1   Creating Persistent Database Tables

Let's start by creating the table. Our table name will be ZSTUDENT_TBL. We'll also define some basic annotations while creating the table, such as table category, delivery class, enhancement category, and data maintenance, which will be used by our application.

You can refer field directly to the available data types, or you can create custom data elements and custom domains for use when defining a database table.

Listing 2.1 shows how you can create a standard application database table using generic data types. We can also use standard data elements as well. However, only a few standard data elements exist for cloud-based development, unlike in an on-premise ABAP environment. You can also develop your own data types, called *custom data types*, or import your own data type definitions from your existing developments.

```
@EndUserText.label : 'Student Table'
@AbapCatalog.enhancementCategory : #NOT_EXTENSIBLE
@AbapCatalog.tableCategory : #TRANSPARENT
@AbapCatalog.deliveryClass : #A
@AbapCatalog.dataMaintenance : #RESTRICTED
define table zstudent_tbl {
key client : abap.clnt;
key studentid : abap.int4 not null;
fname : abap.char(40);
lname : abap.char(40);
mname : abap.char(40);
gender : abap.char(1);
birthdate : abap.dats;
commnumber : abap.int4; }
```

Listing 2.1 Defining a Database Table

> **Note**
>
> We described how to navigate to released objects in ADT for Eclipse in Chapter 1. You can refer back to that chapter for instructions on how to do so.

2.3.2   Creating a Custom Class to Insert Values into the Table

Now, we'll define a custom class and interfaces in the public section. Then, in the class implementation, we'll define methods and add code in the method to insert the values into the table. This class contains reference to the interface IF_OO_ADT_CLASSRUN and to the implementation of method IF_OO_ADT_CLASSRUN~MAIN. With this approach, our class is executable and will be runnable, which means you can execute it as a console application and write its output to the console as well.

Listing 2.2 shows how you can create the class for populating the custom database table with data. This class will then output the data into the console to confirm successful execution.

```abap
class ZCL_STUDENT_TBL definition
public
final
create public.
public section.
INTERFACES if_oo_adt_classrun.
protected section.
private section.
endclass.
class ZCL_STUDENT_TBL IMPLEMENTATION.
method if_oo_adt_classrun~main.
data: lt_student type standard table of ZSTUDENT_TBL.
lt_student = value #(
( studentid = '1000101' fname = 'RAKESH' lname = 'KUMAR' gender = 'M'
birthdate = '19900414' commnumber = '101' )
( studentid = '1000102' fname = 'SUNITA' lname = 'RAVAL' gender = 'F'
birthdate = '19900612' commnumber = '102' )
( studentid = '1000103' fname = 'BHAVAN' lname = 'SHARMA' gender = 'M'
birthdate = '19900911' commnumber = '103')
( studentid = '1000104' fname = 'MOHAMMAD' lname = 'KHAN' gender = 'M'
birthdate = '19890612' commnumber = '104')
( studentid = '1000105' fname = 'JENNIFER' lname = 'DESOZA' gender = 'F'
birthdate = '19900403' commnumber = '105')
( studentid = '1000106' fname = 'PRIYANK' lname = 'CHATURVEDI' gender
= 'M' birthdate = '19911206' commnumber = '106')
( studentid = '1000107' fname = 'RISHABH' lname = 'BHAT' gender = 'M'
```

```
birthdate = '19911125' commnumber = '107')
(studentid = '1000108' fname = 'MUKESH' lname = 'SAHANI' gender = 'M'
birthdate = '19891010' commnumber = '108')
( studentid = '1000109' fname = 'KETKI' lname = 'BHAT' gender = 'F'
birthdate = '19900501' commnumber = '109')
( studentid = '1000110' fname = 'KANHA' lname = 'PANDE' gender = 'M'
birthdate = '19910923' commnumber = '110') ).
delete from ZSTUDENT_TBL.
insert ZSTUDENT_TBL FROM TABLE @lt_student.
clear lt_student. "clear internal table
select *
from ZSTUDENT_TBL
into table @lt_student .
if sy-subrc is initial. "check if internal table is not initial
out->write( 'data inserted successfully!').
out->write( lt_student ).
endif.
endmethod.
ENDCLASS.
```

Listing 2.2 Class to Insert Entries into a Custom Database Table

> **Note**
>
> According to the new coding standards, values can be assigned directly to a field; you don't need to define work areas and append values to internal tables.

2.3.3   Creating CDS Views on the Table

The data model for an OData service must be defined in CDS. In the CDS layer, you can use and manipulate data that is persisted in the database. To make data available in the ABAP application server, CDS views use SQL queries to project persisted data to the ABAP layer. This approach is necessary when creating an OData service so that the data ready to be consumed.

Start with opening the context menu and choose **New • Other • ABAP Repository Object • Core Data Services • Data Definition**. The data definition creation wizard, shown in Figure 2.17, will be launched.

As in the table definition shown in Listing 2.1, notice how annotations play an important role while defining CDS views. You'll need to understand the required annotations, which depend on the type of application and accordingly must used while defining the CDS view.

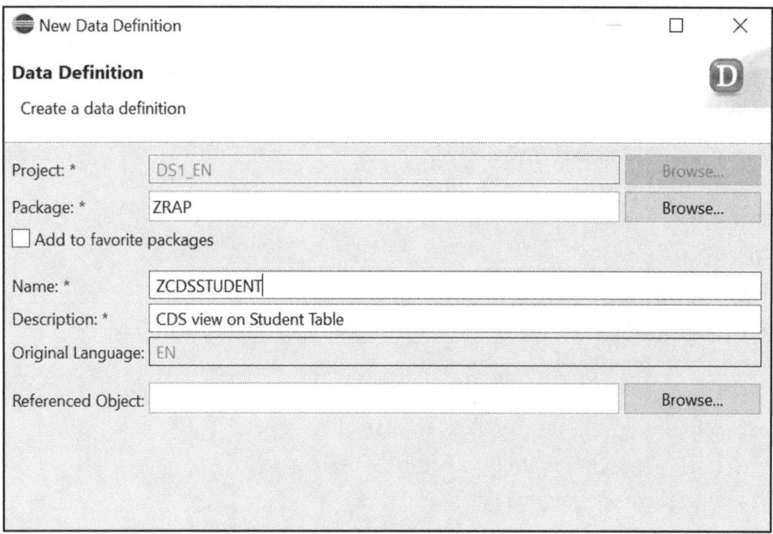

Figure 2.17 Data Definition

> **Note**
>
> More information about CDS annotations can be found in the ABAP keyword documentation at *http://s-prs.co/v523602*. Keep this documentation always handy since you'll use it frequently.

Listing 2.3 shows how to define the student view for our example.

```
@AbapCatalog.sqlViewName: 'ZCDS_STUDENT_TBL'
@AbapCatalog.compiler.compareFilter: true
@AbapCatalog.preserveKey: true
@AccessControl.authorizationCheck: #CHECK
@EndUserText.label: 'CDS view on Student Table'
define view ZCDSSTUDENT as select from ZSTUDENT_TBL {
@UI.hidden: true
key studentid,
@Search.defaultSearchElement: true @UI:{lineItem: [{position: 10, label:
'Fname'}],
identification: [ {position: 10, label: 'fname'}],
selectionField: [ {position: 10}]}
fname,
@Search.defaultSearchElement: true
@UI: {lineItem: [ {position: 20, label: 'Lname'}],
identification: [ {position: 20, label: 'Lname'}],
selectionField: [ {position: 20}]}
lname,
```

```
@Search.defaultSearchElement: true
@UI: {lineItem: [ {position: 30, label: 'Gender'}],
identification: [ {position: 30, label: 'Gender'}],
selectionField: [ {position: 30}]}
gender,
@Search.defaultSearchElement: true
@UI: {lineItem: [ {position: 40, label: 'DOB'}],
identification: [ {position: 40, label: 'DOB'}],
selectionField: [ {position: 40}]}
birthdate,
@Search.defaultSearchElement: true
@UI: {lineItem: [ {position: 50, label: 'commnumber'}],
identification: [ {position: 50, label: 'Lname'}],
selectionField: [ {position: 50}]}
commnumber
}
```

Listing 2.3 Student CDS View

As shown in Listing 2.3, the following annotations enable specific features:

- `@Search` enables the search feature on CDS view elements.

- `@UI` enables the semantic view on data, using specific patterns.

On completion of the view definition, save and activate it. Execute the view by pressing the F8 function key to check the data in the preview tool, as shown in Figure 2.18.

StudentId	FName	LName	MName	Sex	DOB	commnumber
1,000,101	RAKESH	KUMAR		M	1990-04-14	101
1,000,102	SUNITA	RAVAL		F	1990-06-12	102
1,000,103	BHAVAN	SHARMA		M	1990-09-11	103
1,000,104	MOHAM...	KHAN		M	1989-06-12	104
1,000,105	JENNIFER	DESOZA		F	1990-04-03	105
1,000,106	PRIYANK	CHATURV...		M	1991-12-06	106
1,000,107	RISHABH	BHAT		M	1991-11-25	107
1,000,108	MUKESH	SAHANI		M	1989-10-10	108
1,000,109	KETKI	BHAT		F	1990-05-01	109
1,000,110	KANHA	PANDE		M	1991-09-23	110

Figure 2.18 Student Data

To preview the contents of the CDS view, right-click on the view definition, then select the **Data Preview** option, as shown in Figure 2.19.

Now that you've created and tested the CDS view, let's move to the next step—defining and exposing the OData service.

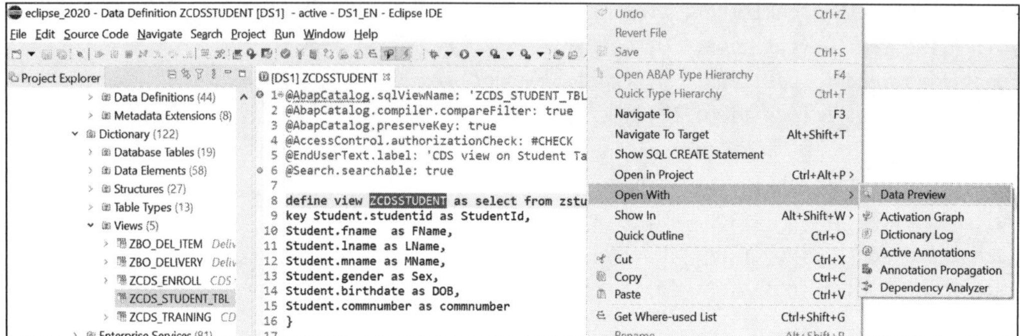

Figure 2.19 CDS View: Data Preview

2.3.4   Defining and Exposing an OData Service

In this section, we'll expose our CDS view as an OData service, which will enable UIs to query the data and consume it.

To create a service definition and expose the CDS view, open the context menu and select **New · Other · ABAP Repository Object · Business Services · Service Definition**, which will launch the service definition creation wizard shown in Figure 2.20.

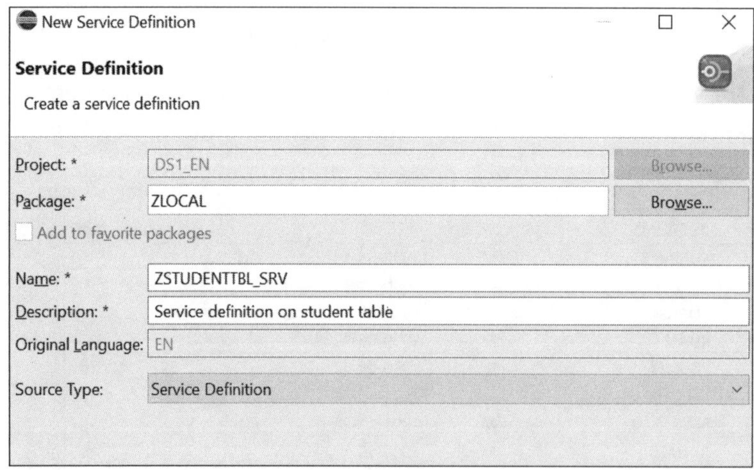

Figure 2.20 Service Definition for ZSTUDENTTBL_SRV

Maintain the required fields, and as a result, you'll get generated source code similar to the code shown in Listing 2.4.

```
@EndUserText.label: 'Service definition on student table'
define service ZSTUDENTTBL_SRV {
expose ZCDSSTUDENT as STUDENT;
}
```

Listing 2.4 Service Definition

With this step, you've created and exposed the OData service in order to publish it. The next step is creating the service binding.

2.3.5   Creating a Service Binding

A service binding implements the protocol that is used for the OData service. A service binding uses a service definition that projects the data models and their related behaviors to the service. To launch the service binding creation wizard, open the context menu and select **New** · **Other** · **ABAP Repository Object** · **Business Services** · **Service Definition**.

This step will create the service binding on the defined service (for student data, in this example). On creating the service binding, you'll see the screen shown in Figure 2.21.

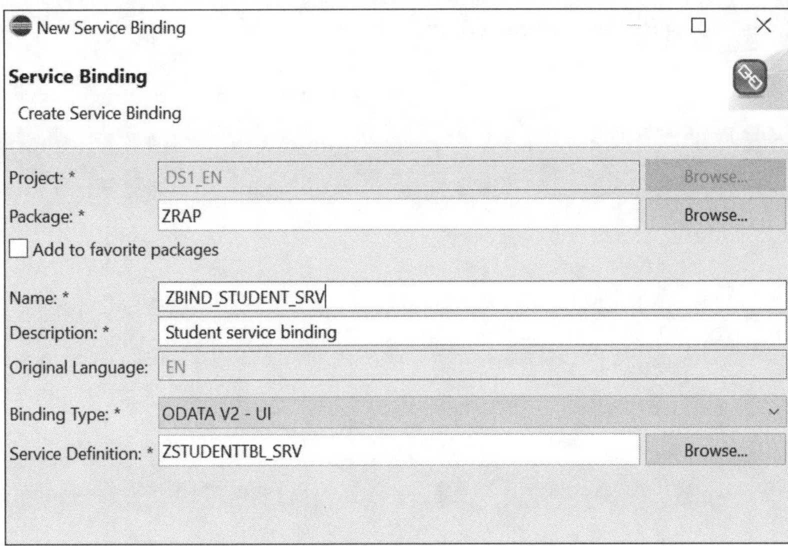

Figure 2.21 Service Binding for ZBIND_STUDENT_SRV

Now, activate the service. As part of activation, a service URL will be generated. For our example service, the relative path is:

/sap/opu/odata/sap/ZBIND_STUDENT_SRV

The next step is to bind the OData service against a protocol and publish it locally. The contents of the service can be previewed in any browser. To preview the data, click on the **Service URL** link, available on **Service Binding** screen, as shown in Figure 2.22.

This action will allow you to preview the contents, as shown in Figure 2.23.

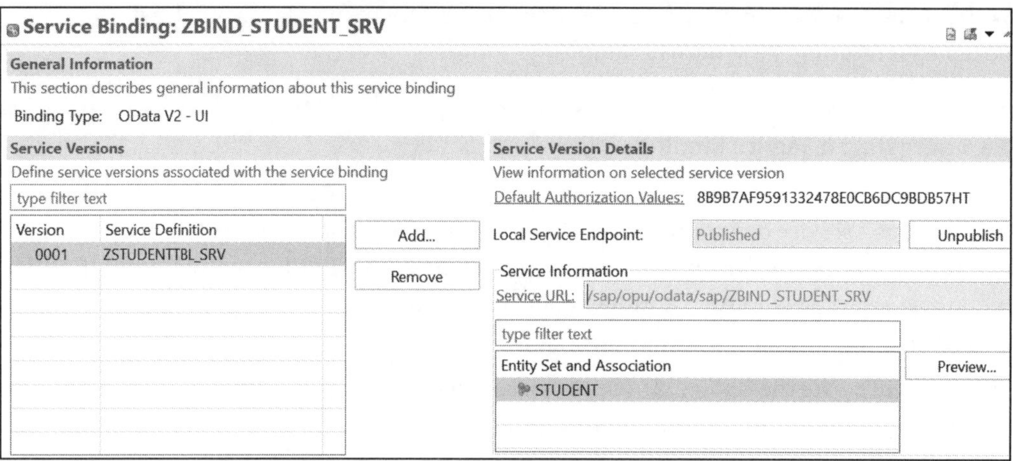

Figure 2.22 Service Binding for the Student Service

Student Id	FNAME	LNAME	MNAME	DOB	GENDER
1	RAKESH	KUMAR	L	Apr 14, 1990	M
10	KANHA	PANDE	A	Sep 23, 1991	M
2	SUNITA	RAVAL	P	Jun 12, 1990	F
20	Mark	Brinka	A	Jun 19, 1992	M
3	BHAVAN	SHARMA	O	Sep 11, 1990	M
4	MOHAMMAD	KHAN	M	Jun 12, 1989	M
5	JENNIFER	DESOZA	S	Apr 3, 1990	F
6	PRIYANK	CHATURVEDI	Q	Dec 6, 1991	M
7	RISHABH	BHAT	M	Nov 25, 1991	M

Figure 2.23 Student Data: Preview Elements App

Note

The data is presented in the Preview Elements app, which displays the selected entity set or association. (This preview app is limited in functionality and scope.)

In our example service binding, we used the UI binding type. Another option for service bindings is a web API. While the UI binding type, as the name suggests, exposes an OData service for UIs, the web API option exposes a service to be consumed by a third-party consumer, which might be an SAP or non-SAP application. Even though the web API will not understand annotations, data can still be consumed through RESTful APIs that provide data in XML or JavaScript Object Notation (JSON) format.

Now, let's expand our data model with additional CDS views to create the complete application.

Start by creating table ZTRAINING_TBL using the code shown in Listing 2.5. As with creating a table in the SAP GUI, in this case, you must also specify a table category, a delivery class, and the table's extensibility, all through annotations. Then, you'll define all the required fields of the table, including their data types and lengths.

```
@EndUserText.label : 'Training Table'
@AbapCatalog.enhancementCategory : #NOT_EXTENSIBLE
@AbapCatalog.tableCategory : #TRANSPARENT
@AbapCatalog.deliveryClass : #A
@AbapCatalog.dataMaintenance : #RESTRICTED
define table ztraining_tbl {
key client : abap.clnt not null;
key trainingid : abap.char(10) not null;
title : abap.char(50);
courseid : abap.char(10);
startdate : abap.dats;
enddate : abap.dats;
iscomplete : abap.char(1);
capacity : abap.int1;
}
```

Listing 2.5 Table ZTRAINING_TBL

Next, we'll create a CDS view on table ZTRAINING_TBL using the code shown in Listing 2.6 by selecting fields from ZTRAINING_TBL. In our example, we've selected all the fields; if required, you can select specific fields.

```
@AbapCatalog.sqlViewName: 'ZCDS_TRAINING'
@AbapCatalog.compiler.compareFilter: true
@AbapCatalog.preserveKey: true
@AccessControl.authorizationCheck: #CHECK
@EndUserText.label: 'CDS view on training table'
define view ZCDSTRAINING as select from ztraining_tbl as TRAINING
{
key TRAINING.trainingid as TRAININGID,
TRAINING.title as TITLE,
```

```
TRAINING.courseid as COURSEID,
TRAINING.startdate as STDATE,
TRAINING.enddate as ENDDATE,
TRAINING.iscomplete as STATUS,
TRAINING.capacity as CAPACITY }
```

Listing 2.6 CDS View for ZTRAINING_TBL

Now that we've created our table and CDS view, we need to populate the table with data. For this step, we'll again use a custom class, shown in Listing 2.7. By defining this custom class, we can assign values to each field of the table and can create as many records as we need to be fetched in the table. This class has a definition in the public section to define the interface if_oo_adt_classrun to execute the class and display the output. While in the implementation section, we'll actually write the code in the method.

```
Create class to enter records in the TRAINING table.
class ZCL_TRAINING definition
public
final
create public .
public section.
INTERFACES if_oo_adt_classrun.
protected section.
private section.
ENDCLASS.
CLASS ZCL_TRAINING IMPLEMENTATION.
METHOD if_oo_adt_classrun~main.
DATA it_training TYPE standard table of ZTRAINING_TBL.
it_training = value #(
( TRAININGID = 'TRNG1' TITLE = 'S/4 HANA INTRODUCTION'
COURSEID = 'SAPPI1' STARTDATE = '20160509' ENDDATE = '20160513'
ISCOMPLETE = '' CAPACITY = '10' )
( TRAININGID = 'TRNG2' TITLE = 'ABAP ADVANCED CONCEPTS'
COURSEID = 'SAPAD1' STARTDATE = '20160510' ENDDATE =
'20160516'
ISCOMPLETE = '' CAPACITY = '15' )
( TRAININGID = 'TRNG3' TITLE = 'RESTful Progarmming' COURSEID =
'SAPABAP11' STARTDATE = '20160511' ENDDATE = '20160517'
ISCOMPLETE = '' CAPACITY = '20' )
( TRAININGID = 'TRNG4' TITLE = 'SAP ANALYTICS CLOUD'
COURSEID = 'SAPABAP12' STARTDATE = '20160512' ENDDATE =
'20160518'
ISCOMPLETE = '' CAPACITY = '10' ) ).
DELETE FROM ZTRAINING_TBL.
INSERT ZTRAINING_TBL FROM TABLE @it_training.
```

```
* clear the internal tab
clear it_training.
* check the result
SELECT * FROM ztraining_TBL INTO TABLE @it_training.
if sy-subrc is initial.
out->write( sy-dbcnt ).
out->write( 'data inserted successfully!').
out->write( it_training ).
endif.
endmethod.
ENDCLASS.
```

Listing 2.7 Custom Class to Populate Table ZTRAINING_TBL with Data

Using same approach as for the student table, create and expose the OData service using the code shown in Listing 2.8.

```
@EndUserText.label: 'Service definition on training table'
define service Z_TRAINING_SRV
{
expose ZCDSTRAINING as training;
}
```

Listing 2.8 OData Service for ZTRAINING_TBL

Now, bind the service and activate it by clicking on the **Service URL** link, as shown in Figure 2.24.

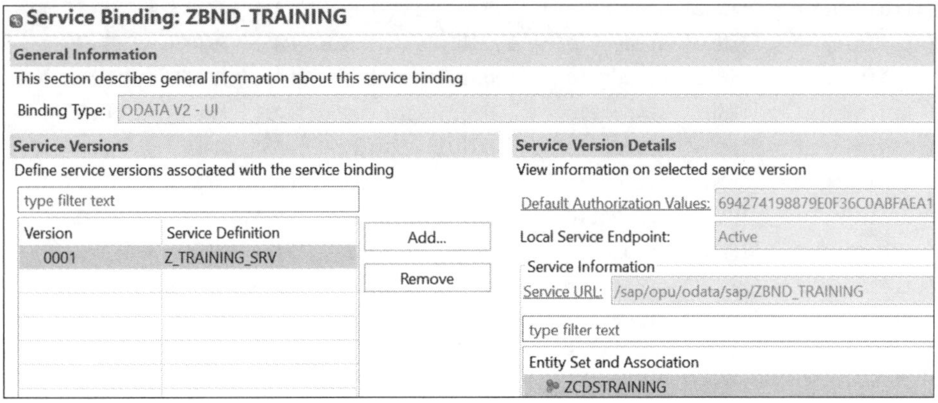

Figure 2.24 Training Service Binding

We'll repeat the same process for table ZENROLL. First, create the table using the code shown in Listing 2.9. The only differences at this stage are the fields—the process of table creation itself is the same.

```
@EndUserText.label : 'Student Enroll'
@AbapCatalog.enhancementCategory : #NOT_EXTENSIBLE
@AbapCatalog.tableCategory : #TRANSPARENT
@AbapCatalog.deliveryClass : #A
@AbapCatalog.dataMaintenance : #RESTRICTED
define table zenroll {
key client : abap.clnt not null;
key trainingid : abap.char(10) not null;
key studentid : abap.char(10) not null;
status : abap.char(4);
marks : abap.int1;
note : abap.char(255);
}
```

Listing 2.9 Creating Table ZENROLL

Next, populate the table with data using a custom class. In Listing 2.10, we've provided only the part of the code with the data. Refer to Listing 2.7 for the complete code and explanations.

```
...
data : lt_enroll type standard table of zenroll.
lt_enroll = value # (
( trainingid = 'TRNG1' studentid = '1000101' status = 'PASS' MARKS = '88.8' )
( trainingid = 'TRNG1' studentid = '1000105' status = 'PASS' MARKS = '92.0' )
( trainingid = 'TRNG1' studentid = '1000103' status = 'PASS' MARKS = '90.0' )
( trainingid = 'TRNG1' studentid = '1000105' status = 'FAIL' MARKS = '68.0' )
( trainingid = 'TRNG2' studentid = '1000102' status = 'PASS' MARKS = '88.8' )
( trainingid = 'TRNG2' studentid = '1000104' status = 'PASS' MARKS = '92.0' )
( trainingid = 'TRNG2' studentid = '1000103' status = 'PASS' MARKS = '90.0' )
( trainingid = 'TRNG2' studentid = '1000105' status = 'FAIL' MARKS = '60.0' )
( trainingid = 'TRNG3' studentid = '1000101' status = 'PASS' MARKS = '88.8' )
( trainingid = 'TRNG3' studentid = '1000102' status = 'PASS' MARKS = '92.0' )
( trainingid = 'TRNG3' studentid = '1000103' status = 'PASS' MARKS = '90.0' )
( trainingid = 'TRNG3' studentid = '1000104' status = 'FAIL' MARKS = '56.0' )
( trainingid = 'TRNG3' studentid = '1000105' status = 'PASS' MARKS = '88.8' )
( trainingid = 'TRNG3' studentid = '1000106' status = 'PASS' MARKS = '92.0' )
( trainingid = 'TRNG4' studentid = '1000109' status = 'PASS' MARKS = '90.0' )
( trainingid = 'TRNG4' studentid = '1000110' status = 'PASS' MARKS = '93.0' ) ).
...
```

Listing 2.10 Populating Table ZENROLL with Data

Now, we'll create a CDS view on table ENROLL, as shown in Listing 2.11. Again, you may refer back to Listing 2.7 for the complete code and explanations.

```
@AbapCatalog.sqlViewName: 'ZCDS_ENROLL'
@AbapCatalog.compiler.compareFilter: true
@AbapCatalog.preserveKey: true
@AccessControl.authorizationCheck: #CHECK
@EndUserText.label: 'CDS view on ENROLL Table'
define view ZCDSENROLL as select from zenroll {
key zenroll.trainingid as TRNGID,
key zenroll.studentid as STUDENTID,
zenroll.status as STATUS,
zenroll.marks as MARKS,
zenroll.note as NOTE
}
```

Listing 2.11 CDS View on Table ZENROLL

Next, let's create and expose the OData service, as shown in Listing 2.12.

```
@EndUserText.label: 'Service definition on Enroll table'
define service ZENROLLTBL_SRV {
expose ZCDSENROLL as ENROLL;
}
```

Listing 2.12 OData Service for ZENROLL

Finally, repeat the same procedure to bind the service and activate it.

With this step, your data model is complete. Now, let's proceed with creating the SAP Fiori application.

2.3.6   Developing the User Interface

Now that we've fulfilled all the prerequisites, developing the UI itself is the final remaining task. For this task, you'll use the SAP Business Application Studio. You can create a new dev space or use an existing one to create a new SAP Fiori elements application in SAP Business Application Studio.

In SAP Business Application Studio, select **File · New Project from Template**, which will open the screen where you can select a template. Select **SAP Fiori elements application**, as shown in Figure 2.25.

> **Note**
>
> Another way to initiate a new project is using a command provided in SAP Business Application Studio for Yeoman UI generators by following the menu path **View · Find Command** and then typing "Yeoman" in the command window.

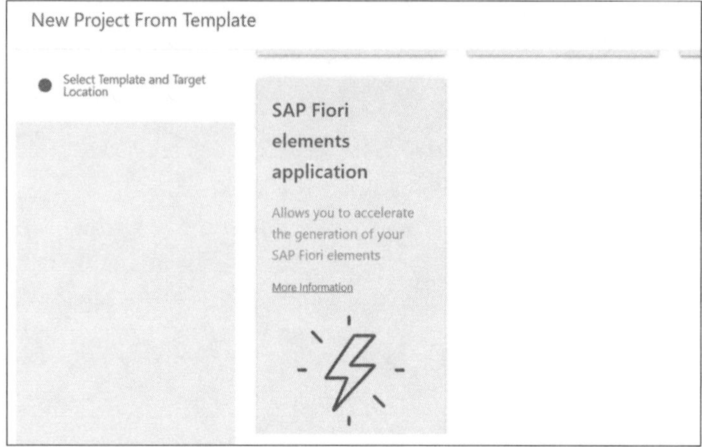

Figure 2.25 New SAP Fiori Application

Select **List Report Object Page** as your SAP Fiori application type, as shown in Figure 2.26.

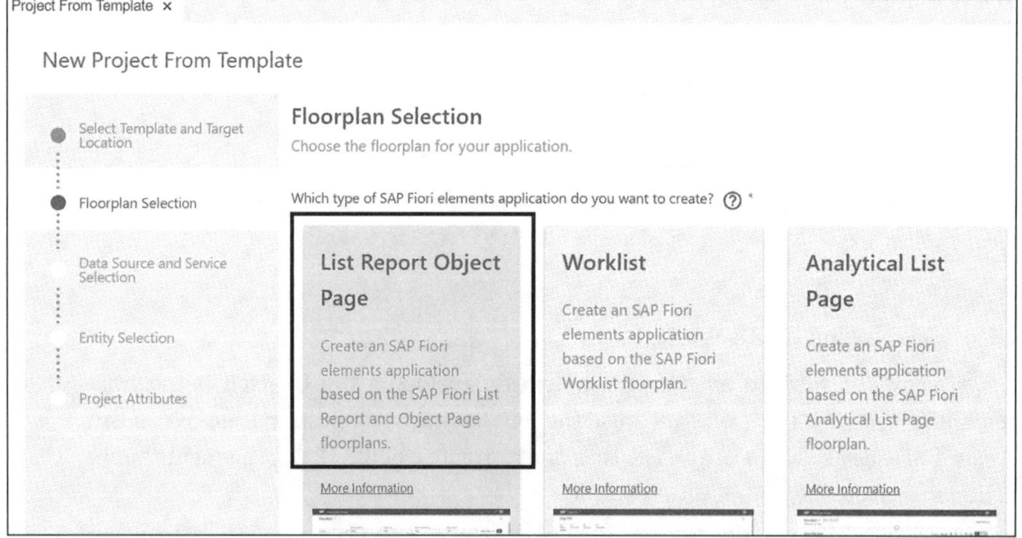

Figure 2.26 Template Selection

Now, you'll be prompted to select the data source. Let's connect to an SAP system, and as our source system, we'll use our SAP BTP, ABAP environment instance. In the **Select a service** dropdown list, shown in Figure 2.27, select the service binding we created in the previous section. To select the data source, choose the **Connect to an SAP System** option.

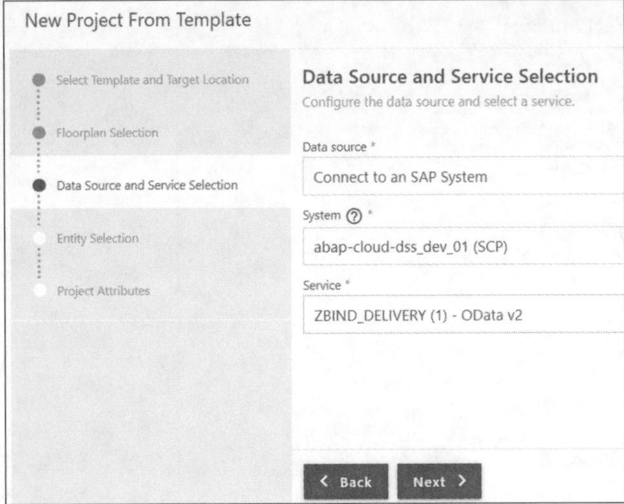

Figure 2.27 Data Source and Service Selection

Now, select the specific entity you want use from the selected data source, as shown in Figure 2.28, and click **Next**.

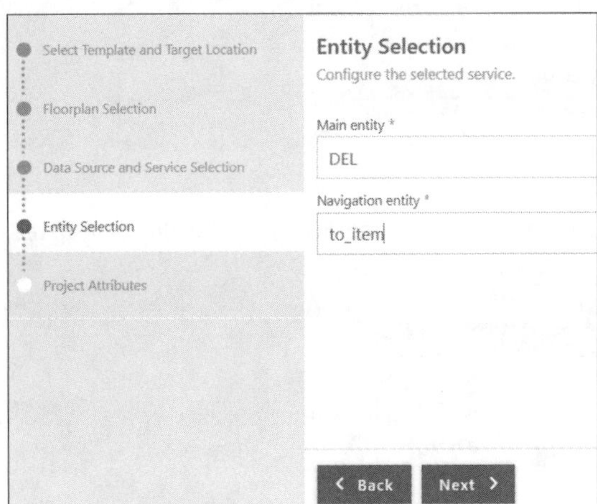

Figure 2.28 Entity Selection

Fill out the project's attributes, as shown in Figure 2.29. Provide the module a name, which should be in lowercase. Note that no special characters are allowed. The application's title can be any suitable text.

When you click on **Finish**, the build process starts. Once completed, you'll be asked if you want to add the project to the current workspace or to create a new workspace, as shown in Figure 2.30.

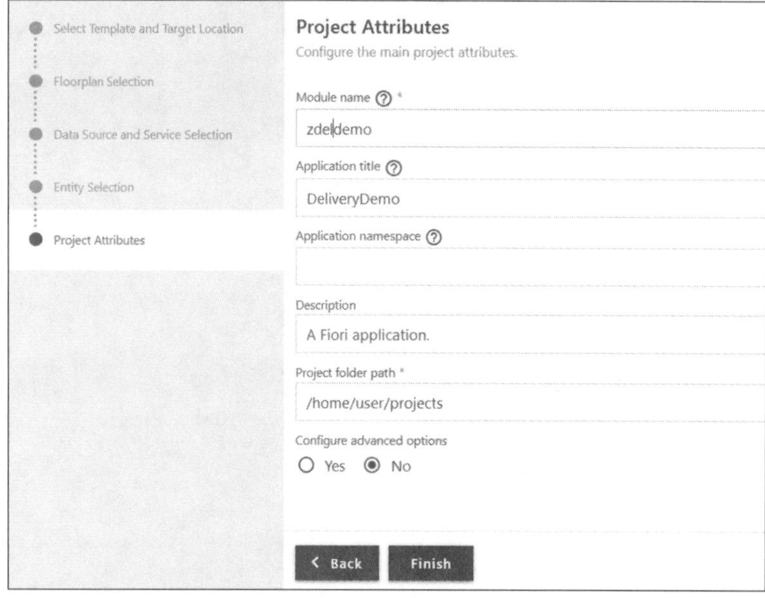

Figure 2.29 Project Attributes

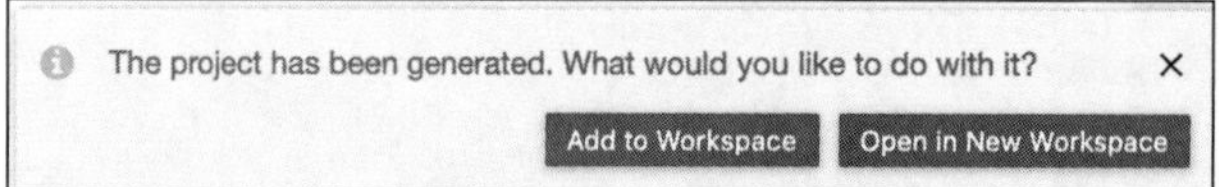

Figure 2.30 Workspace Selection

Now, our application has been generated in the workspace, and you should see it listed in the **Project Explorer** tree, as shown in Figure 2.31.

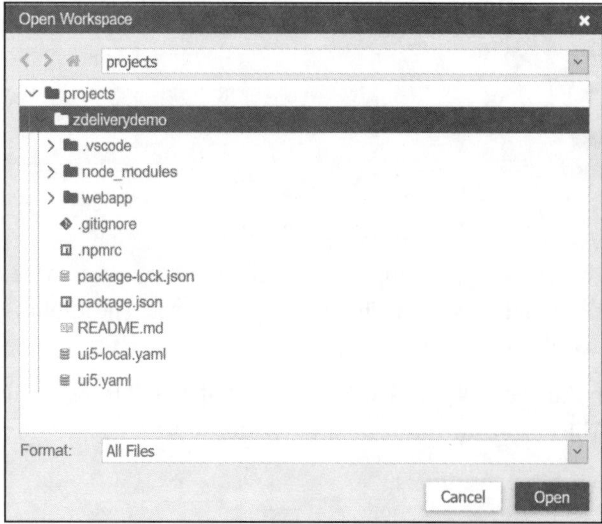

Figure 2.31 Application in Workplace

At this stage, you can preview the application by right-clicking on the project folder and selecting **Preview Application**.

Another option to preview the application is to run the SAP Fiori application directly from SAP Business Application Studio. Click on run button (green triangle) to the left, which will result in the display shown in Figure 2.32.

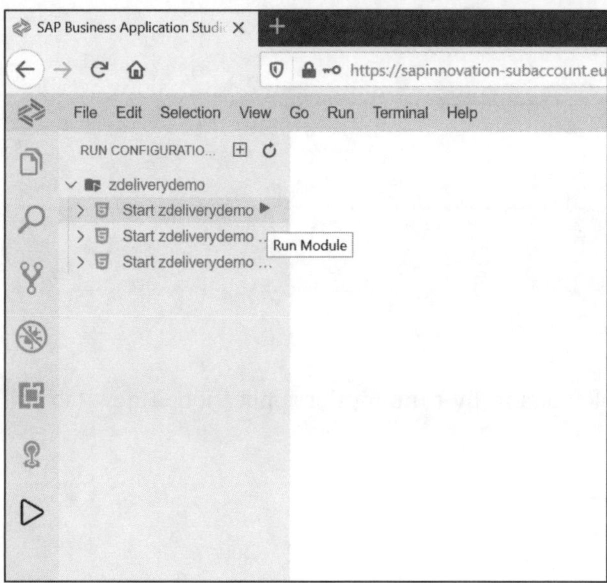

Figure 2.32 Starting an SAP Fiori Application

We'll run the application *zdeliverydemo* by clicking on the **Expose and Open** button, as shown in Figure 2.33.

Figure 2.33 Expose and Open

Then next screen, shown in Figure 2.34, will display the application's list of the files. Click on **index.html** to execute the application.

Name	Size	Modified	Project
annotations/	4.00 KB	1/17/2021, 5:23:20 AM	zdeliverydemo
i18n/	4.00 KB	1/17/2021, 5:23:20 AM	zdeliverydemo
localService/	4.00 KB	1/17/2021, 5:23:20 AM	zdeliverydemo
test/	4.00 KB	1/17/2021, 5:23:20 AM	zdeliverydemo
Component.js	222 Bytes	1/17/2021, 5:23:20 AM	zdeliverydemo
index.html	933 Bytes	1/17/2021, 5:23:20 AM	zdeliverydemo
manifest.json	5.96 KB	1/17/2021, 5:23:20 AM	zdeliverydemo

Figure 2.34 Test Application

Note

You can also right-click on an application in the explorer, select **Open in Terminal**, and enter the command npm start.

The application build is completed, and by running the application, a new tab will appear, as shown in Figure 2.35.

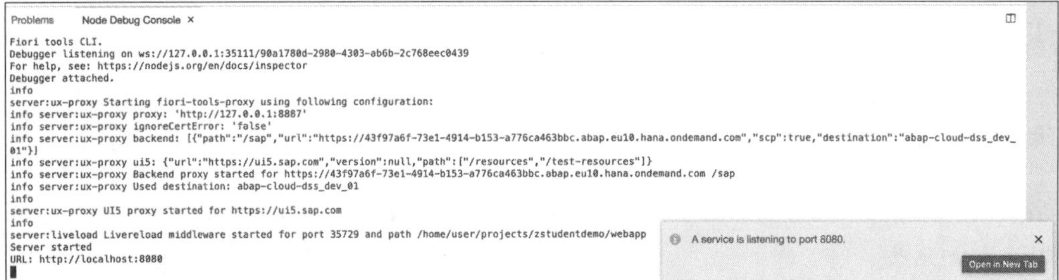

Figure 2.35 Application Build Completed

The screen shown in Figure 2.36 will be provided with the test app.

Name	Size	Modified	Project
annotations/	4.00 KB	12/6/2020, 7:20:27 PM	studentdatademo
i18n/	4.00 KB	12/6/2020, 7:20:27 PM	studentdatademo
localService/	4.00 KB	12/6/2020, 7:20:27 PM	studentdatademo
test/	4.00 KB	12/6/2020, 7:20:27 PM	studentdatademo
Component.js	224 Bytes	12/6/2020, 7:20:27 PM	studentdatademo
manifest.json	5.29 KB	12/6/2020, 7:20:27 PM	studentdatademo

Figure 2.36 Test Application

To run the configuration and test the application, go to the test folder and click on **flpSandbox.html**, as shown in Figure 2.37.

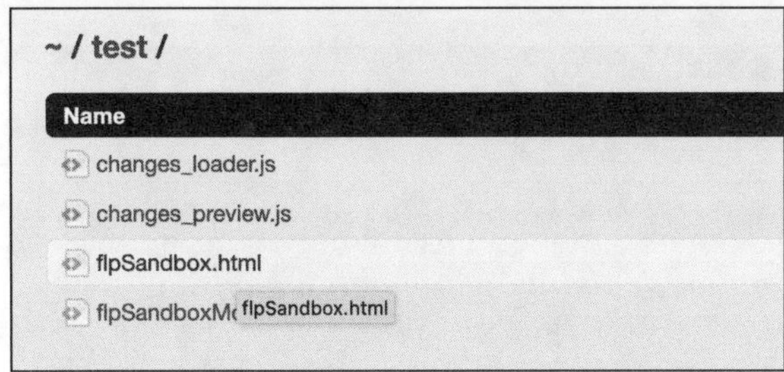

Figure 2.37 Executing the SAP Fiori Application

The SAP Fiori launchpad will open and will contain the application, as shown in Figure 2.38.

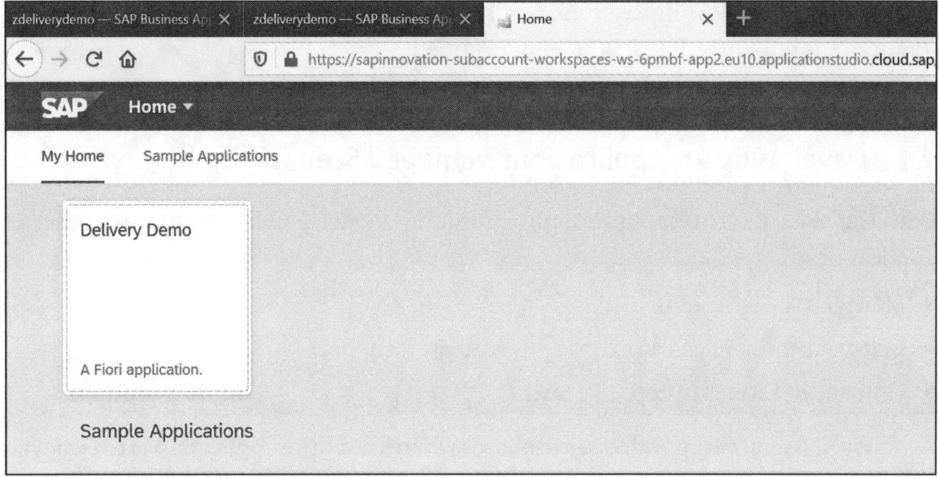

Figure 2.38 SAP Fiori Application

By clicking on the **Delivery Demo** application tile, you'll see the output shown in Figure 2.39.

Since this application is a read-only application, you won't see buttons for creating, editing, or deleting data. You can read only the data from the service that is selected. In the next example, you'll learn how to create, update, and delete records with an SAP Fiori application. The result is an SAP Fiori application created without writing a single line of code, using capabilities of SAP Business Application Studio.

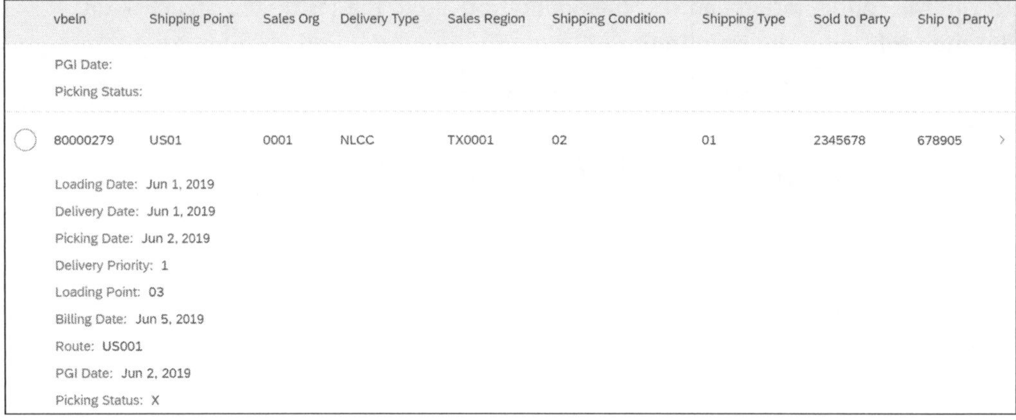

vbeln	Shipping Point	Sales Org	Delivery Type	Sales Region	Shipping Condition	Shipping Type	Sold to Party	Ship to Party
PGI Date:								
Picking Status:								
80000279	US01	0001	NLCC	TX0001	02	01	2345678	678905

Figure 2.39 App Output

We've concluded building our first application in SAP BTP, ABAP environment, with an SAP Fiori UI. In following sections, we'll focus on different scenarios supported for applications built in SAP BTP, ABAP environment.

2.4   Developing an Application: Managed Scenario

With the ABAP RESTful application programming model, you can build three types of applications:

- Managed scenario apps
- Managed/unmanaged scenario apps with draft support
- Unmanaged scenario apps

A managed scenario app addresses use cases where the entire application is to be developed from scratch (greenfield implementations). However, although we're building these applications from scratch, you can reuse behavior components available within the ABAP RESTful application programming model framework itself.

The Business Object Processing Framework (BOPF) implements the interaction phase and the save sequence, while the developer can focus on business logic implementation. In the previous section, we created some tables and built a class to insert data into these tables.

> **Note**
>
> In a managed scenario, the smart ABAP layer takes care of basic scenarios like CRUD operations by using the Business Object Processing Framework (BOPF). Developers don't need to write logic for these operations, but flexibility is still provided to add custom code to meet your business requirements.

The architecture of a managed scenario is shown in Figure 2.40. Notice how the existing functionality can be reused and a new managed implementation built on top. You can reuse standard out-of-the-box operations (CRUD operations); you'll simply specify them in the behavior definition.

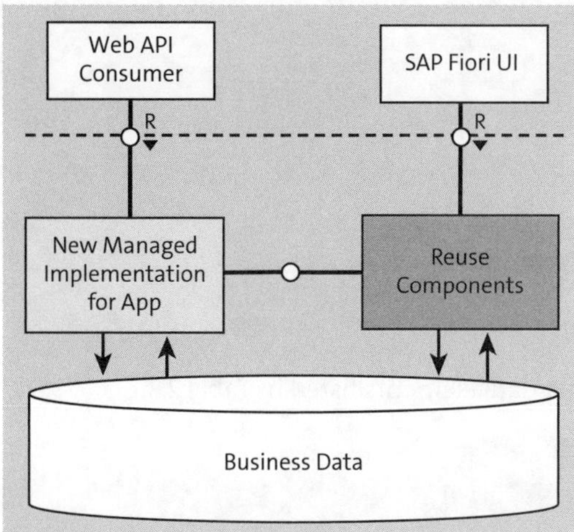

Figure 2.40 Managed Scenario Architecture

Let's look at managed scenarios by exploring two examples:

- Transactional application development
- Screen validation development

In the next section, we'll describe the steps for creating a transactional application in a managed scenario. In the process, you'll also learn about business objects in detail.

2.4.1   Transactional Application Development in a Managed Scenario

First, let's create some persistent database tables. In this example, we'll be using multiple tables, and links between these tables will use foreign keys and associations. Before we start with table creation, we need to create the data elements that will be used in tables. The process of creating data elements is the same as for standard on-premise systems, like SAP ERP and SAP S/4HANA.

In ADT, navigate to **New · Dictionary · Data Element** to create a data element. Figure 2.41 shows the data element maintenance screen in ADT for Eclipse. You'll need to fill out the **Category** and **Data Type** fields at a minimum, but we also recommend providing a field label in the **Long** field.

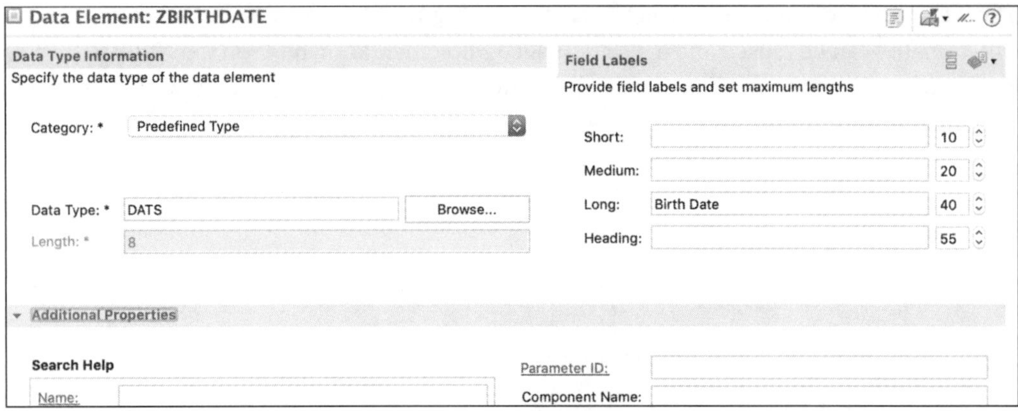

Figure 2.41 Data Element ZBIRTHDATE

Using the same approach, create the other data elements listed in Table 2.3.

Data Element	Description	Data Type
ZSTUID	Student ID	CHAR(5)
ZFNAME	First name	CHAR(40)
ZLNAME	Last name	CHAR(40)
ZMANE	Middle name	CHAR(400
ZGENDER	Gender	CHAR(1)
ZBIRTHDATE	Birth date	DATS
ZCOURSEID	Course ID	CHAR(10)
ZCOURSENAME	Course name	CHAR(40)
ZFLAG	Status	CHAR(1)
ZMARKS	Marks	N(3)
ZTRNGID	Training ID	CHAR(10)
ZST_DT	Start date	DATS
ZEND_DT	End date	DATS
ZCAPACITY	Capacity	N(3)

Table 2.3 Data Element List

After all the data elements are created, we can create the necessary database tables. Let's start with table ZSTUDENT_TABLE, as shown in Listing 2.13. In this table, we'll store basic student information.

```
@EndUserText.label : 'Student Table'
@AbapCatalog.enhancementCategory : #NOT_EXTENSIBLE
@AbapCatalog.tableCategory : #TRANSPARENT
@AbapCatalog.deliveryClass : #A
@AbapCatalog.dataMaintenance : #RESTRICTED
define table zstudent_table {
key client : abap.clnt not null;
key studentid : zstuid not null;
fname : zfname;
lname : zlname;
mname : zmname;
dob : zbirthdate;
gender : zgender;
}
```

Listing 2.13 Table ZSTUDENT_TABLE

The next table is table ZCOURSE_TABLE, as shown in Listing 2.14. In this table, we'll store information about the courses that students might be attending.

```
@EndUserText.label : 'Course Table'
@AbapCatalog.enhancementCategory : #NOT_EXTENSIBLE
@AbapCatalog.tableCategory : #TRANSPARENT
@AbapCatalog.deliveryClass : #A
@AbapCatalog.dataMaintenance : #RESTRICTED
define table zcourse_table {
key client       : abap.clnt not null;
key courseid     : zcourseid not null;
coursename       : zcoursename;
virtualpossible : zflag;
ishandson        : zflag;
totalmarks       : zmarks;
cutofmarks       : zmarks;
start_date       : zst_dt;
end_date         : zend_dt;
iscomplete       : zflag;
capacity         : zcapacity;
}
```

Listing 2.14 Table ZCOURSE_TABLE

Continuing with the data model setup, the remaining table is ZSTUDENT_ENROLL, which we'll create using the code shown in Listing 2.15. In this table, we'll store student enrollment information.

```
@EndUserText.label : 'Student Enroll'
@AbapCatalog.enhancementCategory : #NOT_EXTENSIBLE
@AbapCatalog.tableCategory : #TRANSPARENT
@AbapCatalog.deliveryClass : #A
@AbapCatalog.dataMaintenance : #RESTRICTED
define table zstudent_enroll {
key client      : abap.clnt not null;
key enroll_uuid : sysuuid_x16 not null;
@AbapCatalog.foreignKey.label : 'CourseId'
@AbapCatalog.foreignKey.keyType : #KEY
@AbapCatalog.foreignKey.screenCheck : true
key studentid   : zstuid not null
  with foreign key [0..*,1] zstudent_table
    where studentid = zenroll_table.studentid;
@AbapCatalog.foreignKey.screenCheck : false
key courseid    : zcourseid not null
  with foreign key [0..*,1] zcourse_table
    where courseid = zenroll_table.courseid;
status          : zflag;
marks           : zmarks;
}
```

Listing 2.15 Table ZSTUDENT_ENROLL

The next step is to create interface CDS views on these persistent database tables. As mentioned earlier, basic interface views are at the bottom of a VDM. Only basic interface views can access database tables directly. All other CDS views reuse, and therefore receive their data from, these basic interface views. A basic interface view is supposed to select only from a single database table.

> **Note**
>
> The annotation for a basic view is `@VDM.viewType: #BASIC`.
>
> The annotation for a composite view is `@VDM.viewType: #COMPOSITE`.
>
> The annotation for a projection view is `@VDM.viewType: #CONSUMPTION`.

We'll start with creating the CDS view for the basic student info table using the code shown in Listing 2.16.

```
@AccessControl.authorizationCheck: #CHECK
@EndUserText.label: 'CDS View on student table'
@VDM.viewType: #BASIC
define root view entity ZCDSV_STUDENT_TABLE as select from
zstudent_table
```

```
{
key studentid ,
fname ,
lname ,
mname ,
dob ,
gender
}
```

Listing 2.16 Basic CDS View for ZSTUDENT_TABLE

> **Note**
>
> Note that you must define the CDS data model as a business object when defining a behavior definition that is managed. To represent this data model as a business object, the data model should be defined as a root.

Now, we'll create the other required basic interface views using the code shown in Listing 2.17 and Listing 2.18. This code will create the interface view on the table ZCOURSE_TABLE. This view is defined as the root view, and all the fields from the table are mentioned in this view. This interface view resides on the first level of the table; thus, we'll select all fields in this view from the table.

```
@AccessControl.authorizationCheck: #CHECK
@EndUserText.label: 'CDS View on Course table'
@VDM.viewType: #BASIC
define root view entity ZCDSV_COURSE_TABLE as select from
zcourse_table
{
key courseid ,
  coursename ,
  virtualpossible ,
  ishandson        ,
  totalmarks       ,
  cutofmarks ,
  start_date  ,
  end_date   ,
  iscomplete  ,
  capacity }
```

Listing 2.17 Basic CDS View for ZCOURSE_TABLE

```
@AccessControl.authorizationCheck: #CHECK
@EndUserText.label: 'CDS View on StudentEnroll table'
@VDM.viewType: #BASIC
```

```
Define root view entity ZCDSV_ENROLL_TABLE as select from
zstudent_enroll
association [0..*] to  zcourse_draft as _course on  _course.courseid  =
 $projection.Courseid
association [0..*] to  zstudent_draft as _student on  _student.studentid  =
 $projection.Studentid
{
    key enroll_uuid as EnrollUuid,
    key studentid as Studentid,
    key courseid as Courseid,
    status as Status,
    marks as Marks,
      // associations
    _student,
    _course
}
```

Listing 2.18 Basic CDS View ZSTUDENT_ENROLL

We'll use CDS projection views to expose the general data model for a specific business service. A *general business object* defines all the available entities and elements, whereas a *projection business object* defines the entities and elements that will be used for a specific service.

Let's create some projection CDS views. As mentioned earlier, basic interface views can be used to fetch data from a database. To consume the data in an OData services and for UI development, you must create a projection CDS view on top of an interface view. Create projection views on all four interface views we created earlier, with the code shown in Listing 2.19, Listing 2.20, and Listing 2.21. We opted for projection views because we're projecting fields from this view in the application. In this case, we'll define the label on the fields. For example, the label for the field studentid is Student Id, and its position is 10. This key field will come first. Similarly, all other required fields will be defined in this view.

```
@AccessControl.authorizationCheck: #CHECK
@EndUserText.label: 'Projection view on Student table'
@VDM.viewType: #CONSUMPTION
@UI: { headerInfo: { title.label: 'Student Information',title.type: #STANDARD,
typeName: 'Student ',typeNamePlural: 'Students'}}
define root view entity ZSTUDENT_C as projection on
ZCDSV_STUDENT_TABLE
{
@UI.facet: [{ id: 'studentid',position: 10,label: 'Student Id',type:
#IDENTIFICATION_REFERENCE }]
@UI.lineItem: [{position: 10,type: #STANDARD }]
```

```
@UI.identification: [{position: 10,type: #STANDARD }]
key studentid,
@UI.lineItem: [{position: 20,type: #STANDARD,label: 'FNAME' }]
@UI.identification: [{position: 20,type: #STANDARD,label: 'FNAME' }]
fname,
@UI.lineItem: [{position: 30,type: #STANDARD,label: 'LNAME' }]
@UI.identification: [{position: 30,type: #STANDARD,label: 'LNAME' }]
lname,
@UI.lineItem: [{position: 40,type: #STANDARD,label: 'MNAME' }]
@UI.identification: [{position: 40,type: #STANDARD,label: 'MNAME' }]
mname,
@UI.lineItem: [{position: 50,type: #STANDARD,label: 'DOB' }]
@UI.identification: [{position: 50,type: #STANDARD,label: 'DOB' }]
dob,
@UI.lineItem: [{position: 60,type: #STANDARD,label: 'GENDER' }]
@UI.identification: [{position: 60,type: #STANDARD,label: 'GENDER' }]
gender
}
```

Listing 2.19 Projection CDS View ZSTUDENT_C

```
@AccessControl.authorizationCheck: #CHECK
@EndUserText.label: 'Projection view on Course table'
@VDM.viewType: #CONSUMPTION
@UI: { headerInfo: { title.label: 'Course Details',title.type: #STANDARD,
typeName: 'Course',typeNamePlural: 'Courses'}}
define root view entity ZCOURSE_C as projection on
ZCDSV_COURSE_TABLE
{
  @UI.facet: [{ id: 'courseid',position: 10,label: 'Course Id',type: #
IDENTIFICATION_REFERENCE }]
  @UI.lineItem: [{position: 10,type: #STANDARD }]
  @UI.identification: [{position: 10,type: #STANDARD }]
  key courseid,
  @UI.lineItem: [{position: 20,type: #STANDARD,label: 'coursename' }]
  @UI.identification: [{position: 20,type: #STANDARD,label: 'Course Name' }]
      coursename,
      @UI.lineItem: [{position: 30,type: #STANDARD,label: 'VirtualPossible' }]
      @UI.identification: [{position: 30,type: #STANDARD,label:
'VirtualPossible' }]
      virtualpossible,
      @UI.lineItem: [{position: 40,type: #STANDARD,label: 'IsHandsOn' }]
      @UI.identification: [{position: 40,type: #STANDARD,label: 'IsHandsOn' }]
      ishandson,
      @UI.lineItem: [{position: 50,type: #STANDARD,label: 'Total Marks' }]
```

```
      @UI.identification: [{position: 50,type: #STANDARD,label: 'Total Marks' }]
      totalmarks,
      @UI.lineItem: [{position: 60,type: #STANDARD,label: 'Cut of Marks' }]
      @UI.identification: [{position: 60,type: #STANDARD,label: 'Cut of Marks'
}]
      cutofmarks ,
      @UI.lineItem: [{position: 70,type: #STANDARD,label: 'Start Date' }]
      @UI.identification: [{position: 70,type: #STANDARD,label: 'Start Date' }]
      start_date,
      @UI.lineItem: [{position: 80,type: #STANDARD,label: 'End Date' }]
      @UI.identification: [{position: 80,type: #STANDARD,label: 'End Date' }]
      end_date,
      @UI.lineItem: [{position: 90,type: #STANDARD,label: 'IsComplete' }]
      @UI.identification: [{position: 90,type: #STANDARD,label: 'IsComplete' }]
      iscomplete,
      @UI.lineItem: [{position: 100,type: #STANDARD,label: 'Capacity' }]
      @UI.identification: [{position: 100,type: #STANDARD,label: 'Capacity' }]
      Capacity
}
```

Listing 2.20 Projection CDS View ZCOURSE_C

```
@AccessControl.authorizationCheck: #CHECK
@EndUserText.label: 'Projection view on Enroll table'
@VDM.viewType: #CONSUMPTION
@UI: { headerInfo: { title.label: 'Course Enrollment Details',title.type:
#STANDARD,
typeName: 'Enroll',typeNamePlural: 'Enrolls'}}
define root view entity ZENROLL_C as projection on
ZCDSV_ENROLL_TABLE
{
@UI.facet: [{ id: 'trainingid',position: 10,label: 'Training Id',type:
#IDENTIFICATION_REFERENCE }]
@UI.lineItem: [{position: 10,type: #STANDARD }]
@UI.identification: [{position: 10,type: #STANDARD }]
key trainingid,
@UI.lineItem: [{position: 20,type: #STANDARD,label: 'Course Id' }]
@UI.identification: [{position: 20,type: #STANDARD,label: 'Course Id' }]
key courseid,
@UI.lineItem: [{position: 30,type: #STANDARD,label: 'Status' }]
@UI.idedification: [{position: 30,type: #STANDARD,label: 'Status' }]
status,
@UI.lineItem: [{position: 40,type: #STANDARD,label: 'Marks' }]
```

```
@UI.identification: [{position: 40,type: #STANDARD,label: 'Marks' }]
marks
}
```

Listing 2.21 Projection CDS View ZENROLL_C

Now, create a behavior definition to determine the create, update, and delete functionality on each interface and projection CDS view. Open the context menu and choose
New Behavior Definition to launch the creation wizard. Enter a description and then
select the CDS view and implementation type (in this case, **Managed**), as shown in
Figure 2.42.

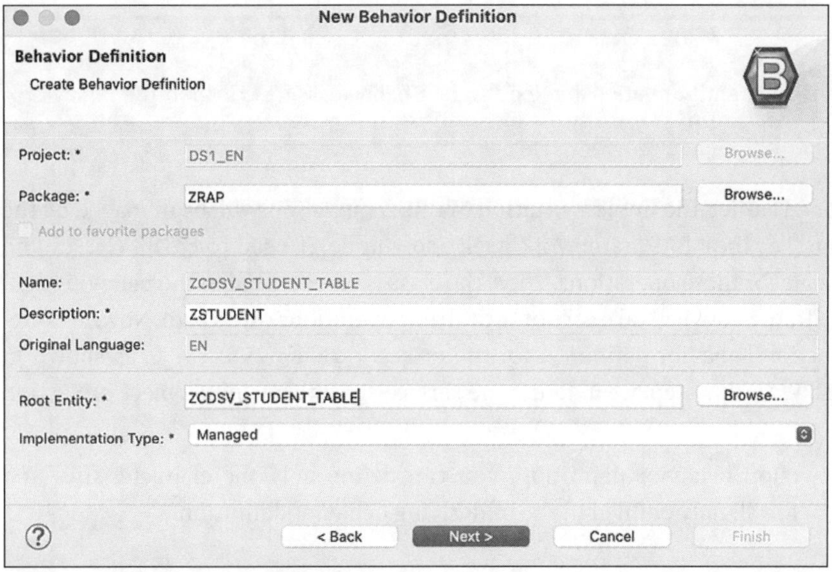

Figure 2.42 New Behavior Definition

You'll now see the source code shown in Listing 2.22, which consists of a header and a
set of definitions. The header defines the business object implementation type—in this
case, the business object implementation type is *managed*. In the definition section,
you'll define the entity's behavior. Enter information about the base table and the CDS
view, together with the standard CRUD operations and other behavior characteristics.

We'll define only the `lock master` characteristic, which means that this entity supports
direct locking. The other option for locking would be in the case of a child entity, which
would have a dependency with its parent entity—`lock dependent by`.

> **Note**
>
> The `lock master` characteristic must be specified for managed scenarios.

```
managed; // implementation in class zbp_cdsv_student_table unique;
define behavior for ZCDSV_STUDENT_TABLE alias STUDENT
persistent table ZSTUDENT_TABLE
lock master
{
create;
update;
delete;
}
```

Listing 2.22 Behavior Definition for ZCDSV_STUDENT_TABLE

> **Note**
>
> If a parent-child entity relationship exists, their behavior is described in the behavior definition.

As mentioned earlier, the implementation of CRUD operations will be managed by the SAP system (i.e., the ABAP framework itself), so you don't need to create classes and write the logic for these operations. Thus, the class is commented in the behavior definition. The framework will take care of these basic operations on its own. Now, let's create a projection behavior definition for the projection view ZSTUDENT_C, as shown in Listing 2.23. With this step, we'll define the service-specific business object projection behavior.

In the projection behavior definition, you can define only the characteristics and options that are already defined in the underlying behavior definition.

```
projection;
define behavior for ZSTUDENT_C alias STUDENT_C
{
use create;
use update;
use delete;
}
```

Listing 2.23 Projection Behavior Definition for ZSTUDENT_C

We'll repeat the same process for the CDS views ZCDSV_ENROLL_TABLE and ZCDSV_COURSE_TABLE and for the projection views ZENROLL_C and ZCOURSE_C, as shown in Listing 2.24, Listing 2.25, Listing 2.26, and Listing 2.27.

```
managed; // implementation in class zbp_cdsv_enroll_table unique;
define behavior for ZCDSV_ENROLL_TABLE alias ENROLL
persistent table ZSTUDENT_ENROLL
lock master
```

```
{
create;
update;
delete;
}
```

Listing 2.24 Behavior Definition for ZCDSV_ENROLL_TABLE

```
projection;
define behavior for ZENROLL_C alias ENROLL_C
{ use create;
use create ;
use delete ;
}
```

Listing 2.25 Projection Behavior Definition for ZENROLL_C

```
managed; // implementation in class zbp_cdsv_course_table unique;
define behavior for ZCDSV_COURSE_TABLE alias COURSE
persistent table ZCOURSE_TABLE
lock master
//authorization master ( instance )
//etag master <field_name>
{
  create;
  update;
  delete;
}
```

Listing 2.26 Behavior Implementation for ZCDSV_COURSE_TABLE

```
projection;
define behavior for ZCOURSE_C alias COURSE_C
{
  use create;
  use update;
  use delete;
}
```

Listing 2.27 Projection Behavior Implementation for ZCOURSE_C

The next step is to define the service. We'll define what data is exposed to the outside world as a business service in the form of an application.

You first must understand the purpose of a service definition. Let's look at a quick example: Imagine you've created 30 database tables and 20 CDS views on these tables, but you only want to expose five CDS views in the cloud, based on your business requirements and needs. You can selectively expose these CDS views by creating a service definition that only exposes the required five CDS views to the outside world. By separating these different layers, plug-and-play flexibility is possible.

To create a new service definition, open the context menu and choose **New Service Definition.** Fill out the wizard and proceed to the source code of the service definition.

In ADT, follow the menu path **File · New · Other**. In the wizard that opens, in the filter section, specify the service and then select **Service Definition**. Create the new service definition by providing a name and a description for the service.

In the source code, enter the lines shown in Listing 2.28.

```
@EndUserText.label: 'Service definition for. Student Information'
define service ZSRV_STUDENT_INFO {
expose ZSTUDENT_C as student;
expose ZCOURSE_C as course;
expose ZENROLL_C as enroll;
}
```

Listing 2.28 Service Definition ZSRV_STUDENT_INFO

In the service definition, we'll expose three projection views, but using aliases instead of technical names (student, course, and enroll).

Now, you'll bind the service. The service binding itself indicates the purpose of the service, so we'll indicate that the service should be exposed to a UI, as shown in Figure 2.43.

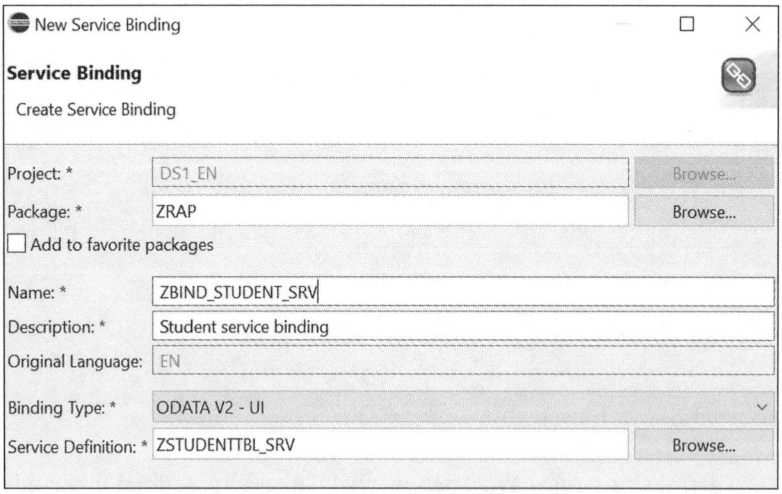

Figure 2.43 Student Service Binding

Now, the remaining step is creating the application in SAP Business Application Studio. The first application we created, in the previous section, only read records from the database. In the new application, we'll add create/update and delete functions for table entries. The steps for creating this application are the same as for the first application, described earlier in Section 2.3.6.

The result, if you use entity ZSTUDENT_C, should look like the screen shown in Figure 2.44.

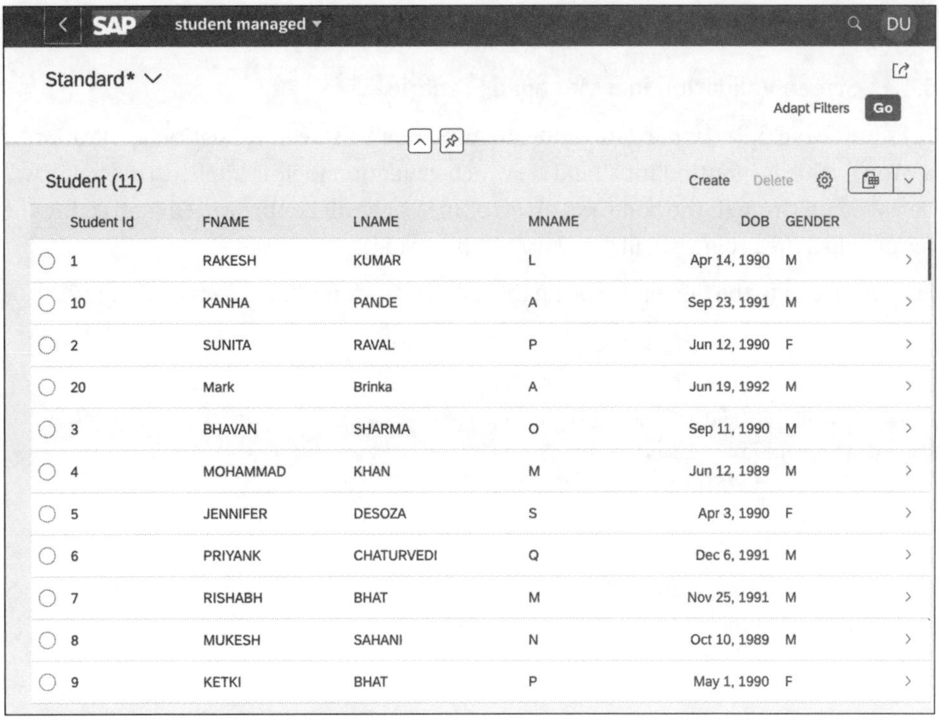

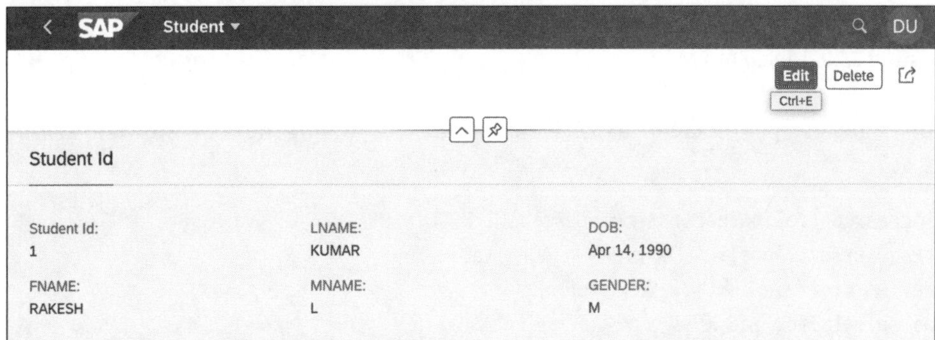

Figure 2.44 Sample Application: Preview

Unlike the first app, you now have option to create new entries. On the detail level, you'll see the screen shown in Figure 2.45.

Figure 2.45 Sample Application: Detail Level

Notice that, in addition to displaying an entry, you also have the option to edit or delete the entry.

At this point, we've completed transactional application development in a managed scenario. Let's now go through steps to create a screen validation in a managed scenario.

2.4.2   Screen Validation in a Managed Scenario

Let's now develop a simple data model to implement a screen validation application. If a wrong value is inputted for a field on which validation logic is applied, that value will be validated through the code and an error message will be thrown. On correcting the input value, the program will allow user to proceed.

First, we'll create the table for storing the validation values, as shown in Listing 2.29.

```
@EndUserText.label : 'Goods Movement for Order'
@AbapCatalog.enhancementCategory : #NOT_EXTENSIBLE
@AbapCatalog.tableCategory : #TRANSPARENT
@AbapCatalog.deliveryClass : #A
@AbapCatalog.dataMaintenance : #RESTRICTED
define table zaufm {
key client : abap.clnt;
key mblnr : abap.char(10) not null;
key mjahr : abap.numc(4) not null;
key zeile : abap.numc(4) not null;
bldat : abap.dats;
budat : abap.dats;
bwart : abap.char(3);
matnr : abap.char(40);
werks : abap.char(4);
lgort : abap.char(4);
}
```

Listing 2.29 Table ZAUFM

In the next step, we'll create a CDS view for table ZAUFM using the code shown in Listing 2.30.

```
@AccessControl.authorizationCheck: #NOT_REQUIRED
@EndUserText.label: 'CDS View on AUFM table'
define root view entity ZCDS_AUFM
as select from zaufm as a
{
@UI.facet: [{ id: 'mblnr',position: 10,label: 'Material Doc Number',type:
#IDENTIFICATION_REFERENCE }]
```

```
@UI.lineItem: [{position: 10,type: #STANDARD }]
@UI.identification: [{position: 10,type: #STANDARD }]
key a.mblnr ,
@UI.lineItem: [{position: 20,type: #STANDARD,label: 'Material Doc Year'
}]
@UI.identification: [{position: 20,type: #STANDARD,label: 'Material Doc
Year' }]
key a.mjahr ,
@UI.lineItem: [{position: 30,type: #STANDARD,label: 'Material Doc Item'
}]
@UI.identification: [{position: 30,type: #STANDARD,label: 'Material Doc
Item' }]
key a.zeile ,
@UI.lineItem: [{position: 40,type: #STANDARD,label: 'Doc Date' }]
@UI.identification: [{position: 40,type: #STANDARD,label: 'Doc Date' }]
a.bldat ,
@UI.lineItem: [{position: 50,type: #STANDARD,label: 'Doc Posting Date' }]
@UI.identification: [{position: 50,type: #STANDARD,label: 'Doc Posting
Date' }]
a.budat ,
@UI.lineItem: [{position: 60,type: #STANDARD,label: 'Movement Type' }]
@UI.identification: [{position: 60,type: #STANDARD,label: 'Movement
Type' }]
a.bwart ,
@UI.lineItem: [{position: 70,type: #STANDARD,label: 'Material Number' }]
@UI.identification: [{position: 70,type: #STANDARD,label: 'Material
Number' }]
a.matnr ,
@UI.lineItem: [{position: 80,type: #STANDARD,label: 'Plant' }]
@UI.identification: [{position: 80,type: #STANDARD,label: 'Plant' }]
a.werks ,
@UI.lineItem: [{position: 90,type: #STANDARD,label: 'Storage Location' }]
@UI.identification: [{position: 90,type: #STANDARD,label: 'Storage
Loacation' }]
a.lgort
}
```

Listing 2.30 CDS View ZCDS_AUFM

We also need a message that will be displayed at screen validation, so we'll need to create a custom message class and add texts to it.

For this activity, create a new object, select a message class, and create a custom class. Open the context menu and select **File · New · Other Repository Object** and then select **Message Class**, as shown in Figure 2.46.

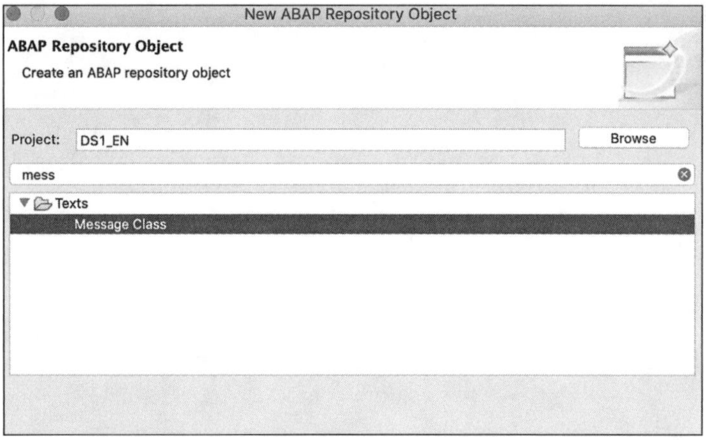

Figure 2.46 Creating a Message Class

A message class can be used to group messages within a development object. After creating a message class, add messages. These messages will notify users about unexpected behaviors, for instance, an error, a specific status, or the result of an action. You may ask why we need a message from the backend when we have an SAPUI5 frontend. In the end, the UI is capable of binding dynamic responses, so whatever happens in the backend can be shown via messages.

Let's proceed with creating a behavior definition of the type managed and uncomment class. As shown in Figure 2.47, specify the package, add a description for the behavior definition, and specify the root entity and implementation type (in this case, **Managed**).

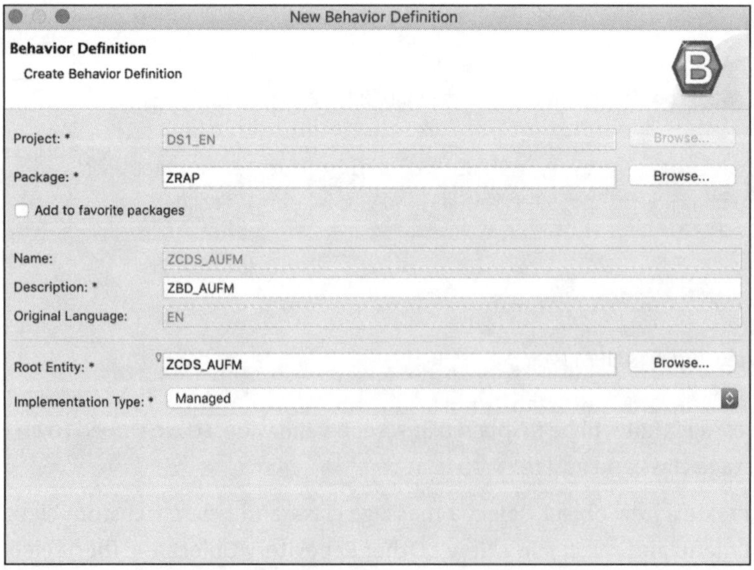

Figure 2.47 Behavior Definition

In previous example, we commented out the class. Now, instead, we'll implement the validation logic, as shown in Listing 2.31.

```
managed implementation in class zbp_cds_aufm unique;
define behavior for ZCDS_AUFM //alias <alias_name>
persistent table ZAUFM
lock master
//authorization master ( instance )
//etag master <field_name>
{
create;
update;
delete;
}
```

Listing 2.31 Behavior Definition

Now, we'll create the behavior implementation for the behavior definition, as shown in Listing 2.32.

```
CLASS zbp_cds_aufm DEFINITION PUBLIC ABSTRACT FINAL FOR
BEHAVIOR OF zcds_aufm.
ENDCLASS.
CLASS zbp_cds_aufm IMPLEMENTATION.
ENDCLASS.
```

Listing 2.32 Behavior Implementation

In the local types section, define the class with the code shown in Listing 2.33.

```
CLASS lhc_ZCDS_AUFM DEFINITION INHERITING FROM
cl_abap_behavior_handler.
PRIVATE SECTION.
METHODS validateAufm FOR VALIDATE ON SAVE
ZCDS_AUFM_INPUT FOR ZCDS_AUFM~validateAufm.
ENDCLASS.
CLASS lhc_ZCDS_AUFM IMPLEMENTATION.
METHOD validateAufm.
read entity ZCDS_AUFM
fields ( MBLNR ) //we'll be validating on document number
with corresponding #( ZCDS_AUFM_INPUT )
result DATA(lt_values).
read table lt_values into data(ls_values) index 1.
if sy-subrc is initial.
append value #( %key = ls_values-%key
%msg = new_message( id = 'ZSY_MSG_CLASS'
```

```
number = '001' //message created in
ZSY_MSG_CLASS
severity = if_abap_behv_message=>severity-error )
) to reported-ZCDS_AUFM.
endif.
ENDMETHOD.
ENDCLASS.
```

Listing 2.33 Behavior Handler Class

In this class, we are implementing the logic to validate field MBLNR (material document number). This class inherits CL_ABAP_BEHAVIOR_HANDLER, and inside the class, you can define methods that will react to specific events. In this case, the triggering event is ON SAVE, which means the handler will be triggered when a user wants to save data. The implemented logic is simple—the content of field MBLNR will be checked, and a message with the severity of the error will be thrown if the field is empty.

Now, let's create a service definition. As shown in Figure 2.48, you'll need to provide a package, a name for the service, a description, and a source type (in this case, **Service Definition**).

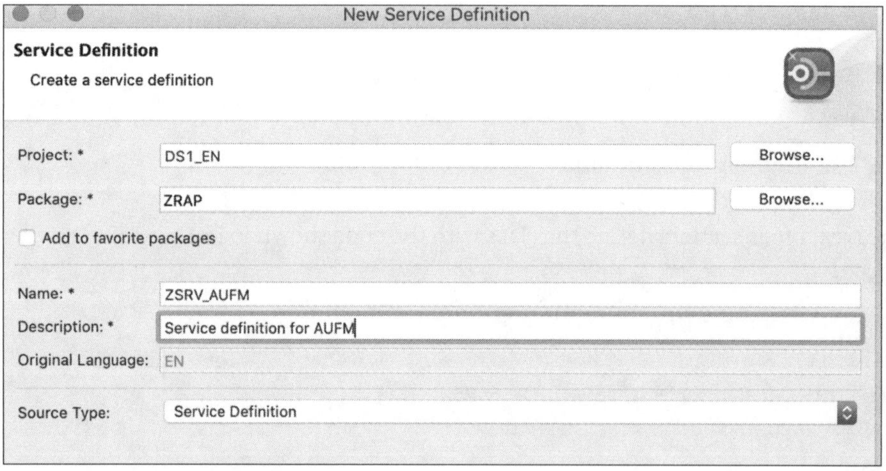

New Service Definition

Service Definition

Create a service definition

Project: * DS1_EN Browse...

Package: * ZRAP Browse...

Add to favorite packages

Name: * ZSRV_AUFM

Description: * Service definition for AUFM

Original Language: EN

Source Type: Service Definition

Figure 2.48 Service Definition

In the service definition, we'll expose the CDS view ZCDS_AUFM, as shown in Listing 2.34.

```
@EndUserText.label: 'Service definition for AUFM'
define service ZSRV_AUFM {
  expose ZCDS_AUFM;
}
```

Listing 2.34 Service Definition ZCDS_AUFM

Now, create the service binding for the service definition, as shown in Figure 2.49.

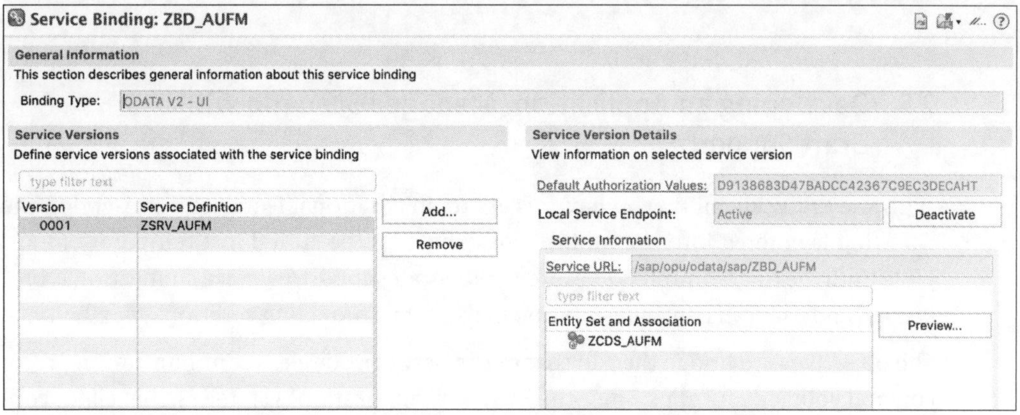

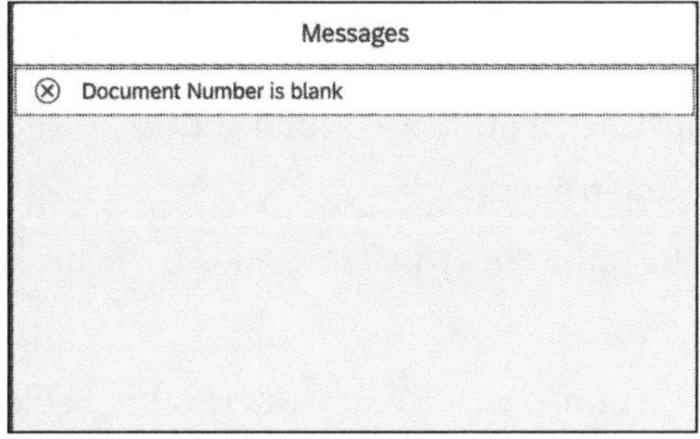

Figure 2.49 Service Binding

With this step, we're ready to test the UI. Simply click the **Service URL** link on the **Service Binding** screen, and you'll see the output shown in Figure 2.50.

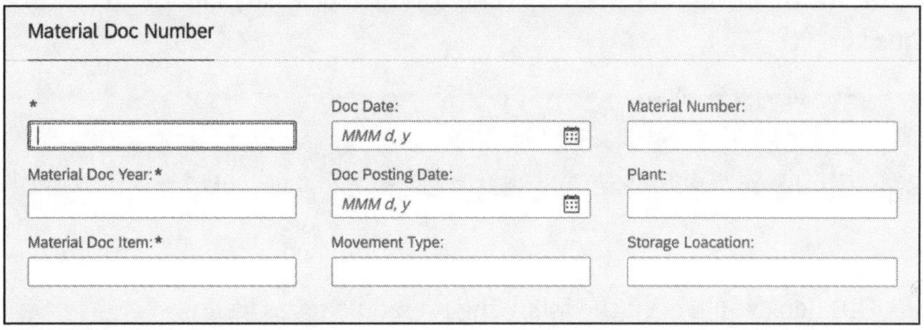

Figure 2.50 Preview UI

If you leave the material number field empty, you'll see the error message shown in Figure 2.51.

Figure 2.51 Error Message

With this step, you've now successfully implemented field validation logic for a screen validation application.

2.5   Developing an Application: Managed Scenario with Draft Support

With the draft capability, you have access to an additional feature for preserving the data that is entered into the input fields, which can be stored in a temporary (draft) table until the final commit is executed. With draft support, even if a connection is broken, an end user can continue entering in data, thus providing additional flexibility.

The objective is to bridge the gap that exists between stateless communication protocols and your stateful apps. Since the REST communication is stateless, that client context is not stored on the server. Instead, this information can be saved in a draft table, which is replica of the active data and is used while in edit mode.

For managed scenarios with draft support, we'll use the data model from our previous example. As an additional feature, we'll be using universally unique identifier (UUID) keys, which means that every business object entity will be easily identifiable with a single unique key.

> **Note**
>
> The usage of UUID keys is optional. You can use the semantic primary keys, as in our example.

Besides the UUID field, you'll need the following prerequisites as well:

- A timestamp field that will indicate when a specific item was last modified. This field is important for ETag master handling. This field is automatically filled in by the runtime managed by the ABAP RESTful application programming model.

- The **Total ETag** field is required to check if the current business object instance was changed in the concurrent clients. The Total ETag field is only required on the `lock master` entity.

- A UUID key in every entity as well as parent UUID fields in each child entity.

- A root UUID field in each grandchild entity and below.

- A semantic ID field to semantically distinguish the entity instance within the business object.

Let's start with table `ZCOURSE_TABLE`. For this example, we'll create a new table, table `ZCOURSE_D`, shown in Listing 2.35, that will contain additional fields for UUIDs as well as a timestamp field that will be used as an ETag, which is mandatory component of the draft capability.

```
@EndUserText.label : 'Course table for DRAFT scenario'
@AbapCatalog.enhancementCategory : #NOT_EXTENSIBLE
@AbapCatalog.tableCategory : #TRANSPARENT
@AbapCatalog.deliveryClass : #A
@AbapCatalog.dataMaintenance : #RESTRICTED
define table zcourse_d {
  key client       : abap.clnt;
  key courseuuid   : sysuuid_x16 not null;
  courseid         : zcourseid not null;
  coursename       : zcoursename;
  virtualpossible  : zflag;
  ishandson        : zflag;
  totalmarks       : zmarks;
  cutofmarks       : zmarks;
  start_date       : zst_dt;
  end_date         : zend_dt;
  iscomplete       : zflag;
  capacity         : zcapacity;
  last_changed_by  : syuname;
  last_changed_at  : timestampl;
}
```

Listing 2.35 Table ZCOURSE_D

The second table in our data model is for enrollments—we'll use table ZENROLL_TABLE as a template and update it as we did for the course table. On top of that, we'll add a foreign key reference that uses UUIDs. The source code for defining table ZENROLL_D is shown in Listing 2.36.

```
@EndUserText.label : 'Enrollment table for DRAFT scenario'
@AbapCatalog.enhancementCategory : #NOT_EXTENSIBLE
@AbapCatalog.tableCategory : #TRANSPARENT
@AbapCatalog.deliveryClass : #A
@AbapCatalog.dataMaintenance : #RESTRICTED
define table zenroll_d {
  key client       : abap.clnt not null;
  key enroll_uuid  : sysuuid_x16 not null;
  courseuuid       : sysuuid_x16 not null;
  studentid        : zstuid not null;
  status           : zflag;
  marks            : zmarks;
  last_changed_by  : syuname;
  last_changed_at  : timestampl;
}
```

Listing 2.36 Table ZENROLL_D

Notice one major difference from our earlier managed scenario without draft support, where we used semantic primary keys and established a corresponding foreign key relationship. In contrast, in this case, we're defining this relationship via UUIDs instead.

Now, moving on to the next step, let's create our CDS view entities. Let's start with the course table's root view, ZCDSV_COURSED, shown in Listing 2.37.

```
@AccessControl.authorizationCheck: #CHECK
@EndUserText.label: 'Course CDS view for draft scenario'
define root view entity ZCDSV_COURSED as select from zcourse_d
    composition [0..*] of ZCDSV_ENROLLD as _Enroll
{
    key courseuuid as Courseuuid,
    courseid as Courseid,
    coursename as Coursename,
    virtualpossible as Virtualpossible,
    ishandson as Ishandson,
    totalmarks as Totalmarks,
    cutofmarks as Cutofmarks,
    start_date as StartDate,
    end_date as EndDate,
    iscomplete as Iscomplete,
    capacity as Capacity,
    @Semantics.user.lastChangedBy: true
    last_changed_by as LastChangedBy,
    //local ETag field --> OData ETag
    @Semantics.systemDateTime.localInstanceLastChangedAt: true
    last_changed_at as LastChangedAt,
    //Associations
    _Enroll
}
```

Listing 2.37 Course Root View

Notice that our annotations define semantics for the fields changedby and changedat — with this definition, we're informing the ABAP RESTful application programming model runtime about the purpose of these fields, so they will be populated automatically. This step is the first step in defining the changedat field as the local ETag field.

You can use the same approach to define annotations for instance creation, as follows:

```
@Semantics.user.CreateBy
@Semantics.systemDateTime.CreatedAt
```

Once the view for enrollment data has been created, add a compositional relationship to the child entity ZCDSV_ENROLLD with the keyword composition.

Now, let's create a view for the enrollment data, as shown in Listing 2.38.

```
@AccessControl.authorizationCheck: #CHECK
@EndUserText.label: 'Enroll CDS view for draft scenario'
define view entity ZCDSV_ENROLLD as select from zenroll_d
//     association to parent ZCDSV_STUDENTD as _
Student on $projection.Studentuuid = _Student.Studentuuid
    association to parent ZCDSV_COURSED as _Course on $projection.Courseuuid = _
Course.Courseuuid

    association [1..1] to ZCDSV_STUDENT_TABLE as _
Student on $projection.Studentid = _Student.studentid
{
    key enroll_uuid as EnrollUuid,
    courseuuid as Courseuuid,
    studentid as Studentid,
    status as Status,
    marks as Marks,
    @Semantics.user.lastChangedBy: true
    last_changed_by as LastChangedBy,
    //local ETag field --> OData ETag
    @Semantics.systemDateTime.localInstanceLastChangedAt: true
    last_changed_at as LastChangedAt,
    //Associations
    _Course,
    _Student
}
```

Listing 2.38 Enrollment View

In the enrollment view, we defined a compositional relationship to the parent entity ZCDSV_COURSED, using the keyword association to parent.

The next step is to create the behavior definition, as shown in Listing 2.39. By opening the context menu of the root view entity (ZCDSV_COURSED), you can create a new behavior definition for the whole business object hierarchy. In this case, we'll choose the implementation type *managed*.

```
managed; // implementation in class zbp_cdsv_coursed unique;

define behavior for ZCDSV_COURSED //alias <alias_name>
persistent table ZCOURSE_D
lock master
//authorization master ( instance )
```

```
etag master LastChangedAt
{
  create;
  update;
  delete;
  association _Enroll { create; }

  field (numbering : managed, readonly) Courseuuid;

  mapping for zcourse_d
  { Courseuuid = courseuuid;
    Courseid = courseid;
    Coursename = coursename;
    Virtualpossible = virtualpossible;
    Ishandson = ishandson;
    Totalmarks = totalmarks;
    Cutofmarks = cutofmarks;
    StartDate = start_date;
    EndDate = end_date;
    Iscomplete = iscomplete;
    Capacity = capacity;
    LastChangedBy = last_changed_by;
    LastChangedAt = last_changed_at; }
}

define behavior for ZCDSV_ENROLLD //alias <alias_name>
persistent table ZENROLL_D
lock dependent by _Course
//authorization dependent by <association>
etag master LastChangedAt
{
  update;
  delete;
  association _Course { }

  field (readonly) Courseuuid;
  field (numbering : managed, readonly) EnrollUuid;

  mapping for zenroll_d
  { EnrollUuid = enroll_uuid;
    Courseuuid = courseuuid;
    Studentid = studentid;
```

```
    Status = status;
    Marks = marks;
    LastChangedBy =last_changed_by;
    LastChangedAt = last_changed_at; }
}
```

Listing 2.39 Behavior Definition

The behavior definition contains two sections:

- The behavior definition for the root entity
- The behavior definition for the child entity

In this example, for simplicity, we won't implement any authorization checks. Also, we won't use any implementation classes, since we won't be implementing the specific behavior. To use implementation classes in a draft scenario, you would need separate classes for each entity.

For definitions specific to this scenario (that is, a managed scenario with draft support), keep in mind the following considerations:

- Either `lock master` or `lock dependent by` must be specified.
- `ETag master` must be defined on each entity to ensure that the ETag check is performed for every entity independently.
- You can use the numbering capabilities of ABAP RESTful application programming model runtime.
- Mapping is required between database table fields and the CDS data model if different names are used.

Instead of projecting the data model and behavior of our business objects and publishing the service, let's first add the draft capabilities. Start with adding the `with draft` statement, as shown in Figure 2.52.

```
  1  managed;
  2  with draft;
  3
⊗ 4⊖ ⸢There is no draft persistency specified for "ZCDSV_COURSED" ("ZCDSV_COURSED" is activated for drafts).
  5  persistent table ZCOURSE_D
⚠ 6  lock master
  7  //authorization master ( instance )
```

Figure 2.52 Adding Draft Capabilities

Immediately, you'll be notified that draft persistency exists, which means that we need to create a draft table. Add the draft table in the source code of the behavior definition. In the popup window shown in Figure 2.53, we'll use a quick fix to generate it.

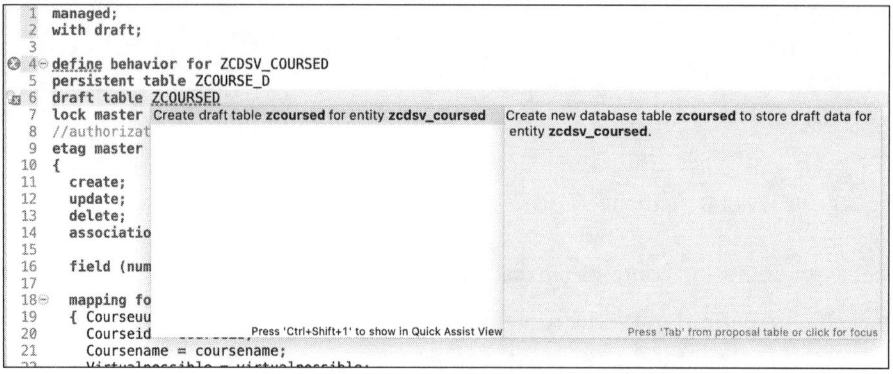

Figure 2.53 Creating a Draft Table

In the popup window that appears, define the new table, as shown in Figure 2.54.

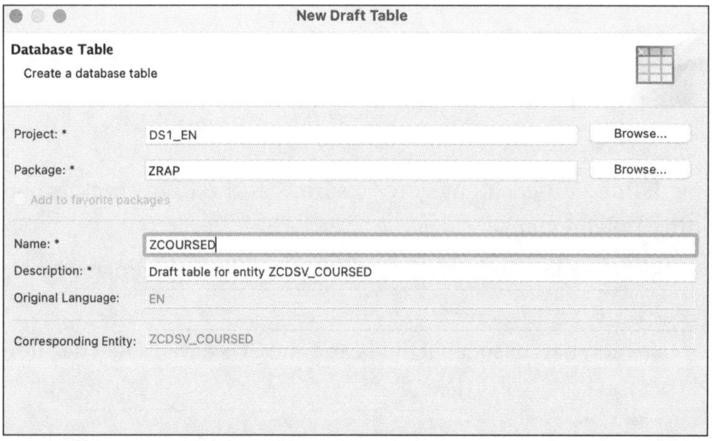

Figure 2.54 New Draft Table

This wizard automatically generates the draft table structure that corresponds to the original table, with the addition of the admin structure %admin, as shown in Listing 2.40.

```
@EndUserText.label : 'Draft table for entity ZCDSV_COURSED'
@AbapCatalog.enhancementCategory : #EXTENSIBLE_ANY
@AbapCatalog.tableCategory : #TRANSPARENT
@AbapCatalog.deliveryClass : #A
@AbapCatalog.dataMaintenance : #RESTRICTED
define table zcoursed {
  key mandt        : mandt not null;
  key courseuuid   : sysuuid_x16 not null;
  courseid         : zcourseid;
  coursename       : zcoursename;
  virtualpossible  : zflag;
  ishandson        : zflag;
```

```
  totalmarks      : zmarks;
  cutofmarks      : zmarks;
  startdate       : zst_dt;
  enddate         : zend_dt;
  iscomplete      : zflag;
  capacity        : zcapacity;
  lastchangedby   : syuname;
  lastchangedat   : timestampl;
  "%admin"        : include sych_bdl_draft_admin_inc;
}
```

Listing 2.40 Draft Table for Entity ZCDSV_COURSED

We'll repeat the same steps to create a draft table for the enrollment entity.

The next step is to add the **Total ETag** field. But first, you'll need to add this field to the root table (changed_at), as shown in Listing 2.41.

```
define table zcourse_d {
  key client        : abap.clnt not null;
  key courseuuid    : sysuuid_x16 not null;
  courseid          : zcourseid not null;
  coursename        : zcoursename;
  virtualpossible   : zflag;
  ishandson         : zflag;
  totalmarks        : zmarks;
  cutofmarks        : zmarks;
  start_date        : zst_dt;
  end_date          : zend_dt;
  iscomplete        : zflag;
  capacity          : zcapacity;
  last_changed_by   : syuname;
  last_changed_at   : timestampl;
  changed_at        : timestampl;
```

Listing 2.41 Table ZCOURSE_D with Total ETag Field

We'll do the same in the root CDS view entity, together with annotation @Semantics.systemDateTime.lastChangedAt:

```
    @Semantics.systemDateTime.lastChangedAt: true
    changed_at as ChangedAt,
```

Update the behavior definition to include the ETag master statement and recreate the draft table, again using the quick fix approach, which appears once you hover your cursor over the overdraft table ZCOURSED declaration, as shown in Figure 2.55.

```
 2  with draft;
 3
 4  define behavior for ZCDSV_COURSED
 5  persistent table ZCOURSE_D
 6  draft table ZCOURSED
 7  lock master    Recreate draft table zcoursed for entity zcdsv_coursed     Recreate draft table zcoursed so that it fits for entity
 8  //authorizat                                                              zcdsv_coursed and reflects recent changes.
 9  etag master
10  {
11    create;
12    update;
13    delete;
14    associatio
15
16    field (num
17
18    mapping fo
19    { Courseuu
20      Courseid                   Press 'Ctrl+Shift+1' to show in Quick Assist View          Press 'Tab' from proposal table or click for focus
21      Coursename = coursename;
22      Virtualpossible = virtualpossible;
23      Ishandson = ishandson;
24      Totalmarks = totalmarks;
25      Cutofmarks = cutofmarks;
26      StartDate = start_date;
27      EndDate = end_date;
28      Iscomplete = iscomplete;
29      Capacity = capacity;
30      LastChangedBy = last_changed_by;
31      LastChangedAt = last_changed_at;
32      ChangedAt = changed_at; }
33  }
```

Figure 2.55 Recreating a Draft Table

With this step, we have completed the behavior definition.

The next step is to project the draft business objects data model. First, create a projection on ZCDSV_COURSED. The easiest approach is right-clicking on the corresponding object and choosing **Define Projection View** in the wizard, as shown in Figure 2.56.

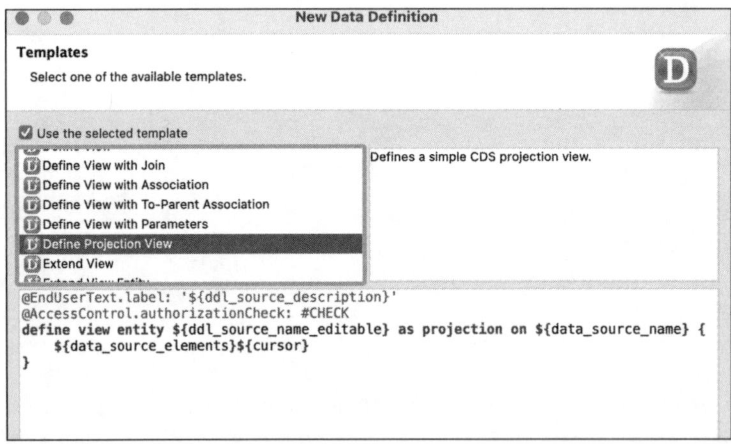

Figure 2.56 New Projection View

While creating the projection view, you can enter the annotation @Metadata.allowExtensions to enable metadata extensions. In metadata extensions, you can use @UI annotations to design the user interface.

As shown in Listing 2.42, we won't use this feature. Instead, we'll only add the root keyword since this is the root entity view.

```
@EndUserText.label: 'Projection for ZCDSV_COURSED'
@AccessControl.authorizationCheck: #CHECK
define root view entity ZCDSV_COURSEDP as projection on ZCDSV_COURSED {
    key Courseuuid,
    Courseid,
    Coursename,
    Virtualpossible,
    Ishandson,
    Totalmarks,
    Cutofmarks,
    StartDate,
    EndDate,
    Iscomplete,
    Capacity,
    LastChangedBy,
    LastChangedAt,
    ChangedAt,
    /* Associations */
    _Enroll : redirected to composition child ZCDSV_ENROLLP}
```

Listing 2.42 Projection View ZCDSV_COURSEDP

Repeat the same process for ZCDSV_ENROLLD to create the projection view ZCDSV_ENROLLP.

Once we're done creating the necessary projection views on our data models, we'll create a projection view for the business object behavior. Open the context menu for the root projection view to open the wizard and proceed with creating the behavior projection, as shown in Figure 2.57.

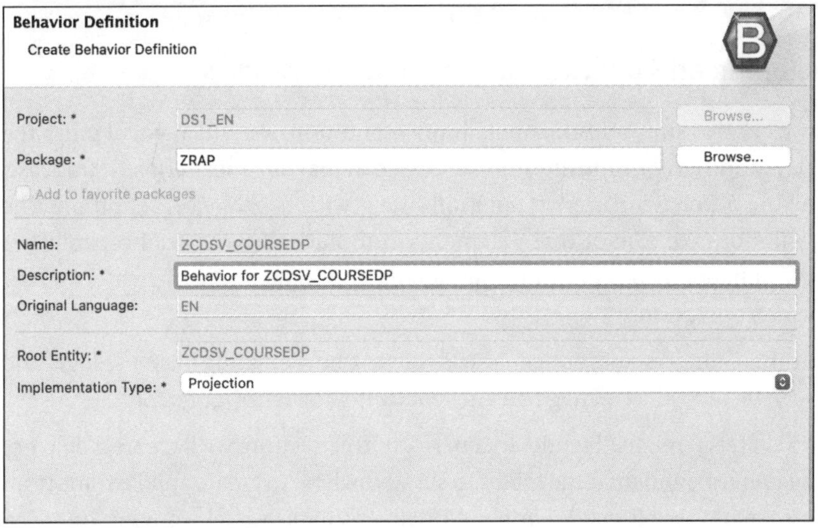

Figure 2.57 Creating a Behavior Projection

The result is a behavior projection similar to the code shown in Listing 2.43.

```
projection;
use draft;

define behavior for ZCDSV_COURSEDP alias Course
use etag
{
  use create;
  use update;
  use delete;

  use association _Enroll { create; with draft; }
}

define behavior for ZCDSV_ENROLLP alias Enroll
use etag
{
  use update;
  use delete;

  use association _Course { with draft; }
}
```

Listing 2.43 Behavior Projection for ZCDSV_ENROLLP

All is left now is to create a service definition and binding—which is the same process as for the managed scenario described in Section 2.4.

2.6   Developing an Application: Unmanaged Scenario

Unmanaged scenarios support brownfield implementations, in which you'll reuse the persistence layer and existing business logic. Unlike in managed scenarios, in this case, you must define a behavior implementation along with a behavior definition. The framework will not take care of these elements automatically. The read capability is implicit and will happen automatically, but you'll need to take care of create, update, and delete (CUD) operations.

The architecture of an unmanaged scenario, depicting how existing business logic and persistence can be reused to manage business data, is shown in Figure 2.58.

Let's look at a delivery-related business scenario. In this section, we'll create a delivery header, delivery items, and material tables to show you how to create applications in an unmanaged scenario. We'll use a simplified delivery data model, copying only the required fields from original tables LIKP and LIPS and from material master table ZMARA.

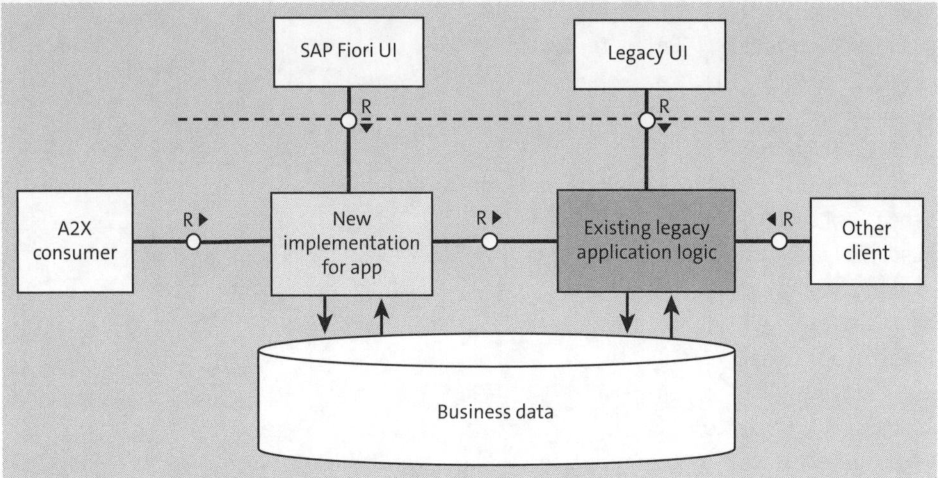

Figure 2.58 Unmanaged Scenario Architecture

This data model (header/item table) is a good way to demonstrate an unmanaged scenario, as well as to demonstrate how to build extensions in SAP BTP, ABAP environment, with data persistence in the platform itself and data replicated from the backend system using APIs.

> **Note**
>
> We'll cover data acquisition from a backend on-premise system using APIs in Chapter 3.

First, we'll create the delivery header table ZLIKP using the code shown in Listing 2.44. This code will generate the table. Key fields are defined with the word key, the table category is #Transparent, and the delivery class is A. The fields of the table are defined with text.label, which will include a short description of the field in addition to specifying the data type and length.

```
@EndUserText.label : 'Delivery Header Data'
@AbapCatalog.enhancementCategory :
#EXTENSIBLE_CHARACTER_NUMERIC
@AbapCatalog.tableCategory : #TRANSPARENT
@AbapCatalog.deliveryClass : #A
@AbapCatalog.dataMaintenance : #RESTRICTED
define table zlikp {
key client : abap.clnt not null;
@EndUserText.label : 'Delivery No.'
key vbeln : abap.char(10) not null;
@EndUserText.label : 'Shipping Point'
vstel : abap.char(4);
@EndUserText.label : 'Sales Organization'
```

```
vkorg : abap.char(4);
@EndUserText.label : 'Delivery Type'
lfart : abap.char(4);
@EndUserText.label : 'Sales Region'
bzrik : abap.char(6);
@EndUserText.label : 'Shipping Condition'
vsbed : abap.char(2);
@EndUserText.label : 'Shipping Type'
vsart : abap.char(2);
@EndUserText.label : 'Sold-to'
kunag : abap.char(10);
@EndUserText.label : 'Ship-to'
kunnr : abap.char(10);
@EndUserText.label : 'Loading Date'
lddat : abap.dats;
@EndUserText.label : 'Delivery Date'
lfdat : abap.dats;
@EndUserText.label : 'Picking Date'
kodat : abap.dats;
@EndUserText.label : 'Delivery Priority'
lprio : abap.numc(2);
@EndUserText.label : 'Loading Point'
lstel : abap.char(2);
@EndUserText.label : 'Billing Date'
fkdat : abap.dats;
@EndUserText.label : 'Route'
route : abap.char(6);
@EndUserText.label : 'PGI Date'
wadat : abap.dats;
@EndUserText.label : 'Picking Status'
kostk : abap.char(1);
}
```

Listing 2.44 Table ZLIKP

The next table, for delivery line items, is shown in Listing 2.45. Again, we're creating a table with significantly fewer fields than the original table LIPS.

```
@EndUserText.label : 'Delivery Item'
@AbapCatalog.enhancementCategory : #NOT_EXTENSIBLE
@AbapCatalog.tableCategory : #TRANSPARENT
@AbapCatalog.deliveryClass : #A
@AbapCatalog.dataMaintenance : #RESTRICTED
define table zlips {
key client : abap.clnt not null;
```

```
@EndUserText.label : 'Delivery No.'
key vbeln : abap.char(10) not null;
@EndUserText.label : 'Delivery Item'
key posnr : abap.numc(6) not null;
@EndUserText.label : 'Material'
matnr : abap.char(40);
@EndUserText.label : 'Quantity'
@Semantics.quantity.unitOfMeasure : 'zlips.meins'
lfimg : abap.quan(15,3);
@EndUserText.label : 'Unit'
meins : abap.unit(3); }
```

Listing 2.45 Table ZLIPS

One more table we need for our data model is the material master table, a reduced version of table MARA, as shown in Listing 2.46.

```
@EndUserText.label : 'Material Master'
@AbapCatalog.enhancementCategory : #NOT_EXTENSIBLE
@AbapCatalog.tableCategory : #TRANSPARENT
@AbapCatalog.deliveryClass : #A
@AbapCatalog.dataMaintenance : #RESTRICTED
define table zmara {
key client : abap.clnt not null;
@EndUserText.label : 'Material'
key matnr : abap.char(40) not null;
@EndUserText.label : 'Material Type'
mtart : abap.char(4);
@EndUserText.label : 'Length'
@Semantics.quantity.unitOfMeasure : 'zmara.meins'
laeng : abap.quan(13,3);
@EndUserText.label : 'Width'
@Semantics.quantity.unitOfMeasure : 'zmara.meins'
breit : abap.quan(13,3);
@EndUserText.label : 'Height'
@Semantics.quantity.unitOfMeasure : 'zmara.meins'
hoehe : abap.quan(13,3);
@EndUserText.label : 'Unit Of Dimension'
meins :abap.unit(3);
}
```

Listing 2.46 Table ZMARA

Now, we're ready to create a CDS view on the delivery header table, table ZBO_DEL, as shown in Listing 2.47.

This CDS view defines the root entity of the data model and represents the root of the compositional hierarchy for the business object to be created. To define a composition relationship from the root to a child entity, the keyword COMPOSITION is used.

With cardinality [0..*], you can express that any number of item instances can be assigned to each delivery instance.

```
@AbapCatalog.sqlViewName: 'ZBO_DELIVERY'
@EndUserText.label: 'Delivery Details'
@AccessControl.authorizationCheck: #NOT_REQUIRED
@AbapCatalog.compiler.compareFilter: true
@AbapCatalog.preserveKey: true
define root view ZBO_DEL as select from zlikp as del
composition [0..*] of ZBO_ITEM as _item
{
key vbeln,
@UI.lineItem: [{ position: 20 }]
@EndUserText.label: 'Shipping Point'
vstel,
@UI.lineItem: [{ position: 30 }]
@EndUserText.label: 'Sales Org'
vkorg,
@UI.lineItem: [{ position: 40 }]
@EndUserText.label: 'Delivery Type'
lfart,
@UI.lineItem: [{ position: 50 }]
@EndUserText.label: 'Sales Region'
bzrik,
@UI.lineItem: [{ position: 60 }]
@EndUserText.label: 'Shipping Condition'
vsbed,
@UI.lineItem: [{ position: 70 }]
@EndUserText.label: 'Shipping Type'
vsart,
@UI.lineItem: [{ position: 80 }]
@EndUserText.label: 'Sold to Party'
kunag,
@UI.lineItem: [{ position: 90 }]
@EndUserText.label: 'Ship to Party'
kunnr,
@UI.lineItem: [{ position: 100 }]
@EndUserText.label: 'Loading Date'
lddat,
```

```
@UI.lineItem: [{ position: 100 }]
@EndUserText.label: 'Delivery Date'
lfdat,
@UI.lineItem: [{ position: 110 }]
@EndUserText.label: 'Picking Date'
kodat ,
@UI.lineItem: [{ position: 120 }]
@EndUserText.label: 'Delivery Priority'
lprio,
@UI.lineItem: [{ position: 130 }]
@EndUserText.label: 'Loading Point'
lstel,
@UI.lineItem: [{ position: 140 }]
@EndUserText.label: 'Billing Date'
fkdat,
@UI.lineItem: [{ position: 150 }]
@EndUserText.label: 'Route'
route,
@UI.lineItem: [{ position: 160 }]
@EndUserText.label: 'PGI Date'
wadat,
@UI.lineItem: [{ position: 170 }]
@EndUserText.label: 'Picking Status'
kostk,
_item }
```

Listing 2.47 CDS View ZBO_DEL

Next is the CDS view for delivery line items, as shown in Listing 2.48.

```
@AbapCatalog.sqlViewName: 'ZBO_DEL_ITEM'
@AbapCatalog.compiler.compareFilter: true
@AbapCatalog.preserveKey: true
@AccessControl.authorizationCheck: #NOT_REQUIRED
@EndUserText.label: 'Delivery Item Details'
define view ZBO_ITEM as select from zlips
association to parent ZBO_DEL as _del
on $projection.vbeln = _del.vbeln
{
key vbeln,
key posnr,
matnr,
lfimg,
```

```
meins,
_del // Make association public
}
```

Listing 2.48 CDS View ZBO_ITEM

Now, we can activate both CDS views. In ADT, right-click on the CDS view code to open the activation menu.

The next step is to create the behavior definition through the wizard (the same one discussed in Section 2.4.1), as shown in Figure 2.59.

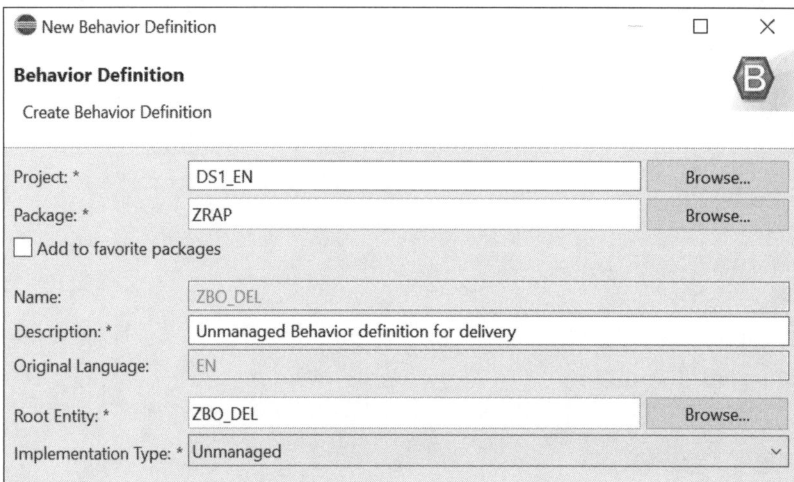

Figure 2.59 New Behavior Definition

In the wizard, select **Unmanaged** from the **Implementation Type** dropdown list. As a result, you'll see the default code shown in Listing 2.49.

```
unmanaged implementation in class zbp_bo_del unique;
define behavior for ZBO_DEL alias Delivery
{
create;
update;
delete;
association _item {
create;
}
}
define behavior for ZBO_ITEM alias Del_Item
{
create;
update;
```

```
delete;
}
```

Listing 2.49 Behavior Definition

We defined our behavior to create a delivery header and item records using CRUD operations. You'll need to modify the code of the behavior definition, as shown in Listing 2.50, specifically including the mandatory fields required to create/update records.

```
implementation unmanaged ;
define behavior for ZBO_DEL alias Delivery
implementation in class zbp_bo_del unique
{
field ( mandatory ) vbeln , vstel, vkorg, lfart , bzrik, vsbed, vsart, kunag ,
kunnr,
lddat, lfdat, kodat, lprio, lstel ,fkdat , route , wadat, kostk ;
create;
update;
delete;
association _item
{
create;
}
mapping for ZLIKP
{
vbeln = vbeln;
vstel = vstel;
vkorg = vkorg;
lfart = lfart;
bzrik = bzrik;
vsbed = vsbed;
vsart = vsart;
kunag = kunag;
kunnr = kunnr;
lddat = lddat;
lfdat = lfdat;
kodat = kodat;
lprio = lprio;
lstel = lstel;
fkdat = fkdat;
route = route;
wadat = wadat;
kostk = kostk;
}
}
```

```
define behavior for ZBO_ITEM alias item
implementation in class zbp_bo_ietm unique
{
field ( mandatory ) vbeln, posnr, matnr , lfimg, meins;
update;
delete;
association _del;
mapping for ZLIPS
{
vbeln = vbeln;
posnr = posnr;
matnr = matnr;
lfimg = lfimg;
meins = meins;
}
```

Listing 2.50 Modified Behavior Definition

In the modified behavior definition, the keyword `association` establishes a connection between the parent and its child entity, which is needed for compositions used in the business object's structure. The use of the keyword `{create;}` means that instances of the associated entity can be created by the source entity.

Another keyword that shown in Listing 2.50 is `mapping`. This keyword defines a mapping between the fields of the database table and the fields of the CDS view.

Now, let's proceed with creating the behavior implementation for this modified behavior definition, as shown in Listing 2.51.

```
CLASS zbp_bo_del DEFINITION PUBLIC ABSTRACT FINAL FOR
BEHAVIOR OF zbo_del.
ENDCLASS.
CLASS zbp_bo_del IMPLEMENTATION.
ENDCLASS.
```

Listing 2.51 Modified Behavior Implementation

For this task, we'll create a local class, as shown in Listing 2.52, that will implement the behavior. In this class, we'll implement CREATE, DELETE, UPDATE, and READ operations, both on the header and at the line-item level.

```
CLASS lhc_Delivery DEFINITION INHERITING FROM
cl_abap_behavior_handler.
PRIVATE SECTION.
METHODS create FOR MODIFY
IMPORTING entities FOR CREATE Delivery.
```

```
METHODS delete FOR MODIFY
IMPORTING keys FOR DELETE Delivery.
METHODS update FOR MODIFY
IMPORTING entities FOR UPDATE Delivery.
METHODS read FOR READ
IMPORTING keys FOR READ Delivery RESULT result.
METHODS cba_item FOR MODIFY
IMPORTING entities_cba FOR CREATE Delivery\_item.
METHODS rba_item FOR READ
IMPORTING keys_rba FOR READ Delivery\_item FULL result_requested
RESULT result LINK association_links.
METHODS set_lock FOR LOCK
IMPORTING keys FOR LOCK Delivery.
ENDCLASS.
```

Listing 2.52 Handler Class Definition

Data from the entity LS_DEL is stored into table ZLIKP upon execution of the CREATE statement, as shown in Listing 2.53.

```
CLASS lhc_Delivery IMPLEMENTATION.
METHOD create.
data: lt_del type table of ZLIKP.
LT_DEL = value #( FOR LS_DEL IN ENTITIES
(
VBELN = LS_DEL-vbeln
VSTEL = LS_DEL-vstel
VKORG = LS_DEL-vkorg
LFART = LS_DEL-lfart
BZRIK = LS_DEL-bzrik
VSBED = LS_DEL-vsbed
Vsart = LS_DEL-vsart
KUNAG = LS_DEL-kunag
KUNNR = LS_DEL-kunnr
LDDAT = LS_DEL-LDDAT
LFDAT = LS_DEL-LFDAT
KODAT = LS_DEL-kodat
LPRIO = LS_DEL-LPRIO
LSTEL = LS_DEL-LSTEL
FKDAT = LS_DEL-fkdat
ROUTE = LS_DEL-ROUTE
WADAT = LS_DEL-wadat
KOSTK = LS_DEL-KOSTK
) ).
```

```
INSERT ZLIKP FROM TABLE @LT_DEL.
ENDMETHOD.
```

Listing 2.53 Method CREATE for LIKP

In the next method, we'll delete data from table ZLIKP, based on marked items (keys). First, we'll populate all delivery document numbers into the corresponding range and then delete all entries from table ZLISKP based on values in the range. This operation is executed with the DELETE statement, as shown in Listing 2.54.

```
METHOD delete.
data: lt_del type RANGE OF CHAR10.
lt_del = VALUE #(
FOR ls_del IN keys
(
sign = 'I'
option = 'EQ'
low = ls_del-vbeln
)
).
DELETE FROM ZLIKP WHERE VBELN IN @lt_del.
ENDMETHOD.
```

Listing 2.54 Method DELETE for LIKP

We'll pick up the values to be updated from the entity LS_DEL and use these values to update table ZLIKP, as shown in Listing 2.55.

```
METHOD update.
data: lt_del type table of ZLIKP.
LT_DEL = value #( FOR LS_DEL IN ENTITIES
(
VBELN = LS_DEL-vbeln
VSTEL = LS_DEL-vstel
VKORG = LS_DEL-vkorg
LFART = LS_DEL-lfart
BZRIK = LS_DEL-bzrik
VSBED = LS_DEL-vsbed
Vsart = LS_DEL-vsart
KUNAG = LS_DEL-kunag
KUNNR = LS_DEL-kunnr
LDDAT = LS_DEL-LDDAT
LFDAT = LS_DEL-LFDAT
KODAT = LS_DEL-kodat
LPRIO = LS_DEL-LPRIO
LSTEL = LS_DEL-LSTEL
```

```
FKDAT = LS_DEL-fkdat
ROUTE = LS_DEL-ROUTE
WADAT = LS_DEL-wadat
KOSTK = LS_DEL-KOSTK
) ).
UPDATE ZLIKP FROM table @lt_del.
ENDMETHOD.
```

Listing 2.55 Method UPDATE for LIKP

The method for reading data can be used for further processing during the interaction phase. We are not going to implement it in our example, as shown in Listing 2.56.

```
METHOD read.
ENDMETHOD.
```

Listing 2.56 Method READ for LIKP

Next, we'll pick up the relevant values from internal table lt_item and insert these values into table ZLIPS, as shown in Listing 2.57.

```
METHOD cba_item.
data: lt_item type table of ZLIPS.
LT_ITEM = value #(
for ls_entity in entities_cba
FOR ls_item IN ls_entity-%target
(
VBELN = ls_item-vbeln
POSNR = ls_item-posnr
MATNR = ls_item-matnr
LFIMG = ls_item-lfimg
MEINS = ls_item-meins
)
).
insert ZLIPS FROM TABLE @LT_ITEM.
ENDMETHOD.
```

Listing 2.57 Method CREATE for ZLIPS

In the same manner as for the read operation for a parent entity, read access for a child entity can be used for further processing during the interaction phase. We're not going to implement it in our example shown Listing 2.58.

```
METHOD rba_item.
ENDMETHOD.
```

Listing 2.58 Method READ for ZLIPS

Finally, let's implement the lock method, as shown in Listing 2.59.

```abap
METHOD lock.
    TRY.
    "Instantiate lock object
    DATA(lock) = cl_abap_lock_object_factory=>get_instance( iv_name = 'ZBO_DEL' ).
      CATCH cx_abap_lock_failure INTO DATA(lr_exp).
        RAISE SHORTDUMP lr_exp.
    ENDTRY.


    LOOP AT lt_del  ASSIGNING FIELD-SYMBOL(<fs_del>).
      TRY.
          "enqueue travel instance
          lock->enqueue(
              it_parameter  = VALUE #( (  name = 'DELIVERY' value =
REF #( <fs_del>-vbeln ) ) )
          ).



        CATCH cx_abap_lock_failure into lr_exp.
          RAISE SHORTDUMP lr_exp.
      ENDTRY.
    ENDLOOP.

  ENDMETHOD.

ENDCLASS.
```

Listing 2.59 Method LOCK

Notice that, in Listing 2.59, failure to issue a lock will result in a short dump.

Other important characteristics that you can add to a behavior implementation to avoid overwriting and to support the locking mechanism include the following:

- `etag master`/`etag dependent by`: ETags can be used for optimal concurrency control by preventing simultaneous updates.

- `lock master`/`lock dependent by`: Defines which entity/entities support direct locking (`lock master`) or depend on the locking status of the parent or root entity (`lock dependent by`). With this characteristic, you can trigger the `enqueue` method of the lock object.

After the behavior implementation, we'll define and expose the service, as shown in Listing 2.60.

```
@EndUserText.label: 'Service definition for delivery'
define service ZSRV_DELIVERY
{
expose ZBO_DEL as DEL;
expose ZBO_ITEM as ITEM;
}
```

Listing 2.60 Defining and Exposing the Service

The final step is to implement the service binding, as shown in Figure 2.60.

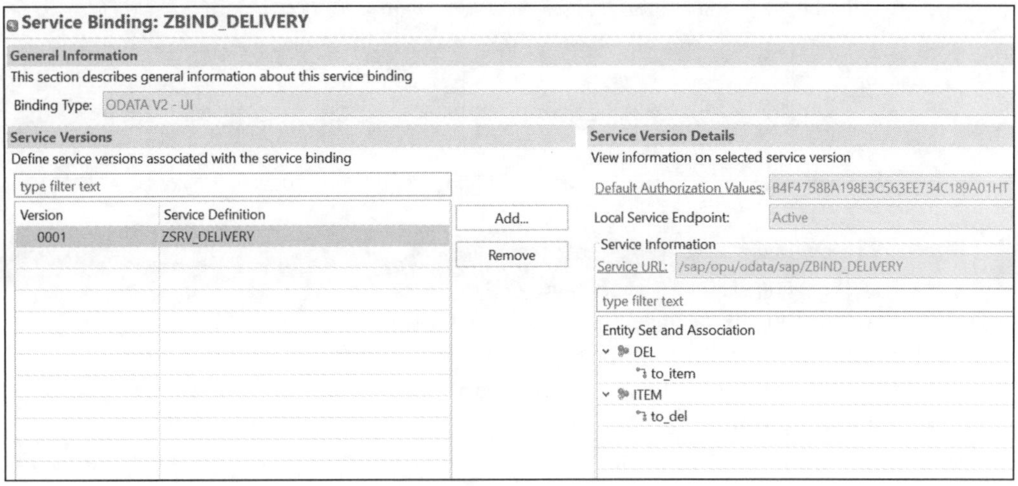

Service Binding: ZBIND_DELIVERY

Figure 2.60 Implementing the Service Binding

The code shown in Listing 2.61 will create records in the delivery header table ZLIKP, delivery item ZLIPS, and material table ZMARA from the Delivery Header Data, Delivery Item Data, and Material Master Data sections of the code, respectively.

```
CLASS z_cl_create_data DEFINITION
  PUBLIC
  FINAL
  CREATE PUBLIC .

  PUBLIC SECTION.
    INTERFACES if_oo_adt_classrun.
  PROTECTED SECTION.
  PRIVATE SECTION.
ENDCLASS.

CLASS Z_CL_CREATE_DATA IMPLEMENTATION.
  METHOD if_oo_adt_classrun~main.
    DATA: it_delv       TYPE TABLE OF zlikp,
```

```abap
            it_delv_item TYPE TABLE OF zlips,
            it_mat       TYPE TABLE of zmara.
***********************Delivery Header Data**************************
    it_delv = VALUE #(
        ( vbeln  = '80000279' vstel = 'US01' vkorg = '0001' lfart = 'NLCC'
          bzrik = 'TX0001' vsbed = '02' vsart = '01' kunag = '2345678' kunnr =
'678905'
          lddat = '20190601' lfdat = '20190601' kodat = '20190602' lprio = '01'
lstel = '03'
          fkdat = '20190605' route = 'US001' wadat = '20190602' kostk = 'X' )
        ( vbeln  = '80000280' vstel = 'CA01' vkorg = '0002' lfart = 'NLCC'
          bzrik = 'TX0001' vsbed = '02' vsart = '01' kunag = '2345678' kunnr =
'678905'
          lddat = '20190601' lfdat = '20190601' kodat = '20190602' lprio = '01'
lstel = '03'
          fkdat = '20190605' route = 'US001' wadat = '20190602' kostk = 'X' )
          ).
**Update or Insert the table entries
    MODIFY zlikp FROM TABLE @it_delv.
    out->write( 'Data inserted into LIKP table successfully!').
**********************************************************************
***********************Delivery Item Data***************************
it_delv_item = VALUE #(
    ( vbeln = '80000279' posnr = '000001' matnr = 'XX-F01' lfimg = 25 meins =
'PC')
    ( vbeln = '80000279' posnr = '000002' matnr = 'YY-F01' lfimg = 50 meins =
'PC')
    ( vbeln = '80000280' posnr = '000001' matnr = 'XX-F01' lfimg = 10 meins =
'PC')
    ( vbeln = '80000280' posnr = '000002' matnr = 'YY-F01' lfimg = 12 meins =
'PC')
    ( vbeln = '80000280' posnr = '000003' matnr = 'ZZ-F01' lfimg = 30 meins =
'PC')
).
**Update or Insert the table entries
    MODIFY zlips FROM TABLE @it_delv_item.
    out->write( 'Data inserted into LIPS table successfully!').
***********************************************************************
***********************Material Master Data****************************
it_mat = value #(
**Trading Goods - Non Packaging materials
    ( matnr = 'XX-F01' mtart = 'HAWA' laeng = 10 breit = 10 hoehe = 10 meins =
'CM' )
    ( matnr = 'YY-F01' mtart = 'HAWA' laeng = 8 breit = 8 hoehe = 8 meins =
```

```
'CM' )
    ( matnr = 'ZZ-F01' mtart = 'HAWA' laeng = 15 breit = 10 hoehe = 12 meins =
'CM' )
**Packaging materials
    ( matnr = 'PACK-16' mtart = 'VERP' laeng = 30 breit = 30 hoehe = 30 meins =
'CM' )
    ( matnr = 'PACK-17' mtart = 'VERP' laeng = 25 breit = 25 hoehe = 25 meins =
'CM' )
    ( matnr = 'PACK-21' mtart = 'VERP' laeng = 20 breit = 20 hoehe = 20 meins =
'CM' )
).
**Update or Insert the table entries
    MODIFY zmara FROM TABLE @it_mat.
    out->write( 'Data inserted into MARA table successfully!').
*********************************************************************************
  ENDMETHOD.
ENDCLASS.
```

Listing 2.61 Data Load Custom Class

Now, let's develop a sample application in SAP Business Application Studio, by following the same steps as in Section 2.3.6. The application output is shown in Figure 2.61.

Delivery		
Delivery: 80000281	Shipping Type: 01	Delivery Priority: 1
Shipping Point: CA01	Sold to Party: 2345678	Loading Point: 03
Sales Org: 0002	Ship to Party: 678905	Billing Date: Oct 20, 2020
Delivery Type: NLCC	Loading Date: Oct 17, 2020	Route: US001
Sales Region: TX001	Delivery Date: Oct 17, 2020	PGI Date: Oct 19, 2020
Shipping Condition: 02	Picking Date: Oct 17, 2020	Picking Status: X

Figure 2.61 Delivery App Output

2.7   Backend Service Development

In previous sections of this chapter, we covered several different scenarios that all had OData services exposed for SAP Fiori. Now, let's expose the service in such a way to make the service available to another ABAP program, or any other program. A service exposed in this way is called a web API.

To create a web API, we'll be using our existing service for this application. The only difference is that the service must now be bound to an OData protocol and published as a web API.

As shown in Figure 2.62, select the **OData V2 - Web API** option instead of the **OData V2 - UI** option. After selecting/creating the transport request, you'll be navigated to service binding management screen, as shown in Figure 2.63.

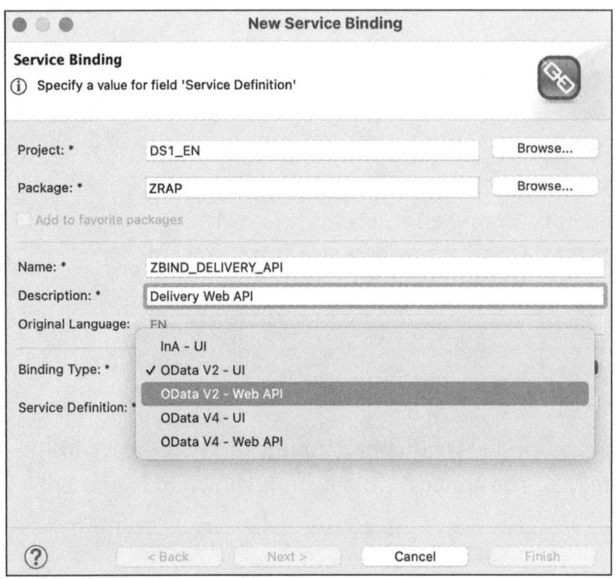

Figure 2.62 Service Binding as a Web API

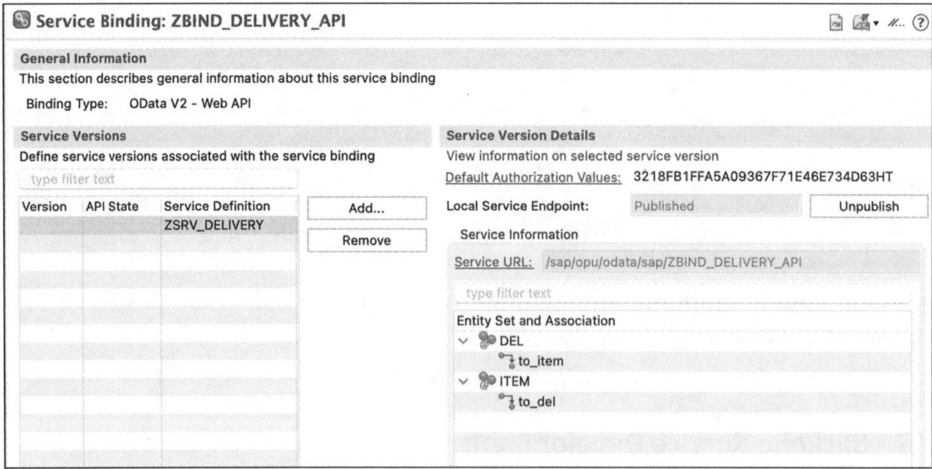

Figure 2.63 Web API Binding

Activate and publish the binding, as described earlier in Section 2.3.5, and now our service is ready for consumption from external applications.

2.8   Setting Up an API Release State

The concept of reusability is important because it helps reduce cost, effort, and time of development. Also, reusability increases the productivity, maintainability, portability, and reliability of software products.

In SAP BTP, before you can make objects available in any of your other software components, you must *release* these objects. Once an object is released in another software component, the two software components become dependent on each other.

Several different contracts exist for releasing objects. Depending on the underlying release contract, released objects must adhere to different stability criteria. Only objects that have the **API State** tab visible can be released as APIs. You can review the various types of API states at the following link: *http://s-prs.co/v523603*. The states listed in the SAP documentation will be available to objects.

Data definition objects, for example, CDS views and source code (i.e., for a class or function module), can be released. Right-click the object to be released and select **API State · Add Use System-Internally (Contract C1)...**, as shown in Figure 2.64.

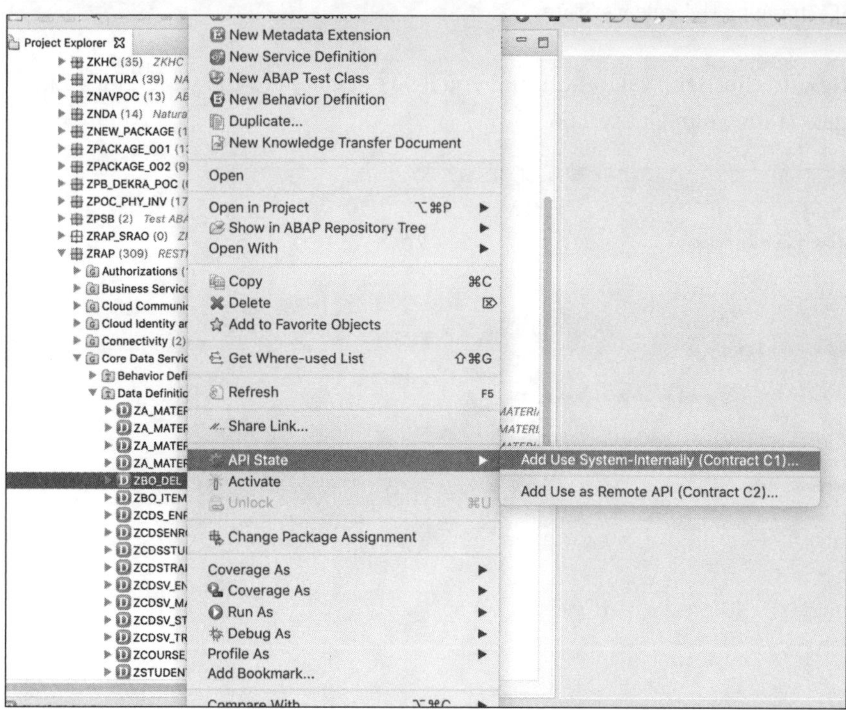

Figure 2.64 API Release

A wizard will open. In our example, we initiated a release for object ZBO_DEL with the **Add Use System-Internally (Contract C1)...** option. On initiating the release of the object, the **Release State** will be set as **RELEASED**, as shown in Figure 2.65.

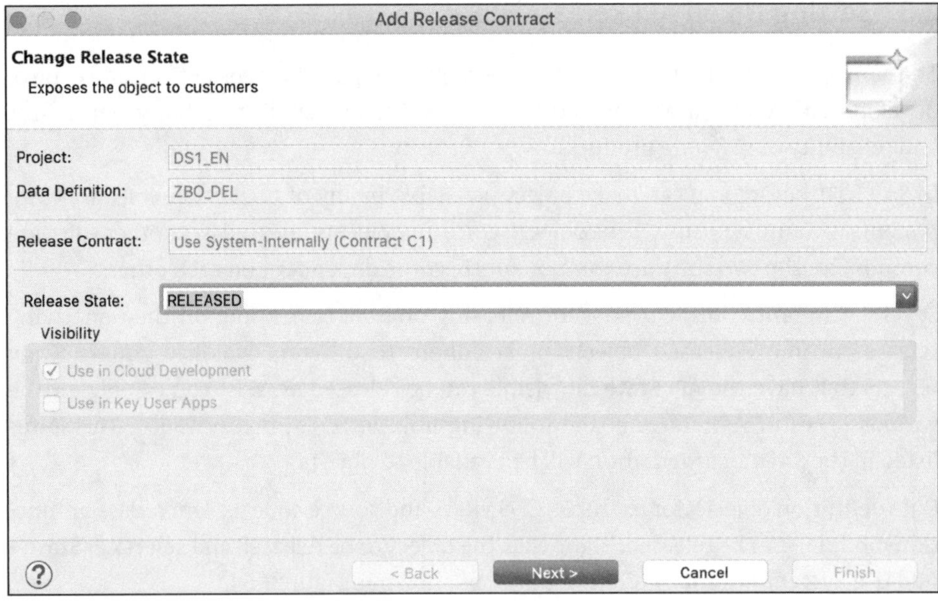

Figure 2.65 Changing the Release State

Click on **Next** to check the validation, and you'll be navigated to the screen for validating changes, as shown in Figure 2.66.

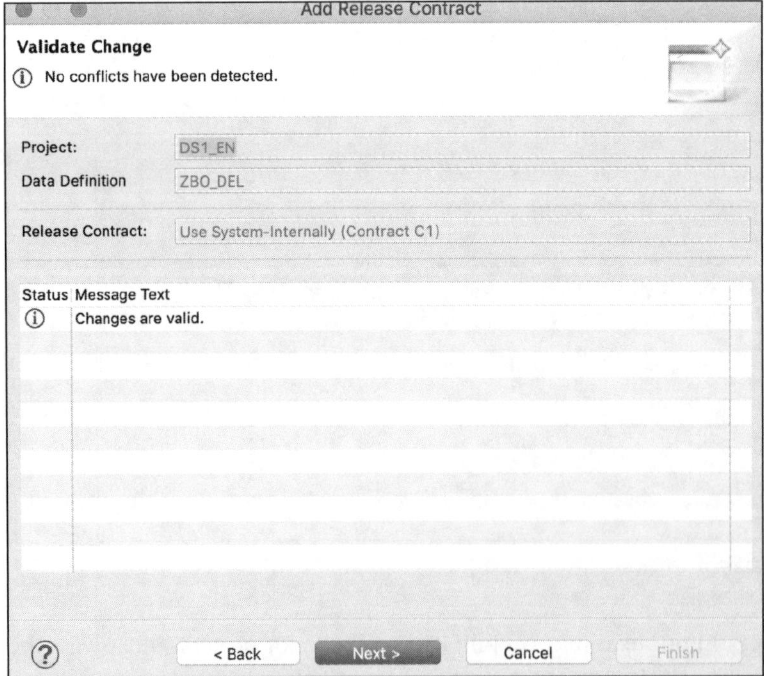

Figure 2.66 Validating Changes

Once you've released the object successfully in the selected contract, you can display, change, or delete the released state of the object. As an alternative, objects can also be released as remote APIs by selecting the **Add Use as Remote API (Contract C2)…** option instead. API release can be applied to remote-enabled function modules and classes as well.

For the list of released objects, navigate to the **Released Objects** folder in the **Project Explorer** view, as shown in Figure 2.67.

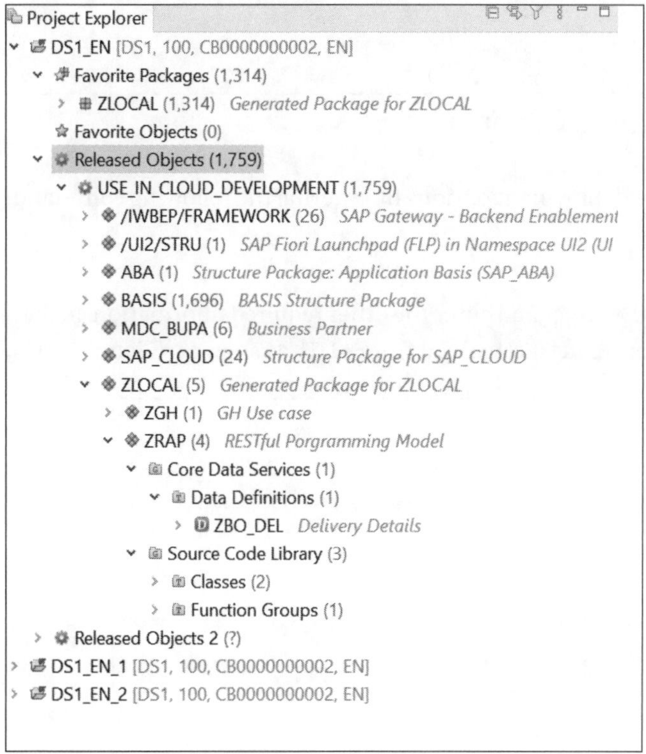

Figure 2.67 Released Objects

2.9   Deploying an Application to SAP BTP, ABAP Environment

In this section, we'll go through the step-by-step process for deploying an application to SAP BTP, ABAP environment.

We'll start with the SAP Fiori application we developed in Section 2.3.6. Launch SAP Business Application Studio and start the development space, which we created in Section 2.3.6. Select the project that you want to deploy in the cloud environment.

The first step is to open a terminal by right-clicking on the desired project and selecting the **Open in Terminal** option.

In the terminal window that opens, at the bottom, type in the following command:

```
npx fiori add flp-config
```

SAP Fiori launchpad content will be added, as shown in Figure 2.68.

```
user@workspaces-ws-6pmbf-deployment-6659444bf9-g55vl: ~/projects/zdeliverydemo  ×

Fiori tools CLI.
Adding flp-config to the project.
In order to integrate your application with the Launchpad module using SAP Business Application Studio, follow instructio
ns available on https://help.sap.com/viewer/ad4b9f0b14b0458cad9bd27bf435637d/Cloud/en-US/ba2dde3670c74d359783061442cfbfe9
.html
? Semantic Object zdeliverydemo_app
? Action display
? Title Delivery Demo
? Subtitle (optional)
user: zdeliverydemo $
```

Figure 2.68 Adding SAP Fiori Launchpad Content

The next step is to define the deployment options by entering the following command:

```
npx fiori add deploy-config
```

Then, we'll choose ABAP as the target and enter the other required information, including the package and transport request (obtained from ADT), as shown in Figure 2.69.

```
Adding flp-config to the project.
? Semantic Object zdeliverydemo_app
? Action display
? Title Delivery Demo App
? Subtitle (optional)
user: zdeliverydemo $ To add deploy config details enter npx fiori add deploy-confi
bash: To: command not found
user: zdeliverydemo $ npx fiori add deploy-config
Fiori tools CLI.
Adding deploy-config to the project.
? Please choose the target ABAP
? Is this an SAP Cloud Platform system? Yes
? Destination abap-cloud-dss_dev_01
? Name (Maximum length: 15 characters) ZDELDEMO
? Package ZRAP
? Transport Request Number DS1K900203
? Generate standalone index.html during deployment Yes
Wrote ui5-deploy.yaml.
user: zdeliverydemo $
```

Figure 2.69 Deployment Options

As a result, the *ui5-deploy.yaml* file will be generated.

Finally, we're ready to trigger deployment. Enter the following command:

```
npm run deploy
```

This statement will execute steps related to npm internally. You don't have to execute any steps explicitly, as shown in Figure 2.70.

```
user@workspaces-ws-6pmbf-deployment-66c87c5c65-bpszg: ~/projects/zdeliverydemo  ×

> zdeliverydemo@0.0.1 deploy /home/user/projects/zdeliverydemo
> ui5 build preload --clean-dest --config ui5-deploy.yaml --include-task=generateManifestBundle generateCachebusterInfo

info builder:builder Building project zdeliverydemo not including dependencies...
info builder:builder 🔨 (1/1) Building project zdeliverydemo
info builder:builder application zdeliverydemo 🔨(1/11) Running task escapeNonAsciiCharacters...
info builder:builder application zdeliverydemo 🔨(2/11) Running task replaceCopyright...
info builder:builder application zdeliverydemo 🔨(3/11) Running task replaceVersion...
info builder:builder application zdeliverydemo 🔨(4/11) Running task generateFlexChangesBundle...
info builder:builder application zdeliverydemo 🔨(5/11) Running task generateManifestBundle...
info builder:builder application zdeliverydemo 🔨(6/11) Running task generateComponentPreload...
info builder:builder application zdeliverydemo 🔨(7/11) Running task createDebugFiles...
info builder:builder application zdeliverydemo 🔨(8/11) Running task uglify...
info builder:builder application zdeliverydemo 🔨(9/11) Running task generateVersionInfo...
info builder:builder application zdeliverydemo 🔨(10/11) Running task generateCachebusterInfo...
info builder:builder application zdeliverydemo 🔨(11/11) Running task deploy-to-abap...
```

Figure 2.70 Application Deployment Running

Once the deployment is complete, you'll see the success message shown in Figure 2.71.

```
info builder:custom deploy-to-abap * Updating the Application Index *
info builder:custom deploy-to-abap Messages from its application log:
info builder:custom deploy-to-abap UIAD ZSTUDENTDEMO_UI5R was created
info builder:custom deploy-to-abap Updated UIADs for UI5 Repository ZSTUDENTDEMO, package ZRAP, transport DS1K900203
info builder:custom deploy-to-abap For details see the application log (SLG1) in client 000 for object /UI5/APPIDX .
info builder:custom deploy-to-abap SAPUI5 Application  has been uploaded and registered successfully
info builder:custom deploy-to-abap * Done *
info builder:custom deploy-to-abap App available at https://43f97a6f-73e1-4914-b153-a776ca463bbc.abap-web.eu10.hana.ondemand.com/sap/bc/ui5_ui5/sap/zstudentdemo
info builder:custom deploy-to-abap Deployment Successful.
info builder:builder Build succeeded in 28 s
info builder:builder Executing cleanup tasks...
```

Figure 2.71 Deployment Output

However, although the output shows the URL where our deployed app is accessible, if you try to open it, you'll get a **403 – Forbidden** message. This error occurs because our app has not been assigned to a business catalog.

First, let's create the IAM app. In ADT, right-click on the package and select **New · Other Repository Object**. Select the **IAM App**, as shown in Figure 2.72.

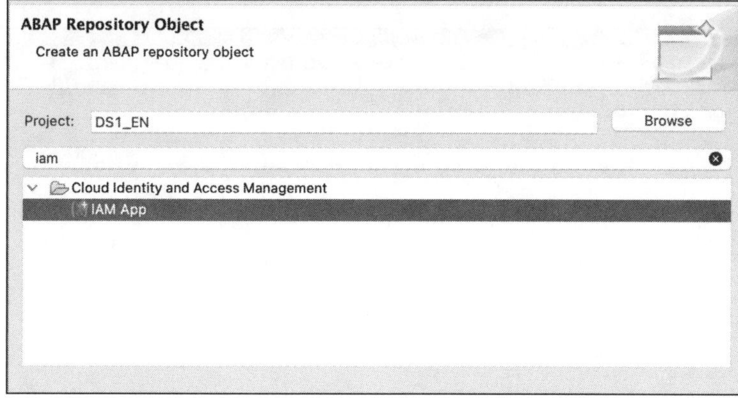

Figure 2.72 New IAM App

Fill out the details for the IAM app and select a transport request. As shown in Figure 2.73, the package should be the custom package in which all objects are stored. In our example, **ZRAP** is the package. You can enter can be any suitable/self-explanatory text in the **Description** field, but you should select **EXT** from the **Application Type** dropdown list.

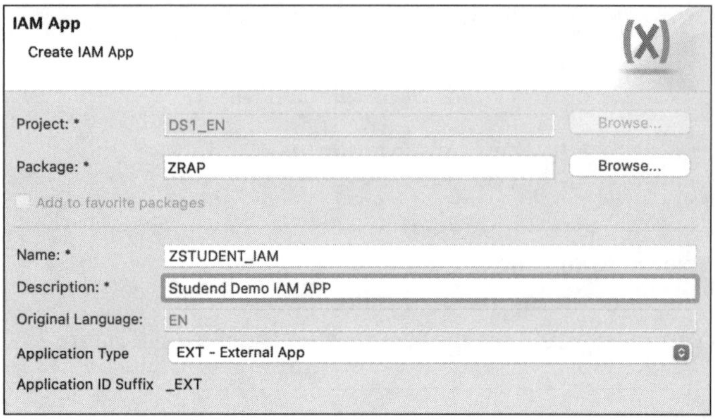

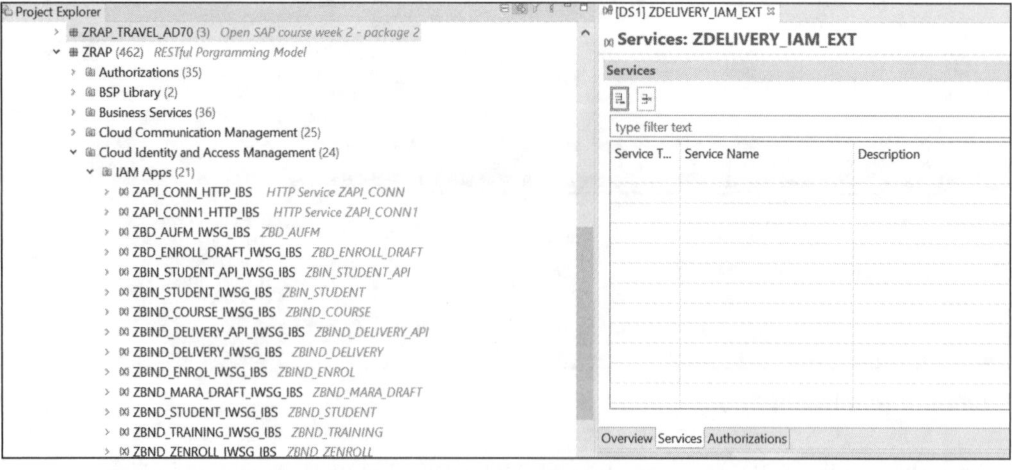

Figure 2.73 IAM App Details

Now that the IAM app has been created, one more task must be performed—adding the service(s) that will be attached to the IAM app. Go to **Project Explorer** • *<your custom package>* (here, **ZRAP**) • **Cloud Identity and Access Management** • **IAM Apps**. Then, select **ZDELIVERY_IAM_EXT** (which is our demo IAM application) and go to the **Services** tab, as shown in Figure 2.74. Next, click on the **Insert** button (leftmost button on the right half of Figure 2.74) which will take you to the next screen, as shown in Figure 2.75.

In Figure 2.75, select a **Service Type** and type in your **Service Name**.

Next, save and publish the IAM app by clicking the **Publish Locally** button, as shown in Figure 2.76.

Figure 2.74 ZDELIVERY_IAM_EXT IAM Application

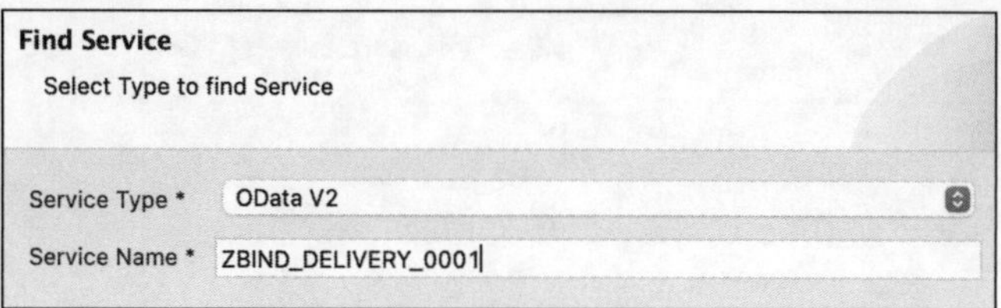

Figure 2.75 Adding a Service

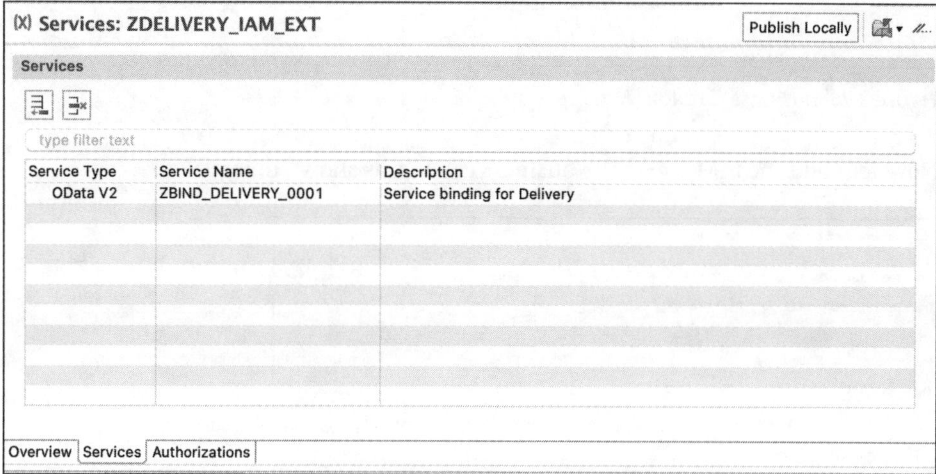

Figure 2.76 IAM App: Services View

Moving on to business catalog—again, right-click on the package and select **New •
Other Repository Object**. In the list of objects, you can filter the list by business catalog,
as shown in Figure 2.77.

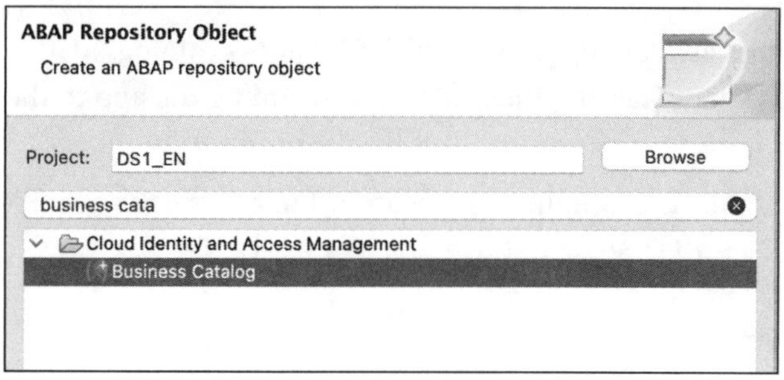

Figure 2.77 Creating a Business Catalog

Enter a name for the business catalog, select a transport request, and proceed to the
next screen. Then, navigate to the **Apps** section, as shown in Figure 2.78.

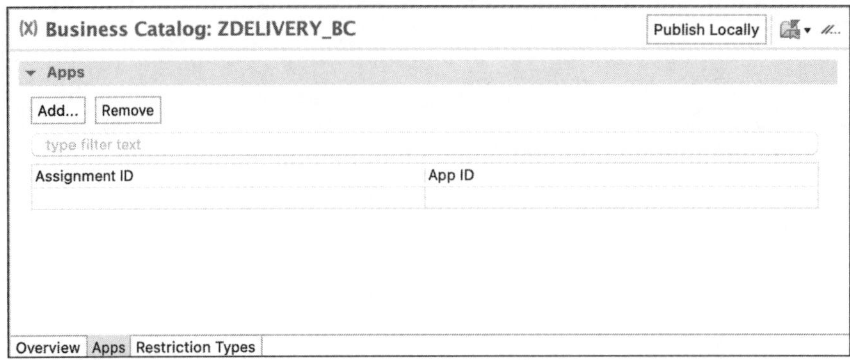

Figure 2.78 Business Catalog: Apps Section

Now, let's add the IAM app to the business catalog, as shown in Figure 2.79.

Figure 2.79 Assigning the IAM App to a Business Catalog

Next, save and publish the business catalog by saving the business catalog. At this point, we have completed the task of creating the IAM app and adding the app to the business catalog.

Now, go back to SAP Business Application Studio and repeat the build. Once the deployment is completed, you should see the message shown in Figure 2.80.

```
info builder:custom deploy-to-abap Transport Request "DS1K900203" has been determined from the corresponding import parameter.
info builder:custom deploy-to-abap The external Code Page Name "UTF8" has been determined from the corresponding import parameter.
info builder:custom deploy-to-abap The acceptance of Unix style end of line markers in text files has been determined from the corresponding import parameter.
info builder:custom deploy-to-abap Unix style end of line markers in text files get accepted.
info builder:custom deploy-to-abap The delta mode has been turned on.
info builder:custom deploy-to-abap Updating existing SAPUI5 ABAP repository ZSTUDENTDEMO
info builder:custom deploy-to-abap UPLOAD FILE   : manifest-bundle.zip (Binary)
info builder:custom deploy-to-abap UPLOAD FILE   : sap-ui-cachebuster-info.json (Text)
info builder:custom deploy-to-abap * Updating the Application Index *
info builder:custom deploy-to-abap Messages from its application log:
info builder:custom deploy-to-abap UIAD ZSTUDENTDEMO_UISR was updated
info builder:custom deploy-to-abap Updated UIADs for UI5 Repository ZSTUDENTDEMO, package ZRAP, transport DS1K900203
info builder:custom deploy-to-abap For details see the application log (SLG1) in client 000 for object /UI5/APPIDX .
info builder:custom deploy-to-abap SAPUI5 Application  has been uploaded and registered successfully
info builder:custom deploy-to-abap * Done *
info builder:custom deploy-to-abap App available at https://43f97a6f-73e1-4914-b153-a776ca463bbc.abap-web.eu10.hana.ondemand.com/sap/bc/ui5_ui5/sap/zstudentdemo
info builder:custom deploy-to-abap Deployment Successful.
info builder:builder Build succeeded in 18 s
info builder:builder Executing cleanup tasks...
user: zstudentdemo $
```

Figure 2.80 Build Completed

If you try to access the URL now, you'll be navigated to the SAP Fiori app, as shown in Figure 2.81.

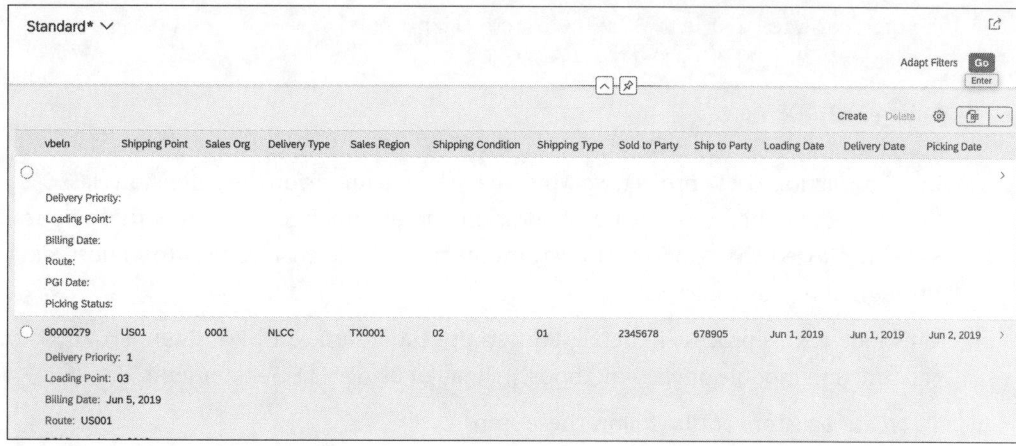

Figure 2.81 Deployed App

Finally, we've deployed the app and made it accessible via a public URL—our application is now deployed to SAP BTP, ABAP environment.

2.10   Custom Entities in the ABAP RESTful Application Programming Model

In this section, you'll learn all about custom entities, including how to create custom entities and consume them in your applications.

As described in earlier sections, our data model is based on CDS in the ABAP RESTful application programming model, on top of which we've defined a behavior definition and a behavior implementation for business logic to enable various operations, for example, CRUD operations. The service definition and service binding expose your CDS data model to the external world.

In the classic approach to ABAP development, you would use Transaction SEGW to build a project in the SAP Gateway service builder. However, in the ABAP RESTful application programming model, no Transaction SEGW SAP Gateway project is used, and thus, custom entities or structures cannot be created directly. All entities are based on CDS, which is based on a SELECT statement on a table.

For example, Listing 2.62 shows sample code outlining how a CDS view can select data from table ZLIKP.

```
@AbapCatalog.sqlViewName: 'ZBO_DELIVERY'
@EndUserText.label: 'Delivery Details'
@AccessControl.authorizationCheck: #NOT_REQUIRED
```

```
@AbapCatalog.compiler.compareFilter: true
@AbapCatalog.preserveKey: true

define root view ZBO_DEL as select from zlikp as del
composition [0..*] of ZBO_ITEM as _item
```

Listing 2.62 CDS View

In a Transaction SEGW project, we would have had to provision the extension class DPC_
EXT to redefine methods in case of additional requirements. To address these issues,
SAP introduced the concept of custom entities, which enables you to expose data
through ABAP custom classes.

Let's look at this process in detail. We can use the interface IF_RAP_QUERY_PROVIDER to
call function module or class methods, instead of using SELECT statements.

To create a custom entity, follow these steps:

1. Create a custom class to implement the query using the code shown in Listing 2.63.

```
class ZCL_CUST_GETLIST definition
      public
 final
create public .
public section.
interfaces if_rap_query_provider.
protected section.
private section.

ENDCLASS.

CLASS ZCL_CUST_GETLIST IMPLEMENTATION.

METHOD if_rap_query_provider~select.

"variables needed to call BAPIs
    DATA lt_cust TYPE STANDARD TABLE OF  ZCE_CUST_VIA_RFC .
    data ls_cust type ZCE_CUST_VIA_RFC.
    DATA lt_result TYPE STANDARD TABLE OF  ZCE_CUST_VIA_RFC .
    DATA ls_product TYPE ZCE_CUST_VIA_RFC .

    "key for BAPI_GET_DETAIL
    TYPES : BEGIN OF cust_rfc_key_type,
            customerid TYPE ZCE_CUST_VIA_RFC-customerid,
          END OF cust_rfc_key_type.
    DATA ls_cust_rfc_key TYPE cust_rfc_key_type.
```

```abap
"select options
    DATA lt_filter_ranges_customerid TYPE RANGE OF ZCE_CUST_VIA_RFC-
customerid.
    DATA ls_filter_ranges_customerid LIKE LINE OF lt_filter_ranges_
customerid.

    " ABAP source code for type definition for BAPIRET2
    TYPES : BEGIN OF ty_bapiret2,
              type      c LENGTH 1,
              id        TYPE c LENGTH 20,
              number    TYPE n LENGTH 3,
              message   TYPE c LENGTH 220,
              logno     TYPE c LENGTH 20,
              logmsgno  TYPE n LENGTH 6,
              messagev1 TYPE c LENGTH 50,
              messagev2 TYPE c LENGTH 50,
              messagev3 TYPE c LENGTH 50,
              messagev4 TYPE c LENGTH 50,
              parameter TYPE c LENGTH 32,
              row       TYPE i,
              field     TYPE c LENGTH 30,
              system    TYPE c LENGTH 10,
            END OF ty_bapiret2.

    "DATA lt_return   TYPE STANDARD TABLE OF bapiret2.
    DATA lt_return   TYPE STANDARD TABLE OF ty_bapiret2.
    "variables generic for implementation of custom entity
    DATA lv_details_read TYPE abap_bool.
    "DATA ls_sel_opt TYPE /iwbep/s_cod_select_option.

*   ensure: in case of a single record is requested (e.g., data for a detail
page),
* only one record is returned and SET_TOTAL_NUMBER_OF_RECORDS = 1
    DATA lv_orderby_string TYPE string.
    DATA lv_select_string TYPE string.
    DATA(lv_abap_trial) = abap_true.

  IF lv_abap_trial = abap_false.

    TRY.
        DATA(lo_rfc_dest) = cl_rfc_destination_provider=>create_by_cloud_
destination(
                                  i_name = |CP8_RFC_ABAP| ).
        DATA(lv_rfc_dest_name) = lo_rfc_dest->get_destination_name( ).
```

```
        CATCH cx_rfc_dest_provider_error INTO DATA(lx_dest).

        CALL FUNCTION 'BAPI_FLCUST_GETLIST'
            DESTINATION lv_rfc_dest_name
              tables
              customer_list  = lt_cust.

          io_response->set_total_number_of_records( lines( lt_cust ) ).
          io_response->set_data( lt_cust ).
      ENDTRY.
    ENDIF.
  ENDMETHOD.
ENDCLASS.
```

Listing 2.63 Custom Class to Implement a Select Query

As shown in Listing 2.63, you'll need to define the class ZCL_CUST_GETLIST. Make sure you define the interface within the public section. Then, you'll need to write the class implementation. In the SELECT method, you must retrieve the destination using cl_rfc_destination_provider and then consume the function module 'BAPI_FLCUST_GETLIST' from that destination.

2. Next, we'll create a custom entity definition. In the **Project Explorer** view, navigate to your custom package (in our example, **ZRAP**), then right-click on the **Dictionary** folder to create a data definition, as shown in Figure 2.82.

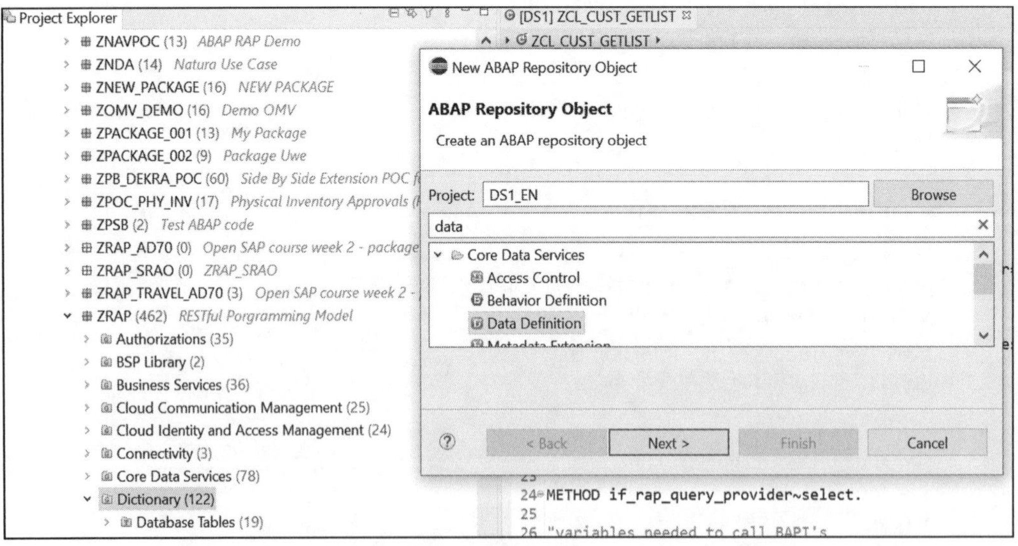

Figure 2.82 New Data Definition

3. As shown in Figure 2.83, for creating the new data definition, you'll need to provide the package, specify a name for the data definition, and add a description.

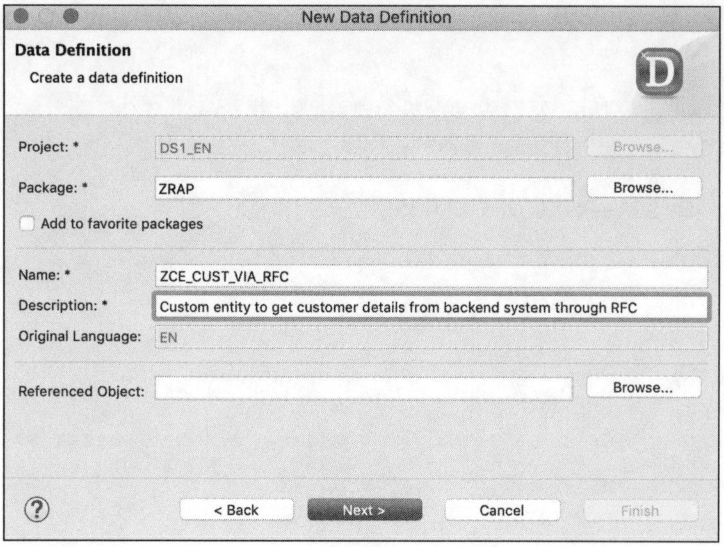

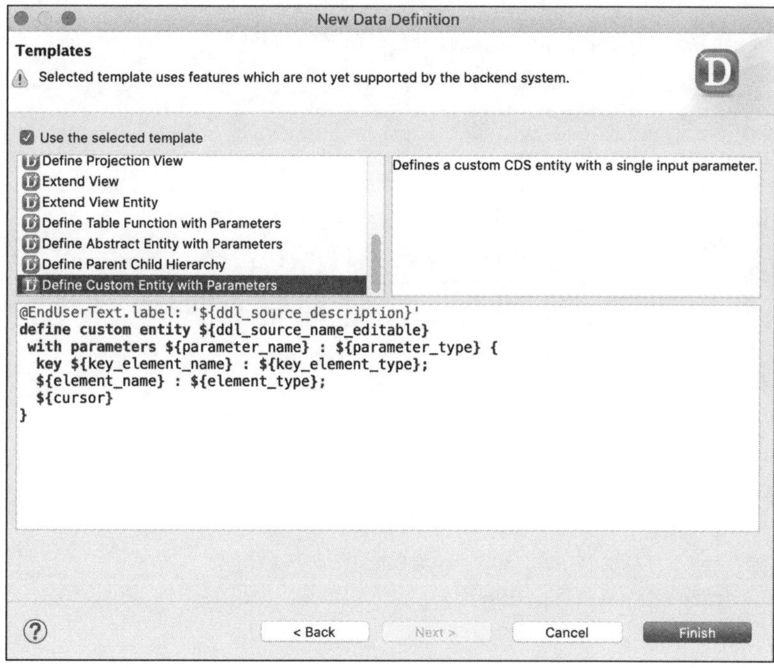

Figure 2.83 Custom Entity Definition

Figure 2.84 shows the available template from the list for the custom entity.

Figure 2.84 Selection of Template for Custom Entity

Now that you've defined the custom class `ZCL_CUST_GETLIST` in Listing 2.63 and created the data definition for `ZCE_CUST_VIA_RFC`, as shown in Figure 2.83, now let's a look at some sample code utilizing these elements and implementing the custom entity. We'll build on the same use case of fetching customer details through RFC, as shown in Listing 2.64.

```
@EndUserText.label: 'Custom entity to get customer details via RFC'
@ObjectModel:{
    query: {
        implementedBy: 'ABAP:ZCL_CUST_GETLIST'
    }
}
define root custom entity ZCE_CUST_VIA_RFC
{

  @UI.facet      : [
          {
              id      :   'Customer',
              purpose:    #STANDARD,
              type    :   #IDENTIFICATION_REFERENCE,
              label   :   'Customer',
              position  : 10 }
        ]
        @UI            : {
    lineItem       : [{position: 10, importance: #HIGH}],
    identification: [{position: 10}],
    selectionField: [{position: 10}]
    }

key CUSTOMERID       : abap.char( 8 );
    TypeCode         : abap.char( 2 );
    @UI              : {
    lineItem       : [{position: 20, importance: #HIGH}],
    identification: [{position: 20}],
    selectionField: [{position: 20}]
    }

    CUSTNAME         : abap.char( 25 );
    @UI              : {
    lineItem       : [{position: 30, importance: #HIGH}],
    identification: [{position: 30}]
    }
    FORM             : abap.char( 15 );
    @UI              : {
    identification: [{position: 40}]
    }
    STREET : abap.char( 30 ) ;
    POBOX  : abap.char( 10 ) ;
```

```
    POSTCODE  : abap.char ( 25 );
    CITY      : abap.char ( 25 ) ;
    COUNTR    : abap.char ( 3 ) ;
    COUNTR_ISO : abap.char ( 2 );
    REGION     : abap.char ( 3 );
    PHONE      : abap.char ( 30 );
    EMAIL      : abap.char ( 40 );
}
```

Listing 2.64 Custom Entity for Displaying Field Contents

As shown in Listing 2.64, you'll need to create an instance of the custom class ZCL_
CUST_GETLIST. You'll then need to lay out the structure of custom entity ZCE_CUST_
VIA_RFC. While defining your custom entities, make sure you use meaningful information in the ID and label to make the business context clear. Note that, in this sample code, CUSTOMERID has been defined as a key for uniquely distinguishing multiple
customer records. Further details of the customer, for instance, name and address,
can be stored in the relevant additional fields added in the structure.

4. Create a service definition and service binding in the same way as when we created
   the data definition in earlier steps. As shown in Figure 2.85, in the **Project Explorer**
   view, go to your custom package and expand the **Business Services** folder. Right-click
   on the **Service Definitions** folder and select **New Service Definition**, which will open
   the screen shown in Figure 2.86.

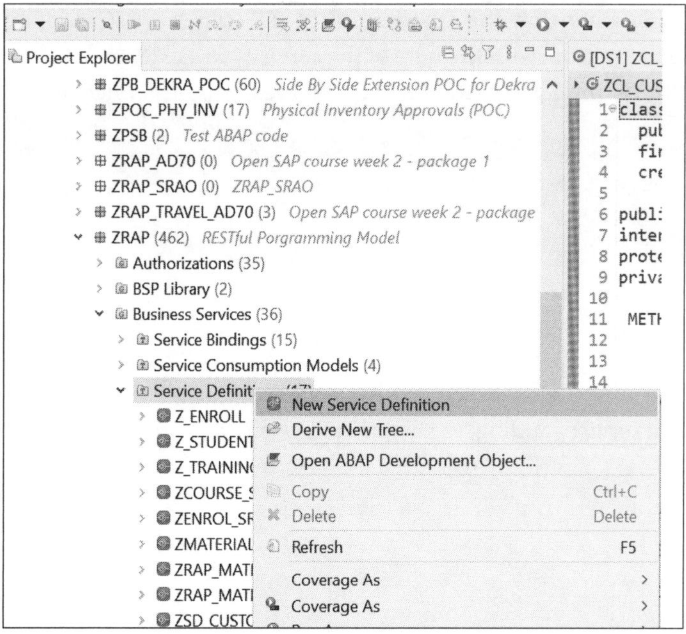

Figure 2.85 New Service Definition

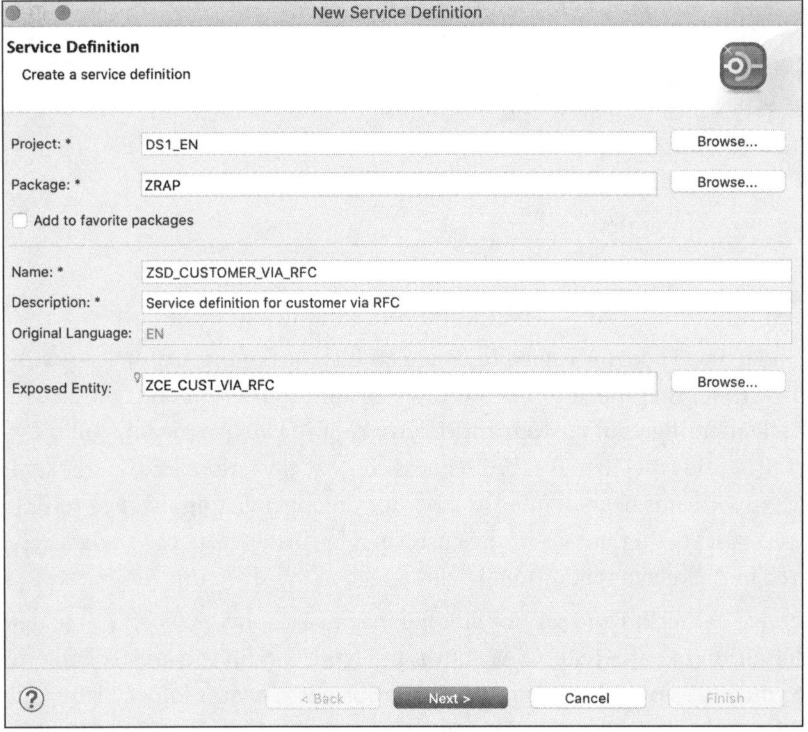

Figure 2.86 New Service Definition, Continued

As shown in Listing 2.65, note that the custom entity ZCE_CUST_VIA_RFC is being exposed for further consumption, with the help of the service ZCE_CUSTOMER_VIA_RFC, as shown in Figure 2.87. Listing 2.65 shows the code for creating the service defini-tion.

```
@EndUserText.label: 'Service definition for customer via RFC'
define service ZSD_CUSTOMER_VIA_RFC {
    expose ZCE_CUST_VIA_RFC;
}
```

Listing 2.65 Service Definition

Once the service is defined, you must bind it so that you can consume the service in UI applications. Create a service binding, as shown in Figure 2.88.

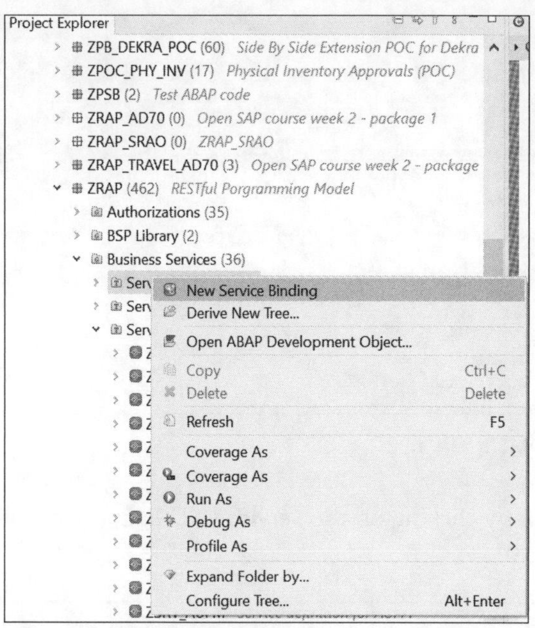

Figure 2.87 New Service Binding

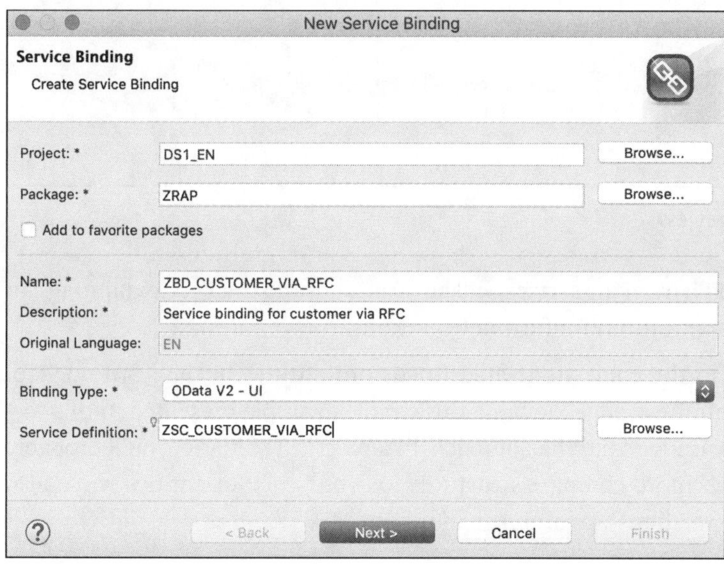

Figure 2.88 New Service Binding, Continued

Activate and publish the service. Once published, a service URL will be generated, as shown in Figure 2.89, that can be utilized in UI applications.

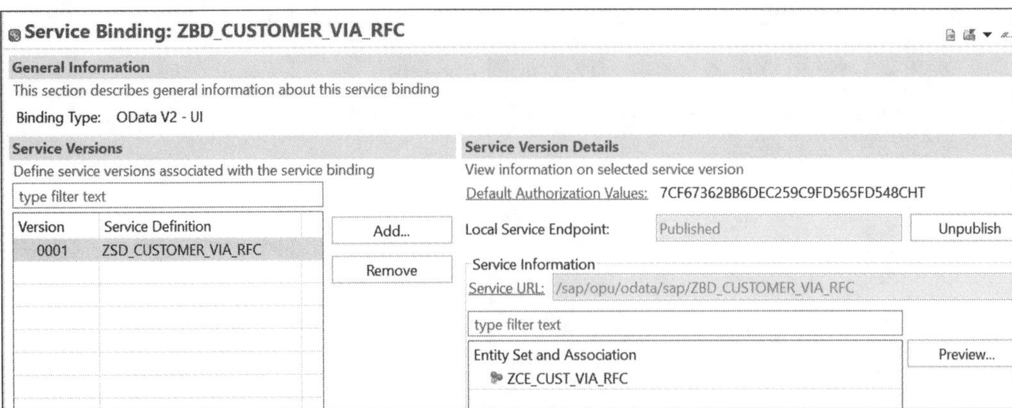

Figure 2.89 Service Binding ZBD_CUSTOMER_VIA_RFC

5. Preview the contents of the service by clicking on the **Service URL** link, which will open the screen shown in Figure 2.90.

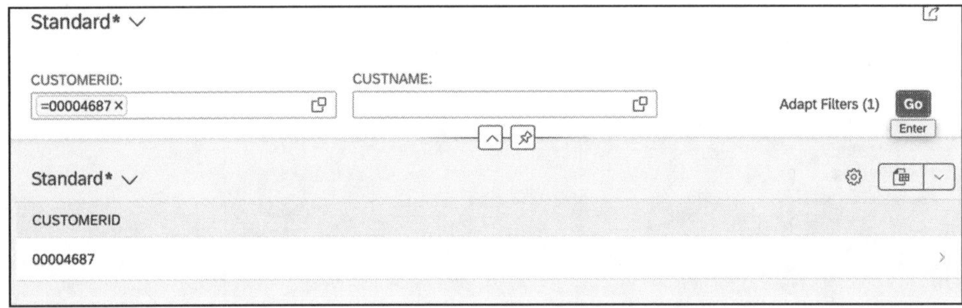

Figure 2.90 Results Preview

Customer details, taken from the service, will be displayed, and this service binding can be used in other UI applications for further operations.

Now, you understand the concepts related to custom entities in the ABAP RESTful application programming model. We hope this simple example presented in this section helps you easily understand the approach. In a practical scenario, you'll probably need to define much more complex structures for your custom entities while also keeping in mind the relationships among various entities.

2.11   Advanced Development Techniques

In this section, we'll cover some advanced techniques that can help you accelerate development in SAP BTP, ABAP environment. Specifically, we'll discuss the XCO library and the RAP generator.

2.11.1   XCO Library

SAP has released the XCO library, which you can use to create custom repository objects. Approximately 44 objects have been released by SAP, in the form of a class and a naming convention for these objects—XCO_CP.

The purpose of the XCO library is to create, update, and delete ABAP repository objects programmatically. APIs are available for the object types shown in Table 2.4.

Object	Description
BDEF	Behavior definitions
CLAS	Classes
DDLS	Data definitions
DDLX	Metadata extensions
DCLS	Access controls
DEVC	Packages
DOMA	Domains
DTEL	Data elements
INTF	Interfaces
MSAG	Message classes
SRVD	Service definitions
SRVB	Service bindings
TABL	Structures and database tables
TTYP	Table types

Table 2.4 Repository Object Types

Let's look at creating custom data element using these classes. In this example, the following XCO classes are called:

1. Class `xco_cp_abap_repository` to get the instance of the custom package in which we'll be creating the data element
2. Class `xco_cp_generation`, a handler instance to create the repository object in the specified transport request
3. Class `xco_cp_abap_dictionary` to create the data element

We'll create class `ZCL_XCO_DATAELEMENT` using the code shown in Listing 2.66.

```abap
class ZCL_XCO_DATAELEMENT definition
  public
  final
  create public.
public section.
  INTERFACES if_oo_adt_classrun.
protected section.
private section.
ENDCLASS.

CLASS ZCL_XCO_DATAELEMENT IMPLEMENTATION.
  METHOD if_oo_adt_classrun~main.
DATA(lo_package) = xco_cp_abap_repository=>package->for( 'ZRAP' ).
** Create the handler instance to create repository objects
    DATA(lo_put_operation) = xco_cp_generation=>environment->dev_system(
      'DS1K900203' "lo_transport_request->value
    )->create_put_operation(  ).
DATA(lo_specification) = lo_put_operation->for-dtel->add_object( 'ZVAX_TP1Q_
3RDPVALUE'
  )->set_package( 'ZRAP'
  )->create_form_specification( ).
lo_specification->set_short_description( '3rd party value' ).
lo_specification->set_data_type( xco_cp_abap_dictionary=>built_in_type->char( 30
) ).
  lo_specification->field_label-short->set_text( '3rd party value' ).
  lo_specification->field_label-medium->set_text( '3rd party value' ).
  lo_specification->field_label-long->set_text( '3rd party value' ).
  lo_specification->field_label-heading->set_text( '3rd party value' ).

DATA(lo_specific) = lo_put_operation->for-dtel->add_object( 'ZVAX_TP1C_VTEXT'
  )->set_package( 'ZRAP'
  )->create_form_specification( ).
lo_specific->set_short_description( 'PH Name' ).
lo_specific->set_data_type( xco_cp_abap_dictionary=>built_in_type->char( 40 ) ).
lo_specific->field_label-short->set_text( 'PH Name' ).
lo_specific->field_label-medium->set_text( 'PH Name' ).
lo_specific->field_label-long->set_text( 'PH Name' ).
lo_specific->field_label-heading->set_text( 'PH Name' ).

DATA(li_specific) = lo_put_operation->for-dtel->add_object( 'ZVAX_TP1C_WNAME'
  )->set_package( 'ZRAP'
  )->create_form_specification( ).
li_specific->set_short_description( 'Plant Name' ).
li_specific->set_data_type( xco_cp_abap_dictionary=>built_in_type->char( 30 ) ).
```

```
li_specific->field_label-short->set_text( 'Plant Name' ).
li_specific->field_label-medium->set_text( 'Plant Name' ).
li_specific->field_label-long->set_text( 'Plant Name' ).
li_specific->field_label-heading->set_text( 'Plant Name' ).

lo_put_operation->execute( ).
endmethod.
ENDCLASS.
```

Listing 2.66 Custom Class for Data Element

The PUT operation is called for the creation and updating of the object according to a provided specification, and the DELETE operation will delete an already existing object.

This class completes the following tasks:

- Define a transport request.

- Create the data element ZVAX_TP1Q_3RDPVALUE in the package ZRAP.

- Create the data element ZVAX_TP1C_VTEXT in the package ZRAP.

- Create the data element ZVAX_TP1C_WNAME in the package ZRAP.

For each data element, you'll define a description, a data type, and texts (short/medium/long/heading).

Once you execute the class, data elements will be generated, and you'll see them available in the system, as shown in Figure 2.91.

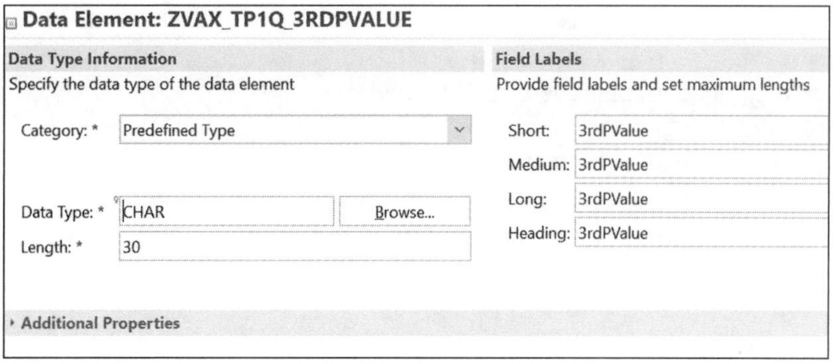

Figure 2.91 XCO Library-Generated Data Element

Now, let's create a custom table with these data elements using the XCO library classes xco_cp_abap_repository and xco_cp_database_table, as shown in Listing 2.67.

```
class ZCL_XCO_TABLE definition
public
final
create public.
```

```abap
public section.
INTERFACES if_oo_adt_classrun.
protected section.
private section.
ENDCLASS.
CLASS ZCL_XCO_TABLE IMPLEMENTATION.
METHOD if_oo_adt_classrun~main.
DATA(lo_package) = xco_cp_abap_repository=>package->for( 'ZRAP' ).
** Get the instance of the transport target.
DATA(lv_transport_target) = lo_package->read(
)-property-transport_layer->get_transport_target( )->value.
** Create the handler instance to create repository objects
DATA(lo_put_operation) = xco_cp_generation=>environment-
>dev_system(
'DS1K900203' "lo_transport_request->value
)->create_put_operation( ) .
DATA(lo_database_table) = lo_put_operation->for-tabl-for-database_table-
>add_object( 'ZVAX_TP1_REPORT'
)->set_package( lo_package->name
)->create_form_specification( ).
lo_database_table->set_short_description( 'Report Protocol' ).
lo_database_table->set_delivery_class(
xco_cp_database_table=>delivery_class->l ).
lo_database_table->set_data_maintenance(
xco_cp_database_table=>data_maintenance->allowed ).
lo_database_table->add_field( 'MANDT'
)->set_key_indicator(
)->set_type( xco_cp_abap_dictionary=>built_in_type->char( 3 )
)->set_not_null( ).
lo_database_table->add_field( 'VKORG'
)->set_key_indicator(
)->set_type( xco_cp_abap_dictionary=>built_in_type->char( 4 )
)->set_not_null( ).
lo_database_table->add_field( 'PRDHA'
)->set_key_indicator(
)->set_type( xco_cp_abap_dictionary=>built_in_type->char( 18 )
)->set_not_null( ).
lo_database_table->add_field( 'VTEXT'
)->set_key_indicator(
)->set_type(
xco_cp_abap_dictionary=>data_element('ZVAX_TP1C_VTEXT')
)->set_not_null( ).
lo_database_table->add_field( 'WERKS'
)->set_key_indicator(
```

```
)->set_type( xco_cp_abap_dictionary=>built_in_type->char( 4 )
)->set_not_null( ).
lo_database_table->add_field( 'NAME1'
)->set_key_indicator(
)->set_type(
xco_cp_abap_dictionary=>data_element('ZVAX_TP1C_WNAME')
)->set_not_null( ).
lo_database_table->add_field( 'VALUE_3RDP'
)->set_type(
xco_cp_abap_dictionary=>data_element('ZVAX_TP1Q_3RDPVALUE') ).
lo_put_operation->execute( ).
endmethod.
ENDCLASS.
```

Listing 2.67 Custom Table Creation from an XCO Class

This code generates the custom table by assigning the delivery class and adding fields to the table.

Once you execute this class, the database table ZVAX_TP1_REPORT will be created. You can see the newly created class in the ABAP Dictionary (DDIC) object list in ADT.

Using a similar approach, you can create other objects as well, using the available XCO library classes. The XCO library is a prerequisite for the RAP generator, which we'll cover in the next section.

2.11.2   RAP Generator

Throughout this chapter, we've been developing all our objects step by step, starting from scratch. Let's revisit the steps for creating any application for any scenario, whether for managed or unmanaged scenarios:

1. Create database tables.

2. Create interface CDS views.

3. Create projection CDS views.

4. Create a service definition.

5. Create a behavior definition.

6. Create a behavior implementation.

7. Consume the OData service in the application.

To speed up all these steps and get up and running with starter business objects, you can use the RAP generator, an open-source project developed by Andre Fischer and maintained by SAP.

The RAP generator picks up tables or CDS views and builds "boilerplate" business objects, leaving it to the developer to implement specific business logic with these generic objects. The RAP generator supports both managed and unmanaged scenarios and can be accessed at *https://github.com/SAP-samples/cloud-abap-rap*. At this GitHub repository, you'll also find a detailed description and instructions as well as some examples.

SAP has developed the custom class `ZCL_RAP_BO_GENERATOR`, which is the base of this concept. This class helps generate an almost complete stack of ABAP RESTful application programming model business objects based on the root, child, and grandchild database structure. For scalable architectures, the object-oriented approach is followed in the RAP generator.

To autogenerate all the necessary repository objects with the help of the RAP generator, follow these steps:

1. You only need to create the database tables, but keep in mind that the table must have a `uuid` field.

   Let's create database tables for sales order headers and sales order items to demonstrate this concept in detail, as shown in Listing 2.68 and Listing 2.69, respectively.

   In this step, you'll create the custom table `ZVBAK` (Sales Order Header Table). The label in this context is the short description of the table, which is defined in the same way as described earlier in Section 2.3.1.

```
@EndUserText.label : 'Sales Order header table'
@AbapCatalog.enhancementCategory : #NOT_EXTENSIBLE
@AbapCatalog.tableCategory : #TRANSPARENT
@AbapCatalog.deliveryClass : #A
@AbapCatalog.dataMaintenance : #RESTRICTED
define table zvbak
  {
    key client : abap.clnt not null;
    key uuid   : sysuuid_x16 not null;
    key vbeln  : abap.char(10) not null;
    erdat      : abap.dats;
    ernam      : abap.char(10);
    auart      : abap.char(4);


  }
```

Listing 2.68 Table ZVBAK

   Then, you'll create custom table `ZVBAP` (Sales Order Item Table).

```
@EndUserText.label : 'Sales Order line items'
@AbapCatalog.enhancementCategory : #NOT_EXTENSIBLE
@AbapCatalog.tableCategory : #TRANSPARENT
```

```
@AbapCatalog.deliveryClass : #A
@AbapCatalog.dataMaintenance : #RESTRICTED
define table zvbap
{
  key client       : abap.clnt not null;
  key uuid         : sysuuid_x16 not null;
  key parent_uuid  : sysuuid_x16 not null;
  key vbeln        : abap.char(10) not null;
  key posnr        : abap.numc(6) not null;
  matnr            : abap.char(10);
  @Semantics.quantity.unitOfMeasure : 'zvbap.zieme'
  zmeng            : abap.quan(13,3);
  zieme            : abap.unit(3);
}
```

Listing 2.69 Table ZVBAP

2. Create a JSON file for these tables and save it on your local system. The JSON file format should resemble the code shown in Listing 2.70. This file must be uploaded to the RAP generator. Through this JSON file, the remaining objects will be created through the RAP generator.

Be sure that this JSON file is in the correct format.

```
{
  "implementationType": "managed_uuid",
  "namespace": "Z",
  "suffix": "_sales",
  "prefix": "RAP_2",
  "package": "ZRAP",
  "datasourcetype": "table",
  "hierarchy": {
    "entityName": "sales",
    "dataSource": "zvbak",
    "objectId": "vbeln",
    "children": [
      {
        "entityName": "SalesItem",
        "dataSource": "zvbap",
        "objectId": "vbeln"
      }
    ]
  }
}
```

Listing 2.70 JSON File

The implementation type for the RAP generator is 'managed_uuid', which is why you need the uuid field as a key field when defining the table.

3. Navigate to the HTTP service, in this case, ZRAP_GENERATOR, in ADT as shown in Figure 2.92.

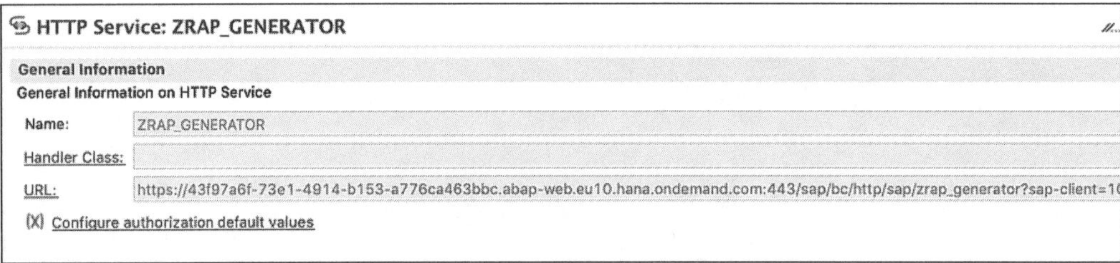

Figure 2.92 ZRAP_GENERATOR

4. Open the URL and, as shown in Figure 2.93, upload the JSON file we created earlier.

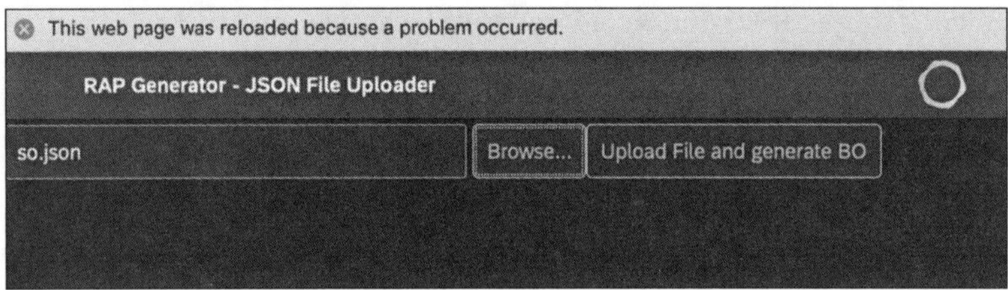

Figure 2.93 RAP Generator: JSON File Uploader

5. Click on the **Upload File and generate BO** button. Now, the RAP generator will create all the necessary repository objects.

The following objects are created through the RAP generator:

- Interface CDS view on table ZVBAK: ZI_RAP_2SALES_SALES
- Interface CDS view on table ZVBAP: ZI_RAP_2SALESITEM_SALES
- Projection view created on the interface view of table ZVBAK: ZC_RAP_2SALES_SALES
- Projection view created on the interface view of table ZVBAP: ZC_RAP_2SALESITEM_SALES
- Behavior definition created for the sales header (i.e., from table ZVBAK): ZI_RAP_2SALES_SALES
- Behavior definition created for the sales item (i.e., from table ZVBAP): ZI_RAP_2SALESITEM_SALES
- Projection of behavior definition created for the sales header (i.e., from table ZVBAK): ZC_RAP_2SALES_SALES

- Projection of behavior definition created for sales items (i.e., from table `ZVBAP`): `ZC_RAP_2SALESITEM_SALES`
- Service definition for sales

As you can see, the RAP generator greatly simplifies the building of the necessary business objects through its auto-generation capabilities.

2.12   Summary

To summarize, the ABAP RESTful application programming model is the preferred modern development platform on the cloud, with its emphasis on the user experience by supporting business users based on their need for only the required functionality.

Performance is crucial for the user experience and is possible with the SAP HANA in-memory database, which offers new features and improves performance significantly. SAP HANA also offers a combination of transactional and analytical capabilities that can be integrated into a single SAP Fiori application.

Besides the user experience, the development experience is also equally important, and the ability to test, support, and document application development is provided in this model. Secure app development and high code quality are supported by ABAP tools.

In next chapter, we'll focus on how SAP BTP, ABAP environment can communicate with external services, both as a provider and a consumer of data.

Chapter 3

Consuming External APIs

In this chapter, you'll learn why ready-to-consume business services are critical for an intelligent solution, and we'll discuss some best practices for connecting on-premise systems or cloud-based systems to business services. We'll also describe some example real-life projects to illustrate which services might be most suitable for your industry or line of business.

Traditionally, businesses have always faced make-or-buy decisions. One popular strategy adopted by successful organizations is a focus on core products that align with the organization's vision, while subcontracting non-core items. A similar concept was inherited by the services industry later, which led to a leap in outsourcing services. This approach paved way for business process outsourcing, knowledge process outsourcing, and many similar service outsourcing ventures. Extending this analogy in business solution development, application programming interfaces (APIs) are the easiest way to consume a business service without having to develop a service from scratch.

Enterprises that use SAP Business Technology Platform (SAP BTP) for ABAP development will need to open their architectures to benefit from the enormous range of APIs available from various service providers. This openness is essential so that the software solutions can be developed efficiently and in a flexible manner. More importantly, this openness reduces overall development costs and lowers the risk of failure since you'll rely on tried and tested APIs instead of custom developments.

In simple terms, an API is a software intermediary that applications (or programs) can use to communicate with each other. You can only leverage the true potential of API communication with proper connectivity and data exchange.

In this chapter, you'll learn about connecting your cloud-based ABAP applications to business systems and additional services. In Section 3.1, we'll walk you through all the steps required to connect on-premise systems, such as SAP S/4HANA or SAP ERP, to external APIs with HTTP and remote function call (RFC) connections. In Section 3.2, you'll also learn how to connect to cloud systems, like SAP S/4HANA, through SAP API Business Hub. We'll also discuss how to consume services that are already part of SAP BTP in Section 3.3 as well as how to consume several powerful, external, non-SAP cloud services in Section 3.4.

3.1   Connecting to On-Premise SAP Systems

Many SAP customers are still using on-premise SAP installations implemented long ago. To get existing business logic into an SAP cloud-based solution, you'll need to establish a connection in the cloud environment following one of two approaches: RFC destinations or HTTP destinations. These destinations must be specified properly to enable communication between cloud-based and on-premise SAP systems.

As a prerequisite, you must establish the connection to the on-premise system using SAP Connectivity service's cloud connector. The cloud connector provides secure communications, requiring only an outbound connection from the on-premise environment to SAP BTP.

> **Note**
>
> You can also use the cloud connector to connect to non-SAP systems and services on-premise.

To connect an on-premise system to the cloud, follow these steps:

1. Set up the cloud connector by installing it.
2. Configure the cloud connector.
3. Grant RFC access in the cloud connector to the on-premise SAP system.
4. For an RFC call, define the destinations in your Cloud Foundry subaccount.
5. Test the cloud connector connection to your on-premise system.

We'll describe all these steps in more detail in the following sections.

3.1.1   Configuring the HTTP and RFC Connection to an On-Premise System

In this section, we'll explain in detail the configuration process for HTTP and RFC connections to on-premise systems.

Cloud Connector Setup

First, we'll need to set up the cloud connector, which can be installed on almost any operating system, including Microsoft Windows, Linux, and macOS. The only prerequisite is that the machine on which the cloud connector is to be installed requires both access to SAP BTP and the on-premise system to which we want to connect. A step-by-step tutorial on how to install the cloud connector is available at *https://developers.sap.com/tutorials/cp-connectivity-install-cloud-connector.html*.

Now that you've installed cloud the connector, we're ready to set up access to our SAP BTP subaccount where SAP BTP, ABAP environment is running. In this stage, prepare the following information about the subaccount:

- **Region**
  The SAP BTP region where our subaccount is running.

- **Subaccount**
  The subaccount ID, which can be obtained by clicking the **Subaccount** tile under the global account and then clicking on the information button.

- **User name and password**
  In this context, the SAP BTP user name and password; this user should belong to the Cloud Foundry subaccount organization.

Note

Always use email IDs for Cloud Foundry subaccounts, unlike in Neo subaccounts, where the S-user ID is used.

Figure 3.1 shows what the filled-out form looks like.

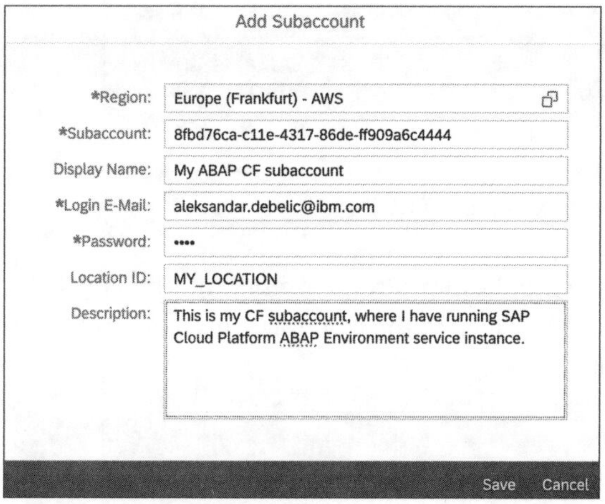

Figure 3.1 Adding a Subaccount

Note

We always recommend using the location ID to easily distinguish specific cloud connector instances when you have multiple cloud connector instances attached to the same SAP BTP subaccount. In fact, this field is mandatory whenever more than one cloud connector is attached to an SAP BTP subaccount.

Although the connectivity option is not visible in a trial SAP BTP subaccount, you can establish connections in a trial subaccount as well.

Now that the subaccount connection is set up, you'll need to establish a connection to the ABAP system, both via HTTP and RFC. A step-by-step guide on setting up the cloud connector for HTTP access is available at *https://developers.sap.com/tutorials/cp-connectivity-create-secure-tunnel.html*. The process for RFC access is similar, but you'll provide an instance number that identifies the corresponding SAP Gateway service. Also, instead a URL path, you'll provide the RFC function module name/prefix.

Figure 3.2 shows several RFC and HTTP connections established with an on-premise system.

Subaccount: SAP Innovation CF

Cloud To On-Premise

ACCESS CONTROL   COOKIE DOMAINS   APPLICATIONS   PRINCIPAL PROPAGATION

Mapping Virtual To Internal System

Status	Virtual Host	Internal Host	Check Result	Protocol	Back-end Type	Actions
☐	sechdb:4300	10.65.160.219:4300	◇ Unchecked	HTTPS	SAP HANA	
☐	sinv-1909v:8101	10.211.40.226:8101	◇ Unchecked	HTTPS	ABAP System	
☐	sinv-1909v:sapgw01	10.211.40.226:sapgw01	◇ Unchecked	RFC	ABAP System	
☐	sinv-copilot:8101	10.65.160.227:8101	◇ Unchecked	HTTPS	ABAP System	
☐	sinv-copilot:sapgw01	10.65.160.227:sapgw01	◇ Unchecked	RFC	ABAP System	

Resources Of sinv-copilot:8101

Status	URL Path	Access Policy	Actions
☐	/	Path And All Sub-Paths	

Figure 3.2 Cloud Connector Configuration Example (HTTPS)

Figure 3.3 shows a list of the Business Application Programming Interfaces (BAPIs), RFCs, and custom function modules to be called in the cloud. The objects included in this list can be consumed in the cloud.

Subaccount: SAP Innovation CF

Cloud To On-Premise

ACCESS CONTROL   COOKIE DOMAINS   APPLICATIONS   PRINCIPAL PROPAGATION

Mapping Virtual To Internal System

Status	Virtual Host	Internal Host	Check Result	Protocol	Back-end Type	Actions
☐	sechdb:4300	10.65.160.219:4300	◇ Unchecked	HTTPS	SAP HANA	
☐	sinv-1909v:8101	10.211.40.226:8101	◇ Unchecked	HTTPS	ABAP System	
☐	sinv-1909v:sapgw01	10.211.40.226:sapgw01	◇ Unchecked	RFC	ABAP System	
☐	sinv-copilot:8101	10.65.160.227:8101	◇ Unchecked	HTTPS	ABAP System	
☐	sinv-copilot:sapgw01	10.65.160.227:sapgw01	◇ Unchecked	RFC	ABAP System	

Resources Of sinv-copilot:sapgw01

Status	Function Name	Naming Policy	Actions
☐	BAPI	Prefix	
☐	RFC	Prefix	
☐	Z	Prefix	

Figure 3.3 Cloud Connector Configuration Example (RFC)

Once the setup is complete on the cloud connector side, you can proceed with setting up destinations on SAP BTP. You can create a destination in one of two ways:

- At the Cloud Foundry subaccount level, under **Connectivity** menu item
- At the Cloud Foundry space level, as a destination service instance

As shown in Figure 3.4, you can navigate to destinations on the Cloud Foundry subaccount level.

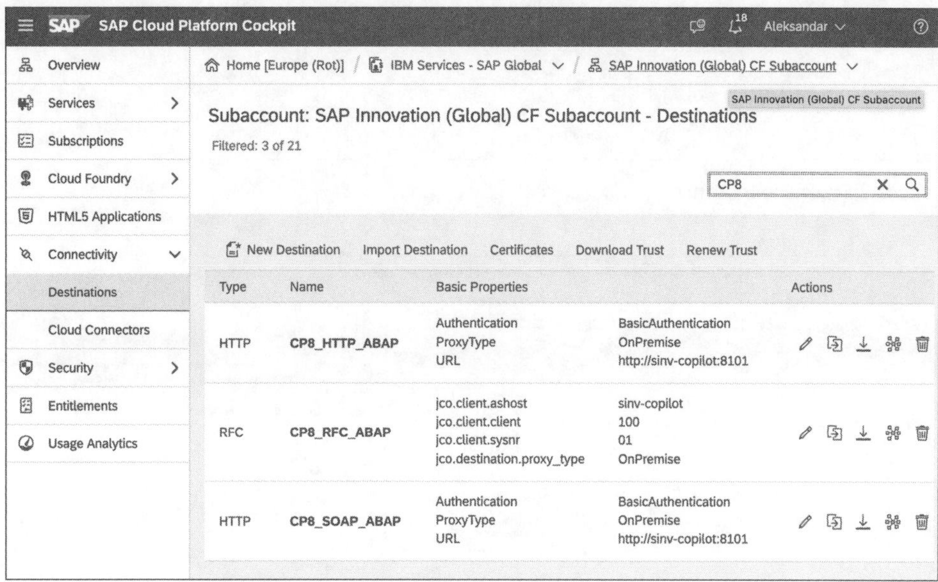

Figure 3.4 Destinations in the Cloud Foundry Subaccount

However, to access destinations defined under the destination service instance within the Cloud Foundry space, you'll need to follow these steps:

1. Access your Cloud Foundry space. This step is optional since you can select service instances directly on the Cloud Foundry subaccount level. However, with this approach, you'll get a list of all service instances in all Cloud Foundry spaces, which can sometimes be difficult to navigate.

2. Select the **Service Instances** menu item, as shown in Figure 3.5. On this screen, you can manage service instances, including creating new service instances or changing/ deleting existing service instances.

3. By clicking on the selection icon (arrow pointing right), the menu on the right will open, as shown in Figure 3.6.

4. Note the text that says **More information is available for this instance**. Click on **See here**, and another browser session will open, where you can manage destinations belonging to a specific destination service, as shown in Figure 3.7.

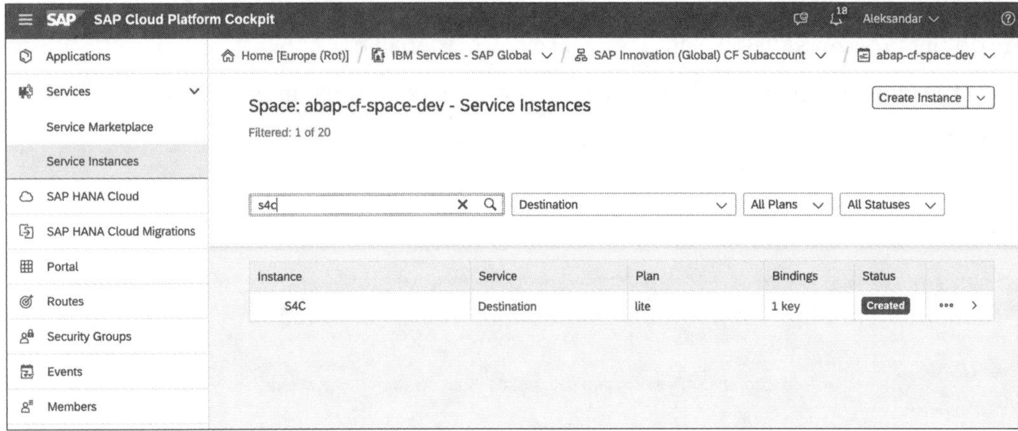

Figure 3.5 Service Instances

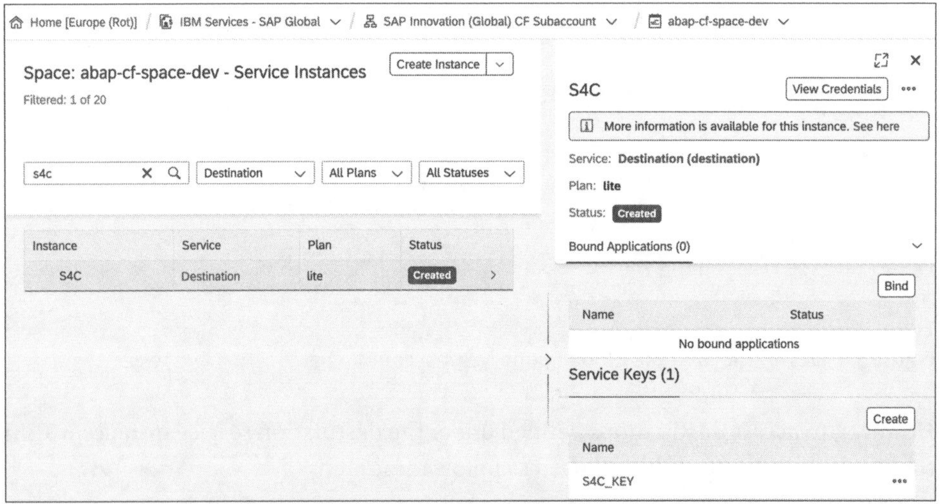

Figure 3.6 Destination Service Instance Details

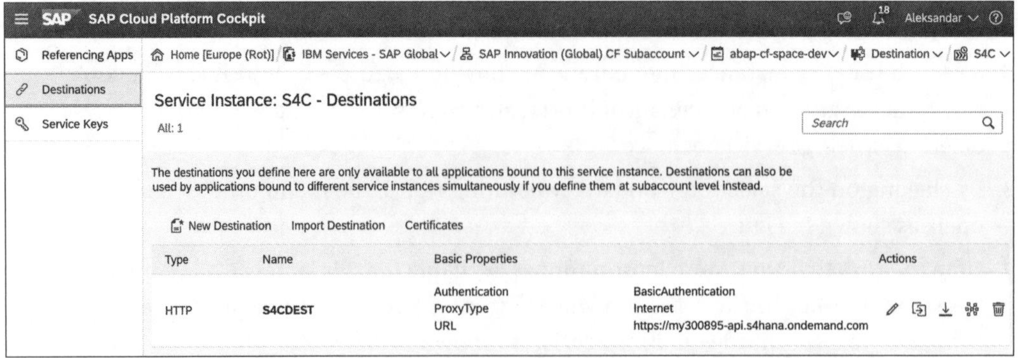

Figure 3.7 Destinations Maintenance

Now that you know how to navigate to a destination for maintenance, let's configure some destinations.

> **Note**
>
> For connections to on-premise systems, you'll use destinations defined on the Cloud Foundry subaccount level.
>
> For other types of connectivity (for example, to SAP S/4HANA Cloud), you'll use destinations defined under the destination service instance.

HTTP Destination Setup

To configure an HTTP destination in the SAP BTP cockpit, follow these steps:

1. Navigate to the relevant destination service instance, for example, the one we created in the previous step.

2. Select the **Destinations** menu item.

3. Select **New Destination**.

4. In the **Destination Configuration** section, shown in Figure 3.8, use the value help to select **HTTP** as **Type**.

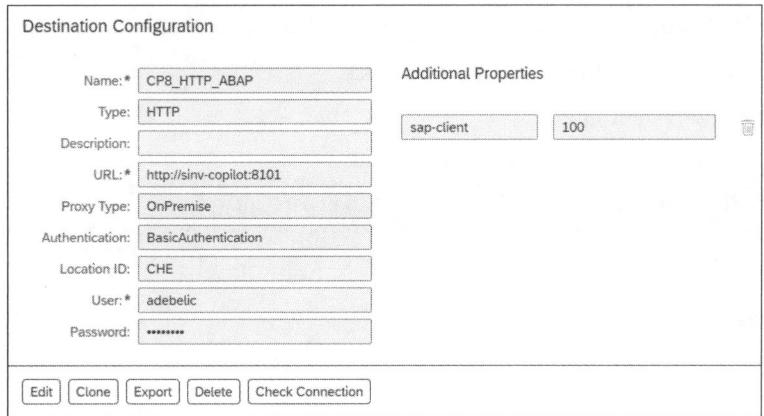

Figure 3.8 HTTP Destination

5. (Optional) If you're using more than one cloud connector in your subaccount, you must enter the location ID of the target cloud connector in the **Location ID** field.

6. For **Proxy Type**, select **OnPremise** from the value help.

7. For **Authentication**, select **BasicAuthentication** or **PrincipalPropagation**.

8. Fill in the required fields and select **Save**.

9. Open Eclipse and create and execute a runnable class.

10. To enable the HTTP communication, use, for example, the API shown in Listing 3.1.

```abap
DATA(lo_destination) = cl_http_destination_provider=>create_by_cloud_
destination(
 i_name = 'ERP_HTTP'
 i_authn_mode = if_a4c_cp_service=>service_specific
 ).
DATA(lo_client) = cl_web_http_client_manager=>create_by_http_destination(
lo_destination ).
DATA(lo_request) = lo_client->get_http_request( ).
DATA(lo_response) = lo_http_client->execute( i_method =
if_web_http_client=>get ).
out->write( lo_response->get_text( ) ).
```

Listing 3.1 Creating an HTTP Destination

In the code shown in Listing 3.1, the created instance `lo_destination` leverages the `cl_http_destination_provider` class. Similarly, the created instance `lo_client` leverages the `cl_web_http_client_manager` class. Then, we'll store the response from the EXECUTE method in the `lo_response` variable and output it.

The next step is to configure the RFC destination in the SAP BTP cockpit.

RFC Destination Setup

To configure an RFC destination in the SAP BTP cockpit, follow these steps:

1. Navigate to the relevant destination service instance.

2. Select the **Destinations** menu item.

3. Create a destination by selecting **New Destination**.

4. For **Type**, select **RFC** from the value help, as shown in Figure 3.9.

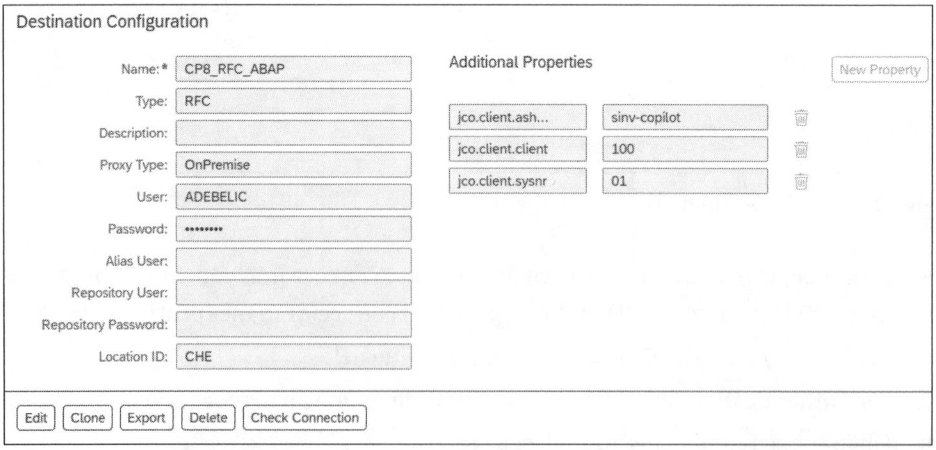

Figure 3.9 RFC Destination

5. (Optional) If you're using more than one cloud connector in your subaccount, you must enter the location ID of the target cloud connector in the **Location ID** field.

6. As authentication type, use basic authentication or set the property `jco.destination.auth_type=PrincipalPropagation`.

7. If you use basic authentication, set a user name and credentials for the destination.

8. To configure the RFC destination, choose one of the following options:

   – For a destination that uses load balancing (system ID and message server), proceed as follows:

     • Select **New Property**, choose **jco.client.r3name** from the value help, and enter the three-letter system ID of your backend system (as configured in the cloud connector) in the **property** field.

     • Create another property, select **jco.client.mshost**, and enter the message server host (as configured in the cloud connector) in the **property** field.

     • Add another property, choose **jco.client.group**, and enter a log group in the **property** field.

     • Create another property, select **jco.client.client**, and enter the three-digit ABAP client number.

   – For a destination without load balancing (application server and instance number), perform these steps:

     • Select **New Property**, choose **jco.client.ashost** from the value help, and enter the application server name of your backend system (as configured in the cloud connector) under the **Additional Properties** heading.

     • Add another property, choose **jco.client.sysnr**, and enter "00," which is the instance number of the application server (as configured in the cloud connector) under the **Additional Properties** heading.

     • Create another property, select **jco.client.client**, and enter the three-digit ABAP client number.

9. Click **Save**.

Call a remote function module with the command `CALL function 'RFC_SYSTEM_INFO' DESTINATION lv_destination`. Then, open Eclipse and create and execute a runnable class.

To execute a remote function module on your on-premise system, use the code shown in Listing 3.2.

```
DATA(lo_destination) = cl_rfc_destination_provider=>create_by_cloud_destination(
  i_name = 'ERP_RFC'
  ).
DATA(lv_destination) = lo_destination->get_destination_name( ).
DATA lv_result type c length 200.
CALL function 'RFC_SYSTEM_INFO'
```

```
DESTINATION lv_destination
 IMPORTING
 rfcsi_export = lv_result.
out->write( lv_result ).
```

Listing 3.2 Executing a Remote Function Module

As shown in Listing 3.2, you'll need to create the instance `lo_destination` leveraging the class `cl_rfc_destination_provider`. Then, you'll store the destination name in `lv_des-tination`. Utilize the function module 'RFC_SYSTEM_INFO' to retrieve and output the resulting details.

3.1.2   Consuming the RFC Function Module

In this section, you'll learn in detail how to call a remote-enabled (RFC-enabled) function module from a backend system, and you'll learn in detail how to consume external APIs in the cloud in the following sections.

We'll start with creating a custom class in our package, in Eclipse. To call the function module from the backend system, first declare a destination as an RFC destination, as shown in Listing 3.3. If you're familiar with SAP ERP, you can call function modules the same way as in that system, using the **Pattern** button on the menu bar, followed by entering the destination name.

```
class ZCL_GET_CUST definition

public
 final
create public .
public section.

INTERFACES if_oo_adt_classrun.
protected section.
private section.

ENDCLASS.

CLASS ZCL_GET_CUST IMPLEMENTATION.
METHOD if_oo_adt_classrun~main.
try.
DATA(lo_rfc_dest) = cl_rfc_destination_provider=>create_by_cloud_destination(
                         i_name = |CP8_RFC_ABAP| ).
DATA(lv_destination) = lo_rfc_dest->get_destination_name( ).
        TYPES : BEGIN OF ty_cust,
                 CUSTOMERID type C LENGTH 8 ,
```

```abap
                CUSTNAME type c length 25 ,
                FORM type c length 15,
                STREET type c length 30,
                POBOX type c length 10,
                POSTCODe type c length 10,
                CITY type c length 25,
                COUNTR type c length 3,
                COUNTR_ISO type c length 2,
                REGION type c length 3,
                PHONE type c length 30,
                EMAIL type c length 40,
          END OF ty_cust.
    DATA: msg TYPE c LENGTH 255.
    DATA lt_cust TYPE STANDARD TABLE OF ty_cust.
    DATA ls_cust TYPE ty_cust.

    CALL FUNCTION 'BAPI_FLCUST_GETLIST'
          DESTINATION lv_destination
            tables
            customer_list  = lt_cust.
      CASE sy-subrc.
        WHEN 0.
          LOOP AT lt_cust INTO ls_cust.
            out->write( ls_cust-customerid && ',' &&  ls_cust-custname && ','
&& ls_cust-street && ',' &&  ls_cust-postcode && ',' && ls_cust-city ).
          ENDLOOP.
        WHEN 1.
          out->write ( |EXCEPTION SYSTEM_FAILURE | && msg ).
        WHEN 2.
          out->write ( |EXCEPTION COMMUNICATION_FAILURE | && msg ).
        WHEN 3.
          out->write( |EXCEPTION OTHERS| ).
      ENDCASE.

CATCH cx_root INTO DATA(lx_root).
      out->write( lx_root->get_text( ) ).
endtry.

endmethod.
ENDCLASS.
```

Listing 3.3 Consuming an RFC Function Module

As shown in Listing 3.3, you'll need to define the class `ZCL_GET_CUST`. Make sure you define the interface within the `PUBLIC SECTION`. Then, you'll need to write the class implementation. In the main method, you'll retrieve the destination name in `lv_destination` and then call/consume RFC `'BAPI_FLCUST_GETLIST'` from that destination. This method will display a list of customers as the output, as shown in Figure 3.10.

```
00004668,Matthias Kramer,Gartenstr. 79,69123,Heidelberg
00004669,Guenther Simonen,Jacobistrasse 219,79104,Freiburg
00004670,Adam Neubasler,Heidelberger Str. 87,69483,Wald-Michelbach
00004671,Stephen Kreiss,Arionweg 30,68723,Schwetzingen
00004672,Kurt Schneider,Waldmann 185,69207,Kurt
00004673,Annemarie Barth,Stauboernchenstrasse 227,67663,Kaiserslautern
00004674,Ulla Marshall,Gruenlingweg 133,69180,Wiesloch
00004675,Andrej Eichbaum,43 Poklukarjeva,1000,Ljubljana
00004676,Ruth Kreiss,Melissenstr. 190,41466,Neuss
00004677,Johann Koller,Ausfallstr. 57,11111,Berlin
00004678,Amelie Deichgraeber,Froschstr. 61,68753,Amelie
00004679,Irmtraut Sudhoff,Max-Planck-Str. 213,63150,Heusenstamm
00004680,Allen Kreiss,6 Sagamore St.,17758,N. Massapequa
00004681,Holm Heller,Muehltalstr. 221,69121,Heidelberg
00004682,Roland Goelke,Gemeindestr. 247,79761,Waldshut
00004683,Achim D´Oultrement,Rankestr. 54,76137,Karlsruhe
00004684,Irmtraut Benjamin,Max-Planck-Str. 15,63150,Heusenstamm
00004685,Anna Detemple,Lerchenstr. 248,86343,Koenigsbrunn
00004686,Anna Buehler,Lerchenstr. 102,86343,Koenigsbrunn
00004687,Laura Ryan,Raupelsweg 27,60118,Mainz
```

Figure 3.10 Output

3.1.3   Consuming the OData Service from the Backend

The Open Data Protocol (OData) was introduced by Microsoft. Later, this protocol became the ISO/IEC-approved OASIS standard, which contains best practices for building and consuming RESTful APIs. You can find more details at *https://www.odata.org/*.

Assuming that an OData service has been created in your backend system (by creating a project through Transaction SEGW or simply added via Transaction /IWFND/MAINT_SERVICE), we'll show you how to consume an OData service from the backend into the cloud through the service consumption model (see Figure 3.11). This model enables communication based on OData and Simple Object Access Protocol (SOAP) client calls to achieve a higher level of abstraction than when using HTTP and RFC while also providing enhanced interfaces with the ABAP RESTful application programming model. To begin, follow these steps:

1. **Create a service consumption model for the OData call**
   Figure 3.11 shows the inputs required for creating a new service consumption model. You'll need to select the package and project where this consumption model is to be stored. The name of the model should follow the project's naming conventions. The **Description** field is a free text field for better documentation.

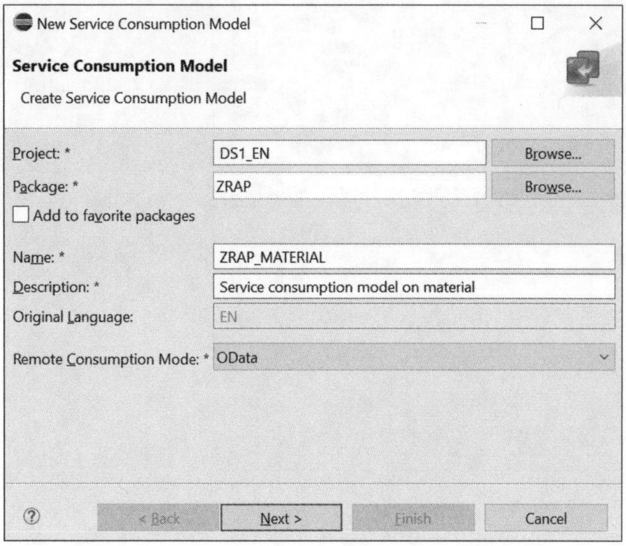

New Service Consumption Model

Service Consumption Model
Create Service Consumption Model

Project: *   DS1_EN   Browse...
Package: *   ZRAP   Browse...
☐ Add to favorite packages

Name: *   ZRAP_MATERIAL
Description: *   Service consumption model on material
Original Language: EN

Remote Consumption Mode: *   OData

< Back   Next >   Finish   Cancel

Figure 3.11 Service Consumption Model

2. **Get the service definition in XML format ($metadata file)**

   The *$metadata* file, shown in Figure 3.12, specifies the data that will be exposed from the backend system. The service definition can be obtained from a service provider, such as an SAP Gateway system or an SAP S/4HANA system, or from a public services inventory, such as the SAP API Business Hub, which lists all published APIs.

   This metadata file can be downloaded from an on-premise system and saved to a local system by executing Transaction SEGW.

```xml
<?xml version="1.0" encoding="UTF-8"?>
<edmx:Edmx xmlns:sap="http://www.sap.com/Protocols/SAPData" xmlns:m="http://schemas.microsoft.com/ado/2007/08/dataservices/metadata"
  xmlns:edmx="http://schemas.microsoft.com/ado/2007/06/edmx" Version="1.0">
  <edmx:Reference xmlns:edmx="http://docs.oasis-open.org/odata/ns/edmx" Uri="http://sinv-
    copilot.isl.edst.ibm.com:8001/sap/opu/odata/IWFND/CATALOGSERVICE;v=2/Vocabularies(TechnicalName='%2FIWBEP%
    2FVOC_COMMON',Version='0001',SAP__Origin='LOCAL')/$value">
      <edmx:Include Alias="Common" Namespace="com.sap.vocabularies.Common.v1"/>
  </edmx:Reference>
  <edmx:Reference xmlns:edmx="http://docs.oasis-open.org/odata/ns/edmx" Uri="http://sinv-
    copilot.isl.edst.ibm.com:8001/sap/opu/odata/IWFND/CATALOGSERVICE;v=2/Vocabularies(TechnicalName='%2FIWBEP%
    2FVOC_CAPABILITIES',Version='0001',SAP__Origin='LOCAL')/$value">
      <edmx:Include Alias="Capabilities" Namespace="Org.OData.Capabilities.V1"/>
  </edmx:Reference>
  <edmx:Reference xmlns:edmx="http://docs.oasis-open.org/odata/ns/edmx" Uri="http://sinv-
    copilot.isl.edst.ibm.com:8001/sap/opu/odata/IWFND/CATALOGSERVICE;v=2/Vocabularies(TechnicalName='%2FIWBEP%
    2FVOC_COMMUNICATION',Version='0001',SAP__Origin='LOCAL')/$value">
      <edmx:Include Alias="Communication" Namespace="com.sap.vocabularies.Communication.v1"/>
  </edmx:Reference>
  <edmx:DataServices m:DataServiceVersion="2.0">
    <Schema xml:lang="en" Namespace="API_MATERIAL_DOCUMENT_SRV" xmlns="http://schemas.microsoft.com/ado/2008/09/edm" sap:schema-
      version="1">
      <EntityType sap:content-version="1" sap:label="API exposure Material Document Header" Name="A_MaterialDocumentHeaderType">
        <Key>
            <PropertyRef Name="MaterialDocumentYear"/>
            <PropertyRef Name="MaterialDocument"/>
        </Key>
        <Property sap:label="Material Document Year" Name="MaterialDocumentYear" sap:display-format="NonNegative" MaxLength="4" Nullable="false"
            Type="Edm.String"/>
        <Property sap:label="Material Document" Name="MaterialDocument" sap:display-format="UpperCase" MaxLength="10" Nullable="false"
```

Figure 3.12 Metadata File

In the service consumption model, you'll need an input file to generate an OData consumption proxy. Figure 3.13 shows how you can select the metadata file, stored in your local drive in the previous step, as the service metadata file.

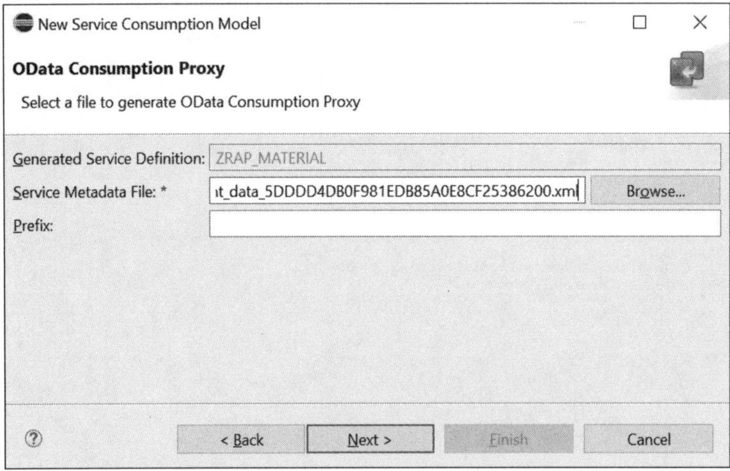

Figure 3.13 OData Service Call

In this step, as shown in Figure 3.13, you can also use a prefix if you plan to import the file several times. This option will help you prevent the framework from automatically creating repository object names when importing the same files (in different clients).

3. **Specify the entity sets to be included with the generation of the service consumption model**
Figure 3.14 shows how to specify the entity sets to be included while creating the service consumption model.

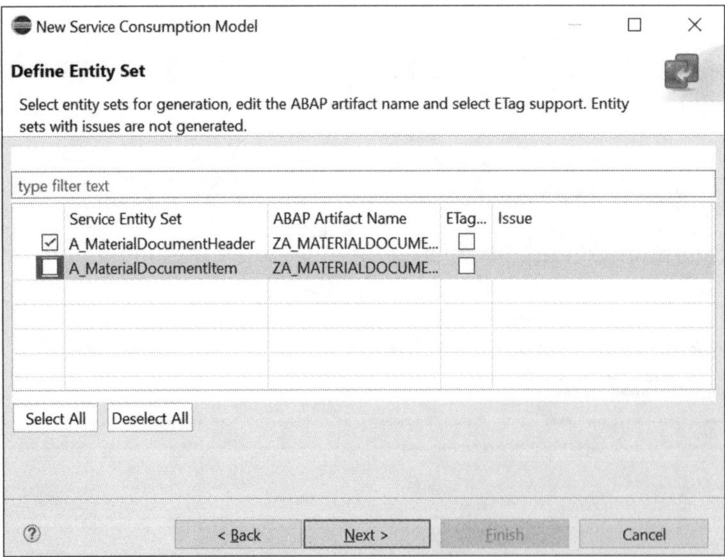

Figure 3.14 Entity Set

Service definitions, abstract entity data definitions, and behavior definitions will be created by a wizard, as shown in Figure 3.15.

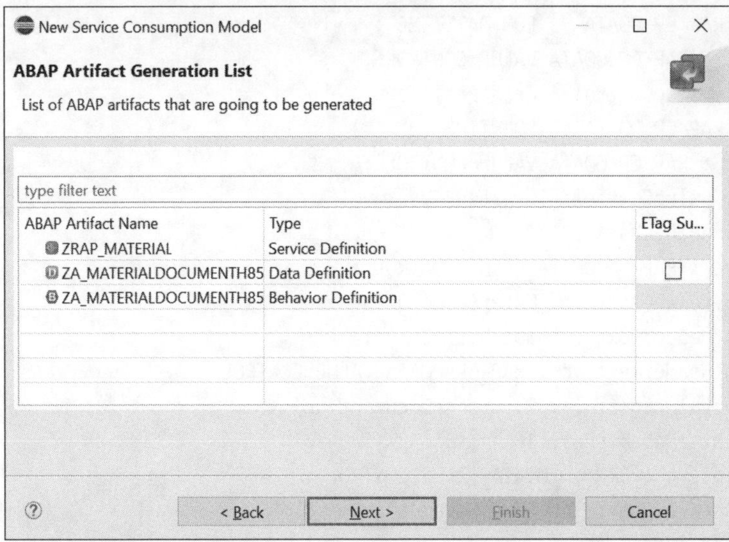

Figure 3.15 ABAP Artifact

Listing 3.4 shows the detailed code that is generated by the wizard tool as part of a service definition.

```
/*Service definition generated by wizard tool*/
@EndUserText.label: 'ZRAP_MATERIAL'
@OData.schema.name: 'API_MATERIAL_DOCUMENT_SRV'
define service ZRAP_MATERIAL {
 expose ZA_MATERIALDOCUMENTH856A03A9E0;
 }

/********* GENERATED on 10/24/
2020 at 09:50:04 by CB0000000002**************/
@OData.entitySet.name: 'A_MaterialDocumentHeader'
@OData.entityType.name: 'A_MaterialDocumentHeaderType'
define root abstract entity ZA_MATERIALDOCUMENTH856A03A9E0 {
key MaterialDocumentYear : abap.numc( 4 ) ;
key MaterialDocument : abap.char( 10 ) ;
@Odata.property.valueControl: 'InventoryTransactionType_vc'
InventoryTransactionType : abap.char( 2 ) ;
InventoryTransactionType_vc : RAP_CP_ODATA_VALUE_CONTROL ;
@Odata.property.valueControl: 'DocumentDate_vc'
DocumentDate : RAP_CP_ODATA_V2_EDM_DATETIME ;
DocumentDate_vc : RAP_CP_ODATA_VALUE_CONTROL ;
@Odata.property.valueControl: 'PostingDate_vc'
```

```
PostingDate : RAP_CP_ODATA_V2_EDM_DATETIME ;
PostingDate_vc : RAP_CP_ODATA_VALUE_CONTROL ;
@Odata.property.valueControl: 'CreationDate_vc'
CreationDate : RAP_CP_ODATA_V2_EDM_DATETIME ;
CreationDate_vc : RAP_CP_ODATA_VALUE_CONTROL ;
@Odata.property.valueControl: 'CreationTime_vc'
CreationTime : RAP_CP_ODATA_V2_EDM_TIME ;
CreationTime_vc : RAP_CP_ODATA_VALUE_CONTROL ;
@Odata.property.valueControl: 'CreatedByUser_vc'
CreatedByUser : abap.char( 12 ) ;
CreatedByUser_vc : RAP_CP_ODATA_VALUE_CONTROL ;
@Odata.property.valueControl: 'MaterialDocumentHeaderText_vc'
MaterialDocumentHeaderText : abap.char( 25 ) ;
MaterialDocumentHeaderText_vc : RAP_CP_ODATA_VALUE_CONTROL ;
@Odata.property.valueControl: 'ReferenceDocument_vc'
ReferenceDocument : abap.char( 16 ) ;
ReferenceDocument_vc : RAP_CP_ODATA_VALUE_CONTROL ;
@Odata.property.valueControl: 'GoodsMovementCode_vc'
GoodsMovementCode : abap.char( 2 ) ;
GoodsMovementCode_vc : RAP_CP_ODATA_VALUE_CONTROL ;
}
```

Listing 3.4 Auto-Generated Service Definition from the Wizard

The results screen of the service consumption model, shown in Figure 3.16, shows an automatically generated code template that can read the entity set.

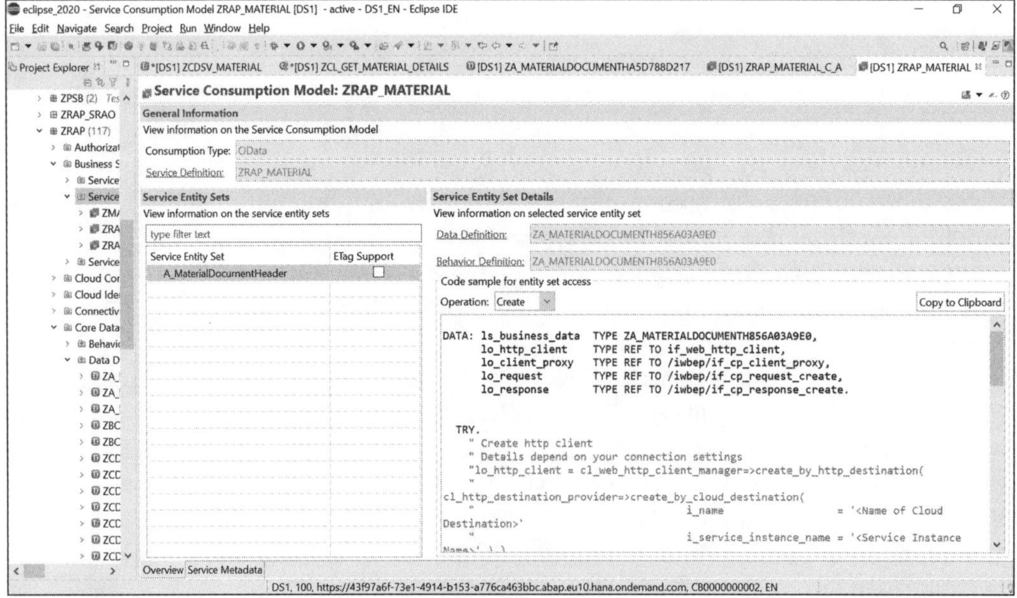

Figure 3.16 Service Consumption Model

4. **Create a custom class to embed this template**

The class shown in Listing 3.5 will produce the console output and list all the materials available from the backend SAP Gateway system.

```
class ZCL_MATERIAL_ODATA_CALL definition
  public
  final
  create public .

public section.
interfaces IF_OO_ADT_CLASSRUN.
protected section.
private section.
ENDCLASS.

CLASS ZCL_MATERIAL_ODATA_CALL IMPLEMENTATION.

method if_oo_adt_classrun~main.
DATA: lt_business_data  TYPE TABLE OF ZA_MATERIALDOCUMENTH856A03A9E0,
      lo_http_client    TYPE REF TO if_web_http_client,
      lo_client_proxy   TYPE REF TO /iwbep/if_cp_client_proxy,
      lo_request        TYPE REF TO /iwbep/if_cp_request_read_list,
      lo_response       TYPE REF TO /iwbep/if_cp_response_read_lst.
  TRY.
  DATA(http_destination) = cl_http_destination_provider=>create_by_cloud_
destination(
        i_name = 'CP8_HTTP_ABAP'
        i_authn_mode = if_a4c_cp_service=>service_specific
        ).
      lo_http_client = cl_web_http_client_manager=>create_by_http_
destination( http_destination ).
    lo_client_proxy = cl_web_odata_client_factory=>create_v2_remote_proxy(
      EXPORTING
        iv_service_definition_name = 'ZRAP_MATERIAL'
        io_http_client             = lo_http_client
        iv_relative_service_root   = '/sap/opu/odata/sap/API_MATERIAL_
DOCUMENT_SRV' ).
    " Navigate to the resource and create a request for the read operation
    lo_request = lo_client_proxy->create_resource_for_entity_set(
'A_MATERIALDOCUMENTHEADER' )->create_request_for_read( ).
    lo_request->set_top( 1 )->set_skip( 0 ).
    " Execute the request and retrieve the business data
    lo_response = lo_request->execute( ).
```

```
      lo_response->get_business_data( IMPORTING et_business_data = lt_business_
data ).

    CATCH cx_web_http_client_error INTO DATA(lx_response).
        out->write( lx_response->get_text(  ) ).

      CATCH cx_http_dest_provider_error INTO DATA(lx_destexception).
        out->write( lx_destexception->get_text(  ) ).

  CATCH /iwbep/cx_cp_remote INTO DATA(lx_remote).
    " Handle remote Exception
    " It contains details about the problems of your http(s) connection

  CATCH /iwbep/cx_gateway INTO DATA(lx_gateway).
    " Handle Exception
ENDTRY.
ENDMETHOD .
ENDCLASS.
```

Listing 3.5 Custom Class for the OData Call

As shown in Listing 3.5, you'll need to define the class `ZCL_MATERIAL_ODATA_CALL`. Make sure you define the interface within the `public` section. Then, you'll need to write the class implementation. In the main method, you'll need to retrieve destination using `cl_http_destination_provider` and then consume OData services from that destination. This example will retrieve material document information.

3.1.4   Consuming the SOAP Web Service

The Simple Object Access Protocol (SOAP) is a messaging protocol that serves as the foundation for the implementation of web services and has been around for more than two decades. Its first release was back in 1998, when it was called "XML-RPC."

SAP introduced support for SOAP web services in 2002, with SAP NetWeaver 6.20 (then called the SAP Web Application Server), which was foundation for SAP R/3 Enterprise 4.70. SOAP web services are supported both in SAP S/4HANA and in all SAP ERP releases, unlike OData, which is not supported in older SAP ERP systems.

To consume SOAP web services from SAP BTP, ABAP environment, you must fulfill two prerequisites:

- **Configuration in the cloud connector**
  An HTTP connection to the on-premise system.

- **Web service configuration**
  A web service configured and exposed via Transaction SOAMANAGER.

For this example, we'll use the *Business User – Read* web service. For more details on this service, refer to *https://api.sap.com/api/QUERYBUSINESSUSERIN/documentation*.

First, download the Web Service Definition Language (WSDL) file via Transaction SOAMANAGER. Locate the web service QUERYBUSINESSUSERIN and open the **Configuration** tab for the service definition, as shown in Figure 3.17.

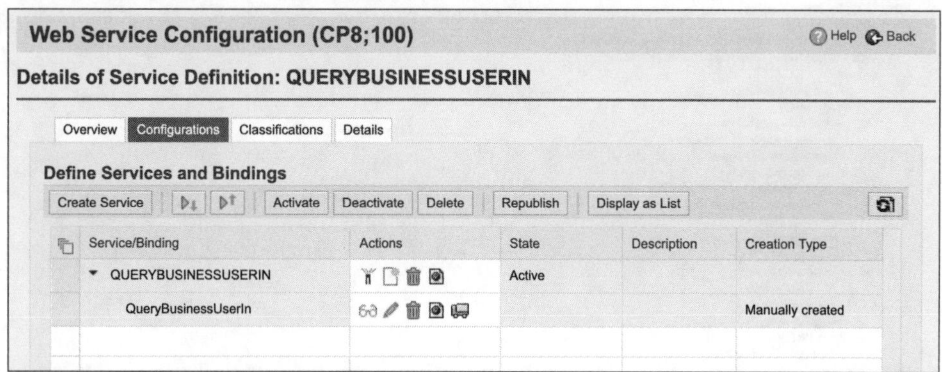

Figure 3.17 Web Service Configuration

By selecting the **Open Binding WSDL Generation** option, a popup window will open. As shown at the bottom of Figure 3.18, locate and copy the WSDL URL. Alternatively, you can simply open the WSDL document directly.

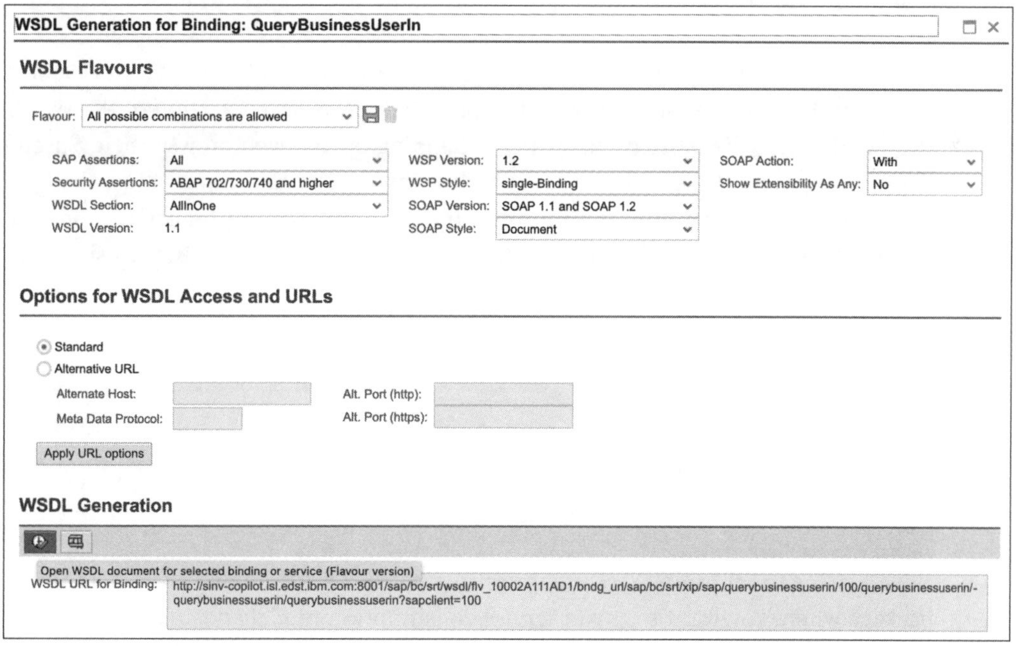

Figure 3.18 Add WSDL URL

The SAP system will ask you to authenticate. After successful authentication, you'll see the WSDL document in the web browser, as shown in Figure 3.19.

```
-<wsdl:definitions targetNamespace="http://sap.com/xi/ABA">
  -<wsdl:documentation>
     <sidl:sidl/>
   </wsdl:documentation>
   <wsp:UsingPolicy wsdl:required="true"/>
  -<wsp:Policy wsu:Id="BN__QueryBusinessUserIn">
   -<wsp:ExactlyOne>
     -<wsp:All>
         <saptrnbnd:OptimizedMimeSerialization wsp:Optional="true"/>
         <wsaw:UsingAddressing wsp:Optional="true"/>
        -<wsp:All>
          -<sp:TransportBinding>
            -<wsp:Policy>
              -<sp:TransportToken>
                 +<wsp:Policy></wsp:Policy>
               </sp:TransportToken>
              -<sp:AlgorithmSuite>
                -<wsp:Policy>
                    <sp:Basic128Rsa15/>
                  </wsp:Policy>
                </sp:AlgorithmSuite>
              -<sp:Layout>
                -<wsp:Policy>
                    <sp:Strict/>
                  </wsp:Policy>
                </sp:Layout>
              </wsp:Policy>
            </sp:TransportBinding>
          </wsp:All>
        </wsp:All>
     -<wsp:All>
         <saptrnbnd:OptimizedXMLTransfer uri="http://xml.sap.com/2006/11/esi/esp/binxml" wsp:Optional="true"/>
         <wsaw:UsingAddressing wsp:Optional="true"/>
        -<wsp:All>
          -<sp:TransportBinding>
            -<wsp:Policy>
              -<sp:TransportToken>
                -<wsp:Policy>
                  -<sp:HttpsToken>
                    -<wsp:Policy>
```

Figure 3.19 WSDL Document

You'll need to download the WSDL document as a text file by selecting the **View Source** option in the web browser, which will display the unformatted version of the WSDL. (This option may be labeled differently depending on the web browser being used.) Copy the content of the web browser window into a text editor and adjust the host name and port so that these details correspond to the cloud connector connection definition. The host name and port information can be located in SAP BTP, as shown in Figure 3.20.

Having completed these steps, the WSDL file is ready.

> **Note**
>
> Make sure that the content is pasted and saved as a plain text file, without any formatting.

Now, moving to Eclipse, we'll create the service consumption model. Right-click on the package where you want to create service consumption model and, in context menu, select **New · Other ABAP Repository Object**. On the next screen, select the **Service Consumption Model** option and click **Next**, as shown in Figure 3.21.

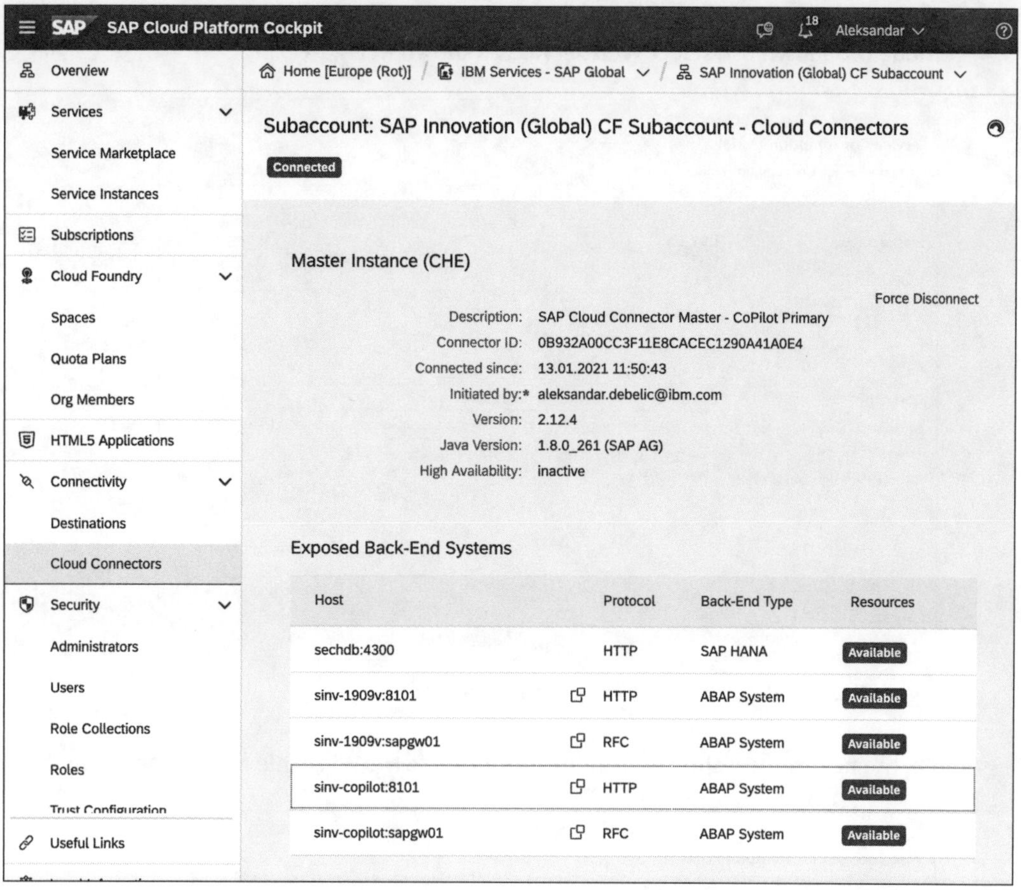

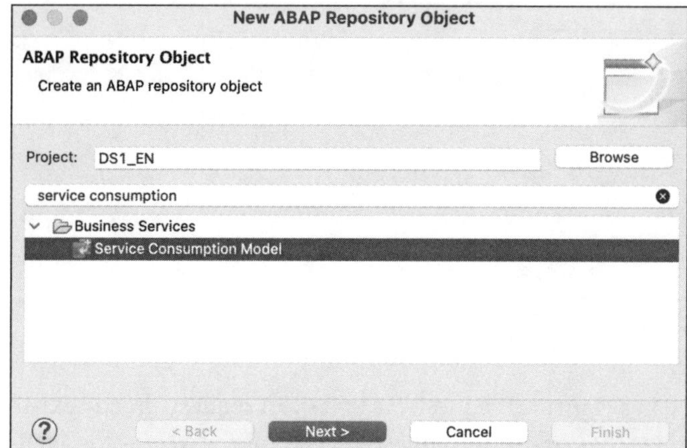

Figure 3.20 Cloud Connector Configuration

Figure 3.21 ABAP Repository Object: Service Consumption Model

On the next screen, enter a name and a description. From the **Remote Consumption Mode** dropdown list, select **Web Service**, as shown in Figure 3.22.

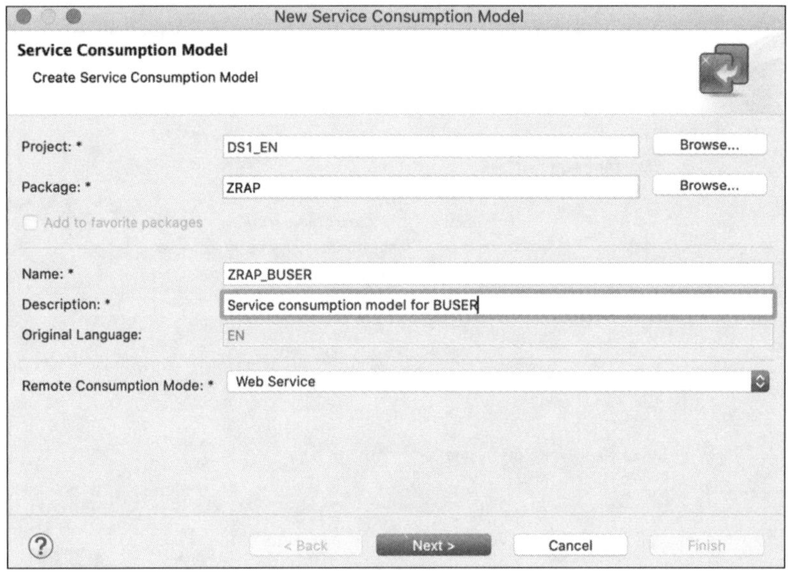

Figure 3.22 Creating a Service Consumption Model

On the next screen, shown in Figure 3.23, provide the WSDL file and define the prefix.

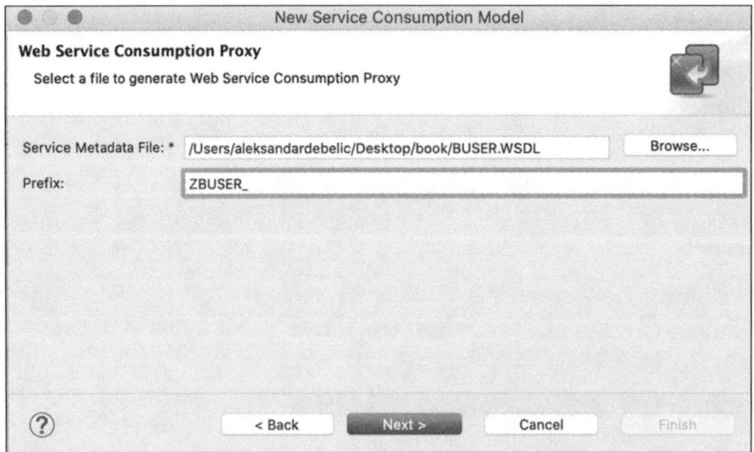

Figure 3.23 Creating a Service Consumption Model, Continued

These changes can be stored in a transport request. If a transport request already exists for this project, you can utilize that transport request. If no transport request is available, then you must create a new transport request.

The service consumption model has now been created, and now, all you need to do is activate it by clicking on the **Activate** button on the menu bar. Now, you'll be presented with the service consumption model maintenance screen, where you'll see properties and metadata as well as a code sample that you can reuse in your development projects, as shown in Figure 3.24.

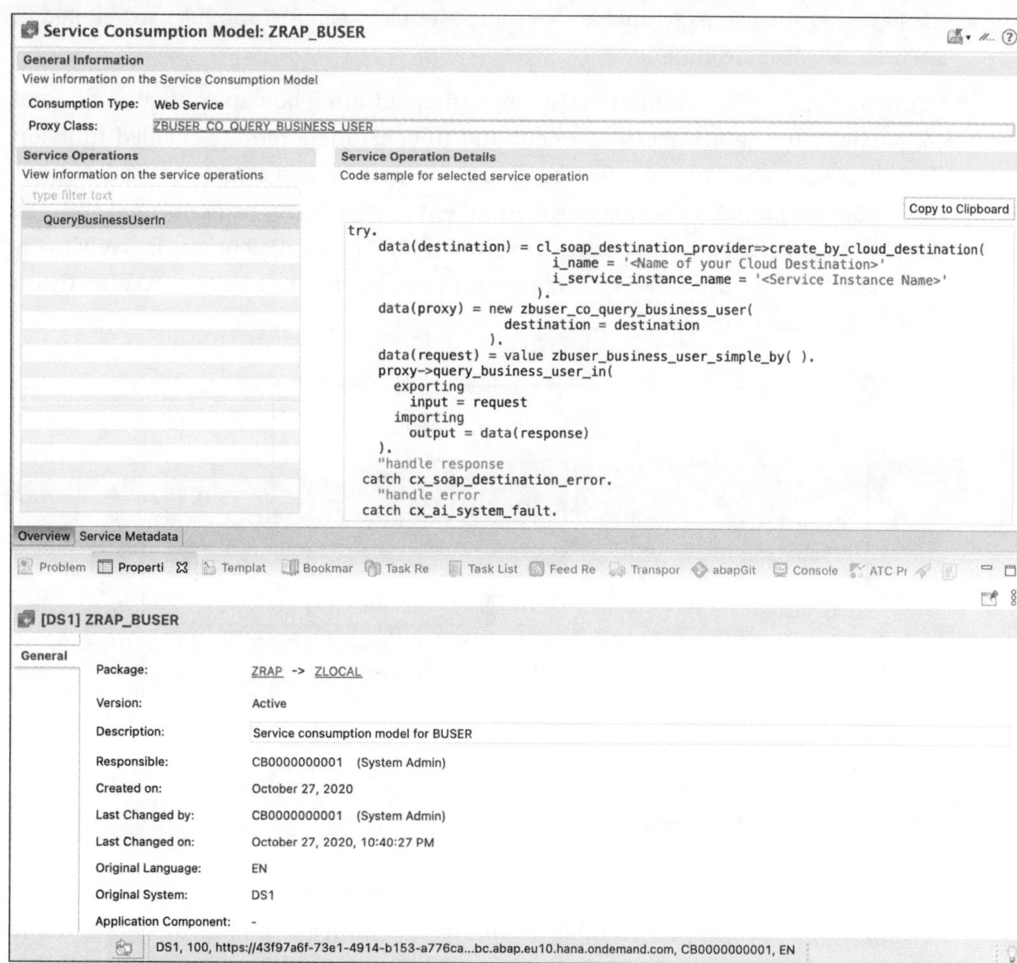

Figure 3.24 Service Consumption Model Maintenance Screen

Alongside the service consumption model, you'll notice that helper classes have been created as well. These proxy classes, shown in Figure 3.25, are required for SOAP service consumption.

Figure 3.25 Generated Proxy Classes

Your new SOAP service is ready for the consumption. Simply use the sample code available in the service consumption model to accelerate development.

3.1.5   Inbound RFC Concepts

Inbound RFC support was introduced in SAP BTP, ABAP environment release 2008. In this section, we'll walk through the process flow to help you understand the prerequisites and required steps to enable external systems to call RFC modules developed in SAP BTP, ABAP environment.

Figure 3.26 shows the inbound RFC architecture, outlining how to configure inbound connectivity to the ABAP environment and to consume a remote-enabled function module from an on-premise ABAP system through the cloud connector. RFC connections from on-premise systems to the ABAP environment can be performed through the cloud connector service channel. From SAP S/4HANA 2019 onwards, WebSocket RFCs, which do not need the cloud connector because they can be opened directly on the Internet, can be used.

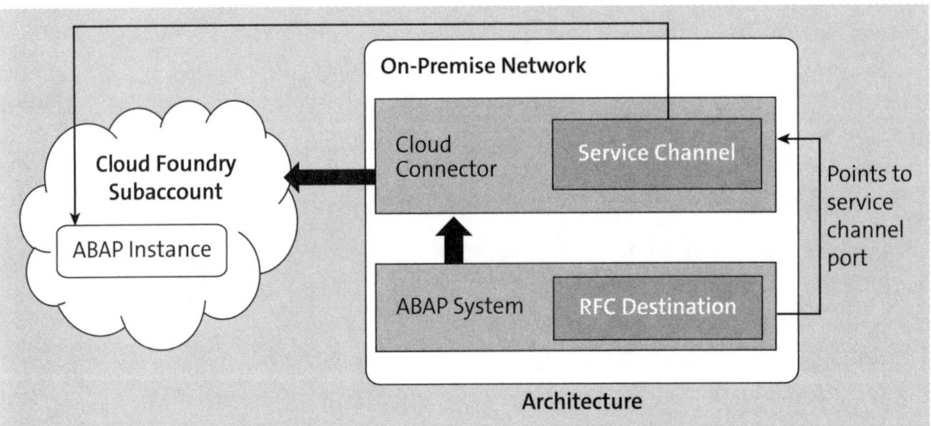

Figure 3.26 Inbound RFC Architecture

Let's now dive more deeply into the technical steps involved in the implementation of the inbound RFC concept. To establish a connection from an on-premise system to the cloud, you must have a service channel in place. Then, you'll need to create a remote-enabled function module. You'll also need a communication scenario, a proper communication arrangement, and a RFC destination already set up. With all these elements in place, finally, you can call the function module to receive the desired output.

Adding a Service Channel

In addition to the standard configuration in the cloud connector for HTTP and RFC connections from SAP BTP to an on-premise system, you must add a service channel with a local instance number. In an on-premise ABAP system, you would thus create an RFC destination of type 3 through Transaction SM59 with the host name of the cloud connector in your on-premise network and the same instance number. The load balancing status should be **NO**.

Now, let's add a service channel to establish the connection from the on-premise system to the cloud. From your subaccount menu, select **On Premise To Cloud**. Then, click the **Add (+)** icon, which will open the screen shown in Figure 3.27.

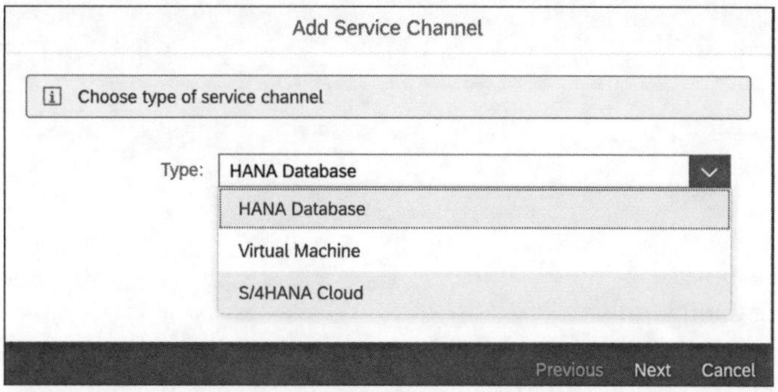

Figure 3.27 Service Channel

On this screen, select **S/4HANA Cloud** from the **Type** dropdown list.

Next, provide the cloud tenant host, local instance number, and connections. Then, select the **Enabled** checkbox, as shown in Figure 3.28. The cloud tenant host information can be obtained from the service dashboard.

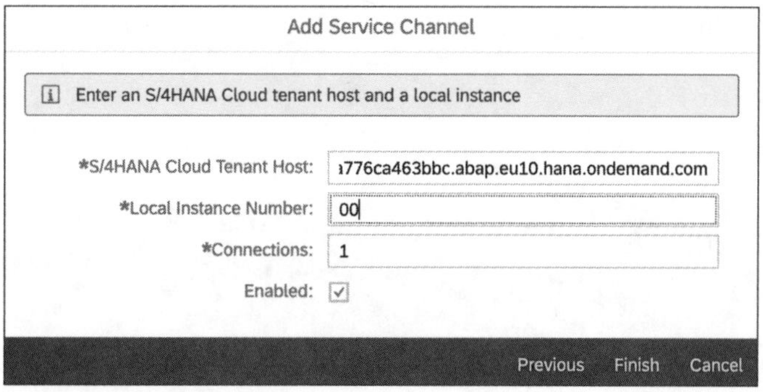

Figure 3.28 Service Channel Details

> **Note**
>
> You can assign a random instance number, and this information will later be used in RFC destination.

This step completes the configuration of the cloud connector. The result looks like the screen shown in Figure 3.29.

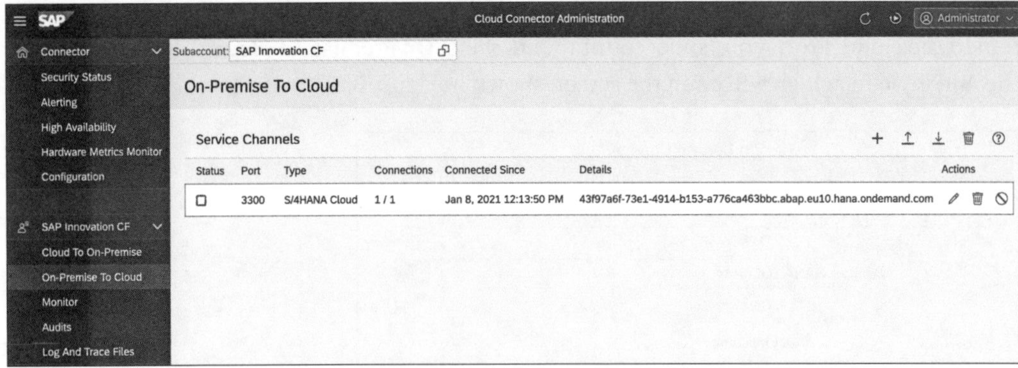

Figure 3.29 Cloud Connector Configuration

A green indicator under **Status** means that the connection with SAP BTP, ABAP environment has been established and the service channel is ready for use.

Creating a Remote-Enabled Function Module

Now, let's use SAP BTP, ABAP environment to create a remote-enabled function module and the communication scenario that will be used for its consumption.

First, create the remote-enabled function module in ABAP Development Tools (ADT) using the code shown in Listing 3.6.

```
FUNCTION ZFM_GET_SO
  IMPORTING
    VALUE(LV_KG) TYPE ZKG
  EXPORTING
    VALUE(LV_POUNDS) TYPE ZLB.


DATA : LV_CON TYPE ZKG.
DATA : GT_LIKP TYPE TABLE OF ZLIKP,
       GS_LIKP TYPE ZLIKP,
       lv_cnt type N,
       CNT TYPE I,
       msg type string .

       LV_CON = '2.204' .
       LV_POUNDS = LV_CON * LV_KG.


ENDFUNCTION.
```

Listing 3.6 Remote-Enabled Function Module Example

As shown in Listing 3.6, this sample code creates a remote-enabled function module in ADT. In our example, for simplicity, the logic converts kilograms to pounds, but any business logic can be similarly embedded.

Figure 3.30 shows in detail how to update the properties of a remote-enabled function module. Make sure you select the **RFC** processing type option from the **Processing Type** dropdown list to make the function module remote-enabled

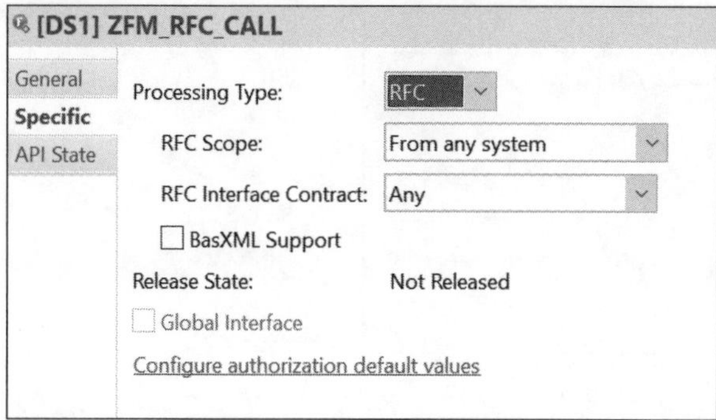

Figure 3.30 Remote Function Module Properties

Creating a Communication Scenario

Now, let's proceed with creating a communication scenario. In ADT, right-click on desired package and select **New Communication Scenario**, as shown in Figure 3.31.

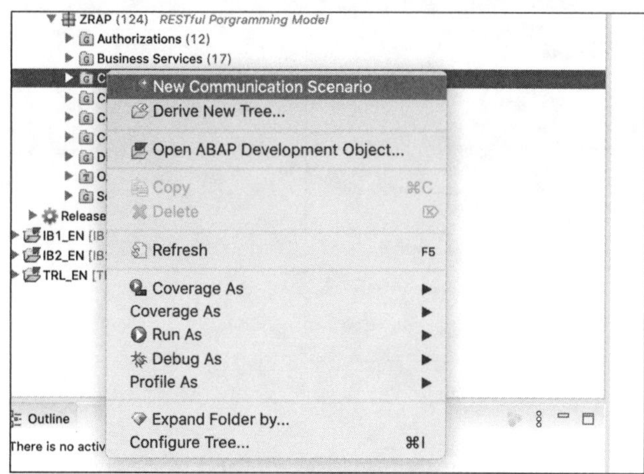

Figure 3.31 Communication Scenario

As shown in Figure 3.32, you'll create the communication scenario in ADT.

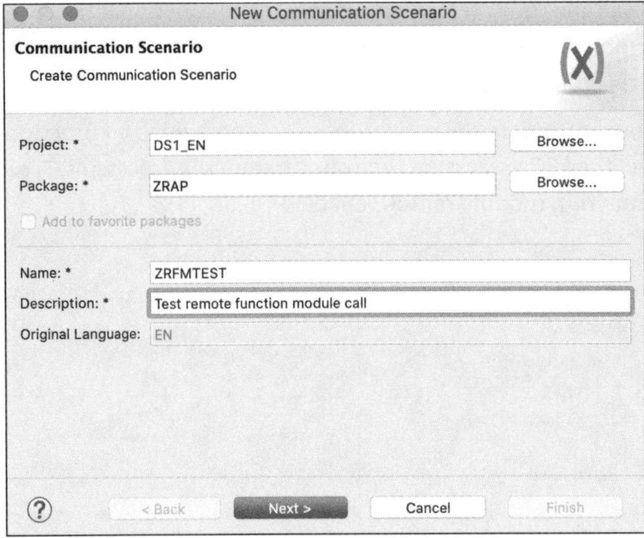

Figure 3.32 Creating a Communication Scenario

Once it's created, on the **Overview** tab you will find the dropdown list for **Allowed Instances**. Choose the **One instance per scenario & communication system** option, as shown in Figure 3.33.

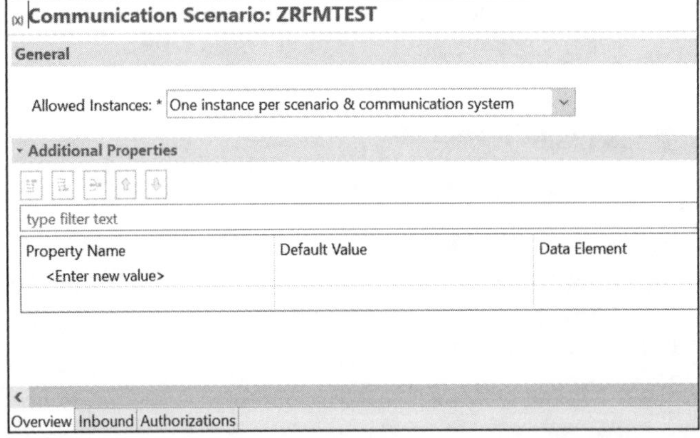

Figure 3.33 Communication Scenario Overview Tab

Select the **Inbound** tab and assign the basic authentication method by selecting the **Basic** checkbox in the **Inbound Settings** section, as shown in Figure 3.34.

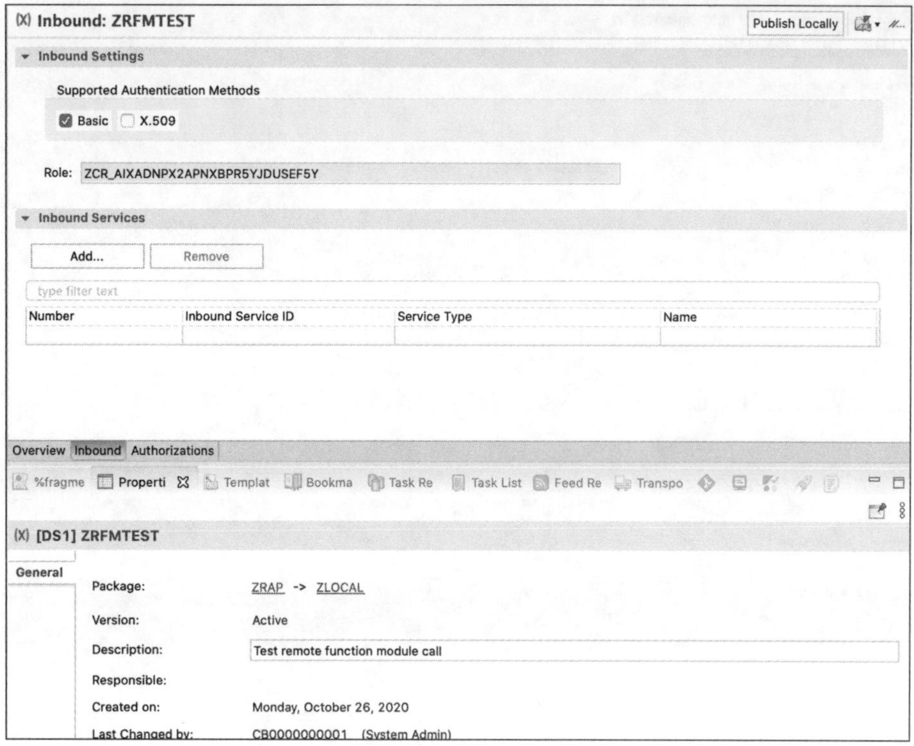

Figure 3.34 Authentication Method Assignment

Now, add the inbound services that you want to include in this communication scenario in the **Inbound Service ID** field, as shown in Figure 3.35.

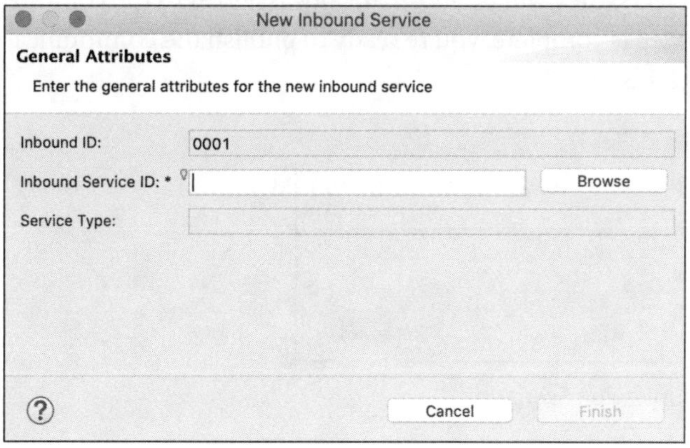

Figure 3.35 General Attributes

Next, assign one or more authorization objects, as shown in Figure 3.36. This step is optional, but if authorization objects need to be assigned, you'll need to provide the authorization objects.

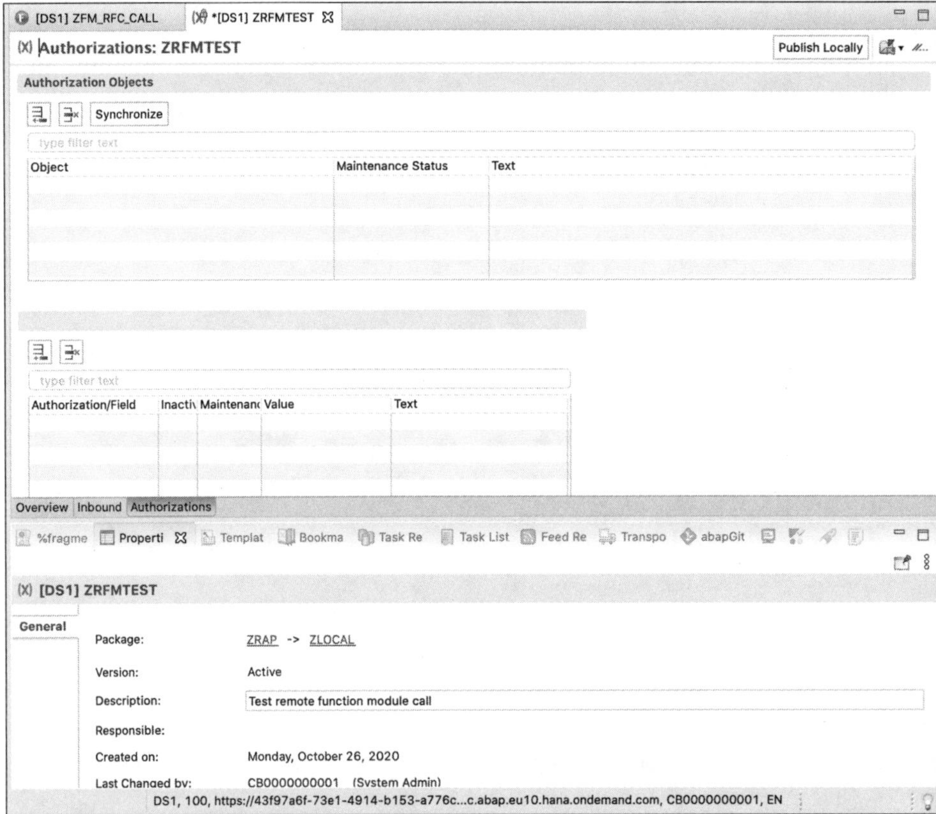

Figure 3.36 Authorization Assignment

If you assign authorization objects, they will be automatically assigned to the communication user later. Once this step is complete, you're ready to publish the communication scenario, as shown in Figure 3.37.

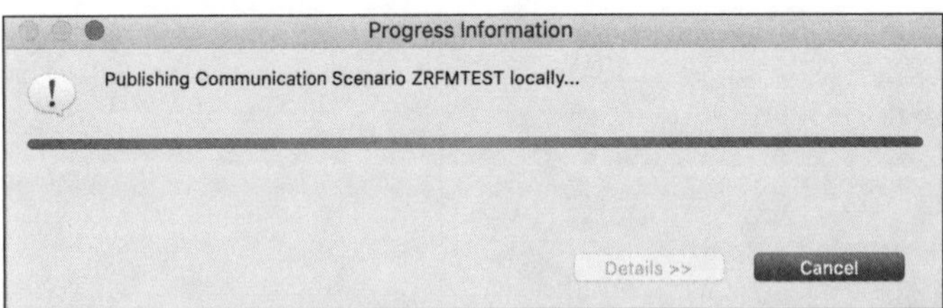

Figure 3.37 Publishing a Communication Scenario

This step concludes the setting up of the communication scenario, which is now ready for use in SAP BTP, ABAP environment.

Creating a Communication Arrangement

Let's now proceed with setting up the SAP BTP, ABAP environment dashboard. In this section, we'll create a communication arrangement via the **Communication Arrangements** tile, shown in Figure 3.38.

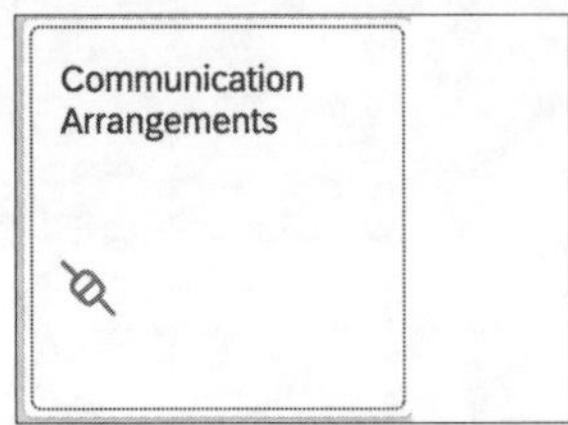

Figure 3.38 Communication Arrangements Tile

But, first, you'll need to complete two prerequisites: creating a communication user and creating a communication system. First, create a communication user via the Create Communication User app. Enter a user name, password, and description, as shown in Figure 3.39.

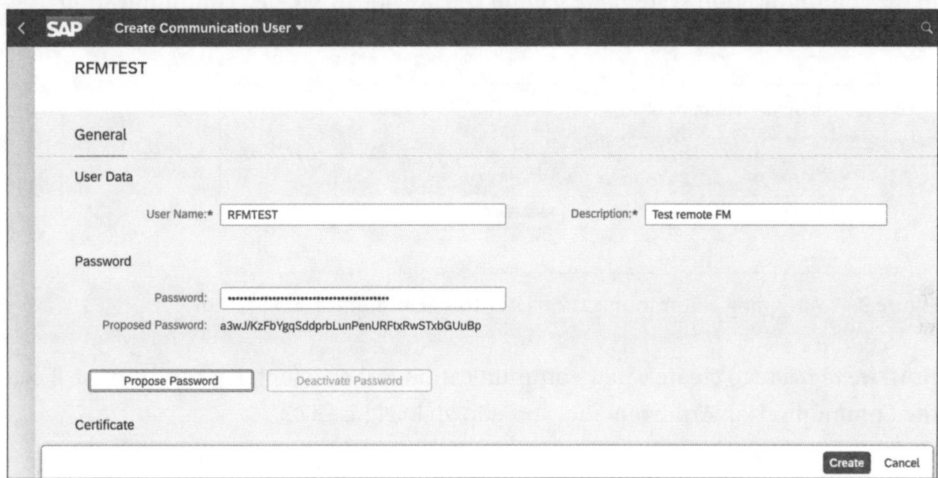

Figure 3.39 Creating a Communication User

The next step is to create a communication system via the Communication Systems app. As shown in Figure 3.40, enter a system ID, name, and business system.

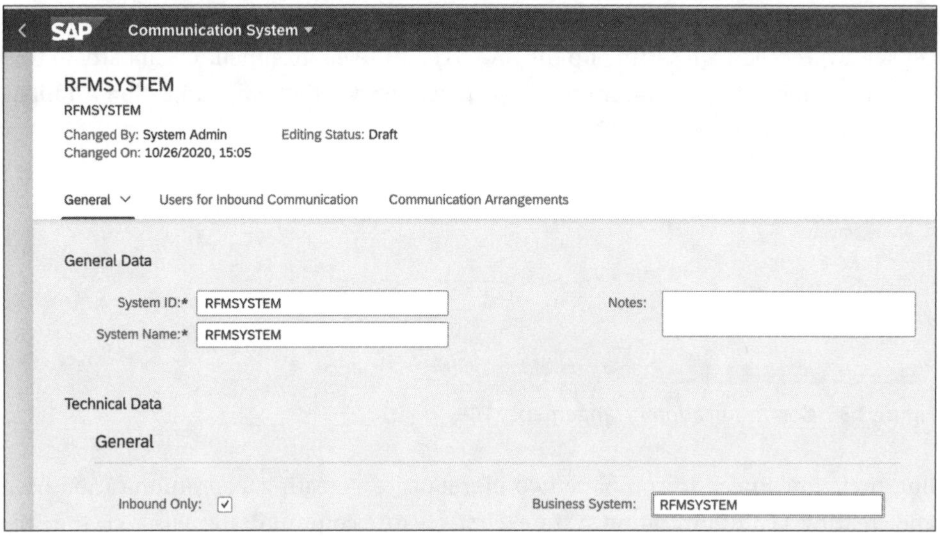

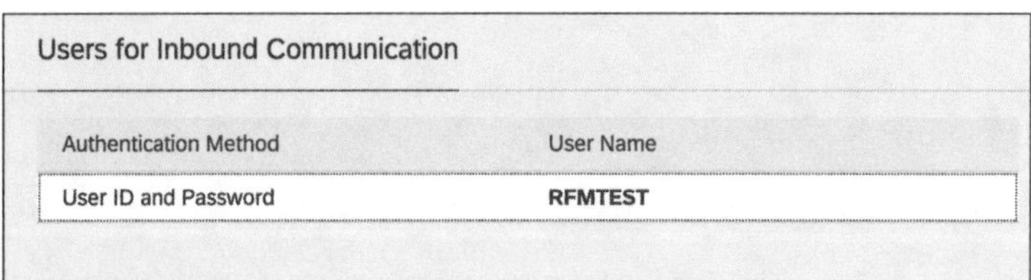

Figure 3.40 Creating a Communication System

In the communication system, we'll enter the previously created communication user in the **Users for Inbound Communication** section, as shown in Figure 3.41.

Figure 3.41 Assigning a Communication User to a Communication System

Now, we're ready to create a new communication arrangement. For this step, we'll use the Communication Arrangements app, shown in Figure 3.42.

To create a communication arrangement, we'll refer to communication scenario we configured previously in ADT.

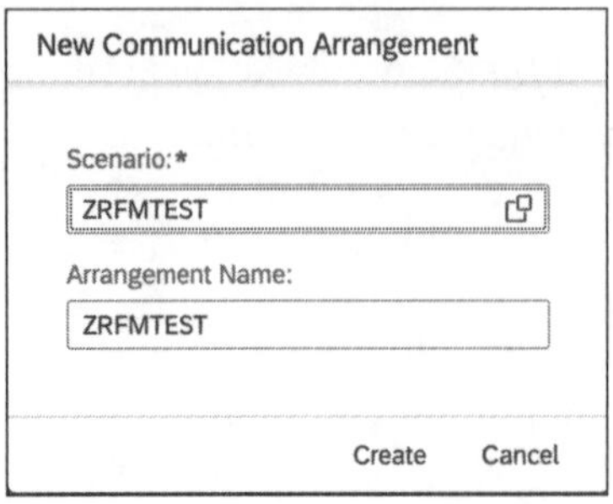

Figure 3.42 New Communication Arrangement Popup Window

Now, maintain the **Communication System** field, and all the relevant data, including communication user and the inbound services, will automatically be populated in the **User Name** field and under the **Inbound Services** heading, respectively, as shown in Figure 3.43.

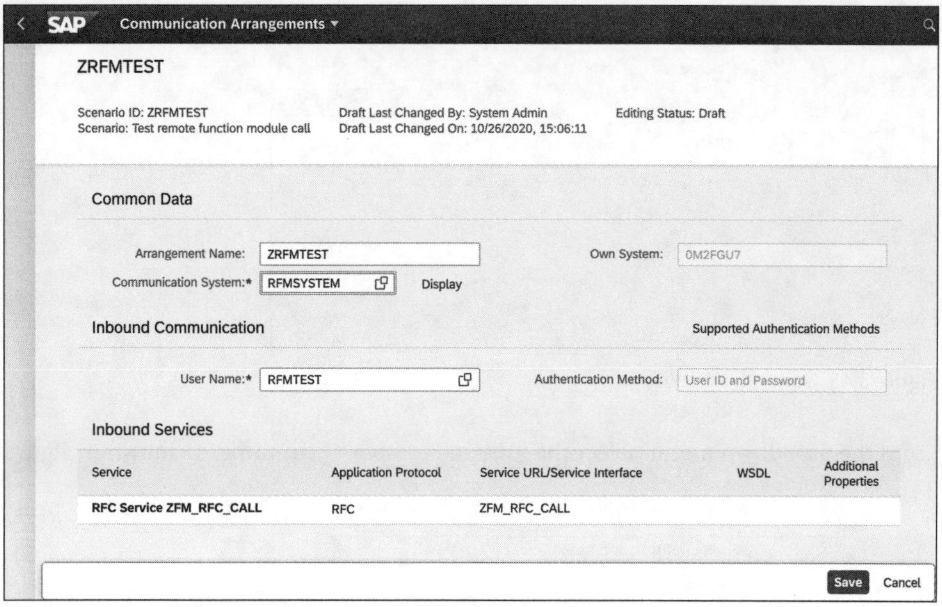

Figure 3.43 Creating a Communication Arrangement

Save your communication arrangement, which is now ready for use.

Let's say we want to add another inbound service for the inbound function module, to be called by on-premise system. On the **Communication Scenario** screen, go to the **Inbound** tab and maintain the fields under the **Inbound Settings** heading, as shown in Figure 3.44.

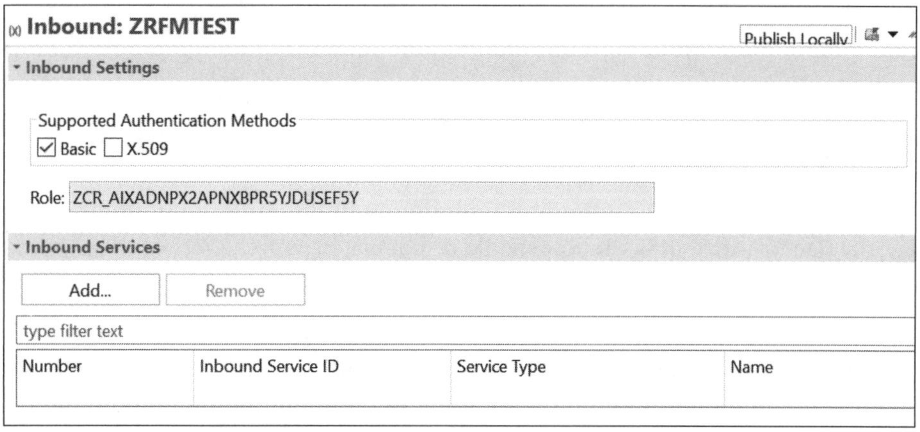

Figure 3.44 Maintaining Inbound Settings

The next step is to add an inbound service by clicking the **Add...** button. In the **General Attributes** window, click **Browse** to reach the screen shown in Figure 3.45.

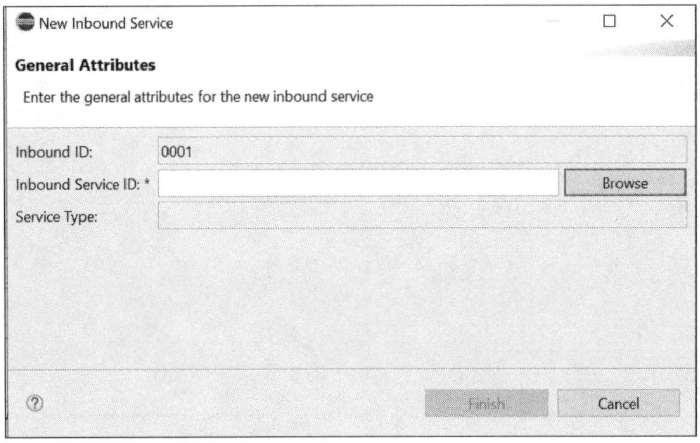

Figure 3.45 captionFigure.

Figure 3.45 General Attributes

From the dropdown menu, select the inbound service by its name, as shown in Figure 3.46.

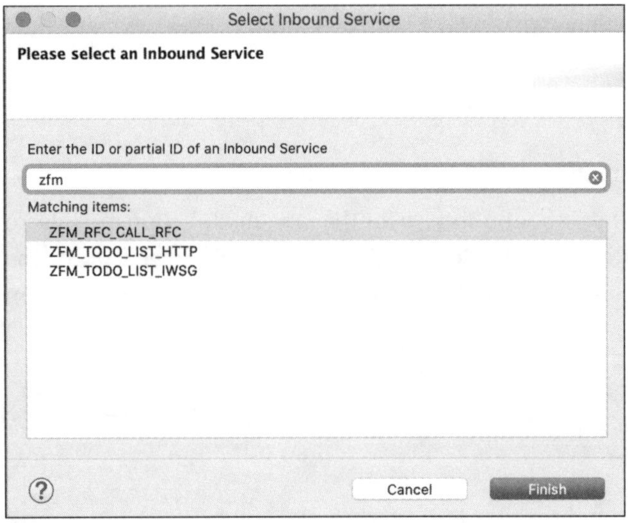

Figure 3.46 Inbound Service

Now, the **General Attributes** window should be populated with the service name in the **Inbound Service ID** field, as shown in Figure 3.47.

Finally, the inbound settings of the communication scenario will be populated with details about the inbound service, as shown in Figure 3.48.

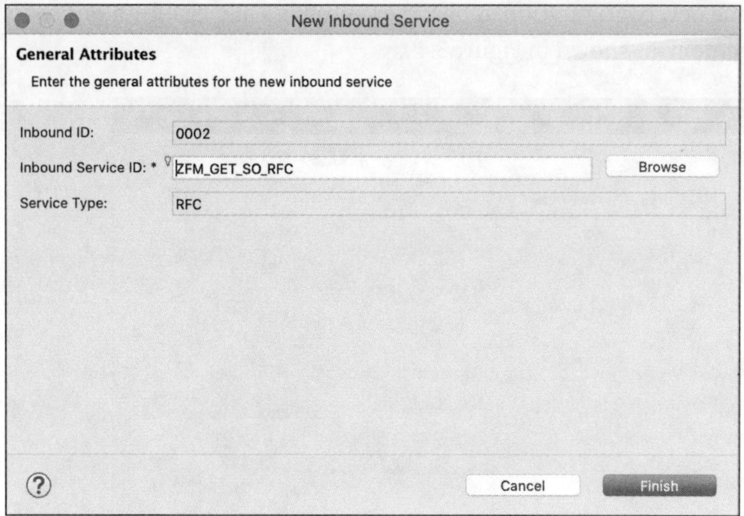

Figure 3.47 General Attributes

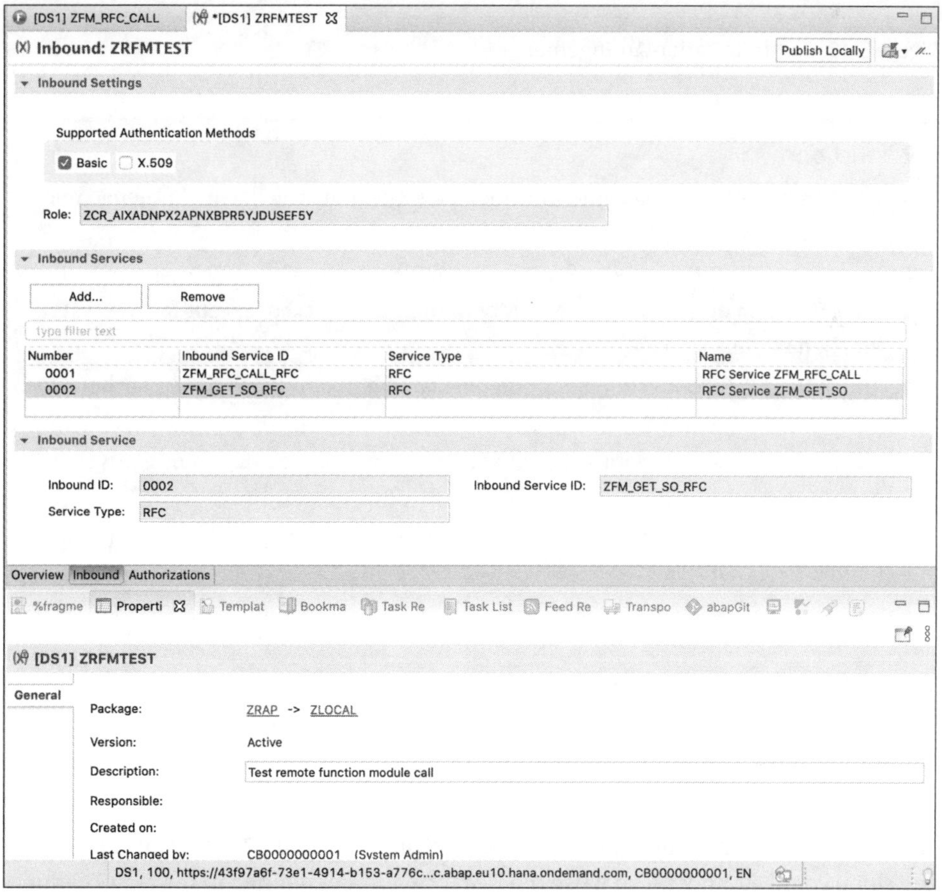

Figure 3.48 Inbound Settings

Once you publish the new configuration, it becomes automatically visible in our communication arrangement, as shown in Figure 3.49.

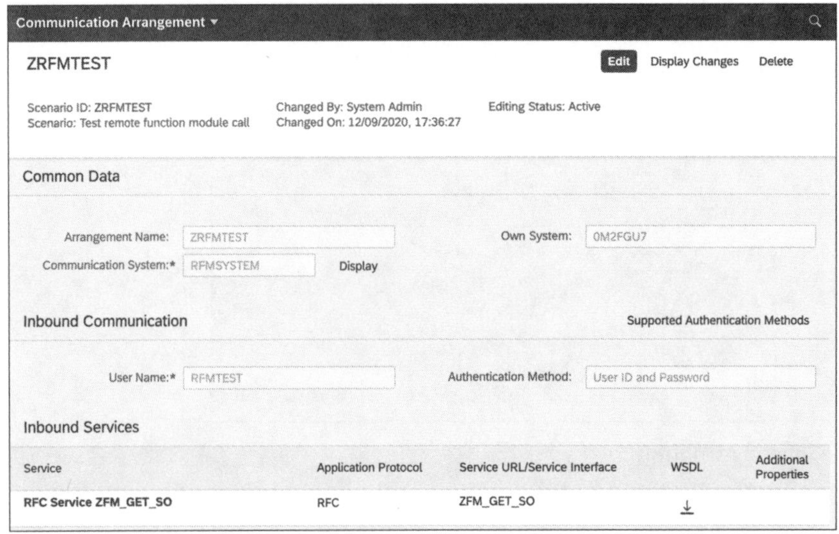

Figure 3.49 Communication Arrangement

> **Restrictions**
>
> Some restrictions regarding communication arrangements you'll need to keep in mind include the following:
>
> - Only remote function modules that are part of a communication scenario can be called from an on-premise system. Most others are blocked, except for the system function module RFCPING.
> - An inbound RFC connection supports only *basic authentication* as the authentication type.
> - The user used for authentication must be a *communication user*; a *business user* is not allowed.
> - The user name you enter in an RFC destination of an on-premise ABAP system is limited to 12 characters.

Setting Up the RFC Destination

Before you can call the service from an on-premise function module, you must set up the RFC destination in Transaction SM59. The destination is of type **3** (**ABAP Connection**).

As shown in Figure 3.50, under the **Technical Settings** tab, in the **Target Host** field, enter the cloud connector host. You'll use the instance number defined in the cloud connector itself, in the previous step, in the **Instance No.** field.

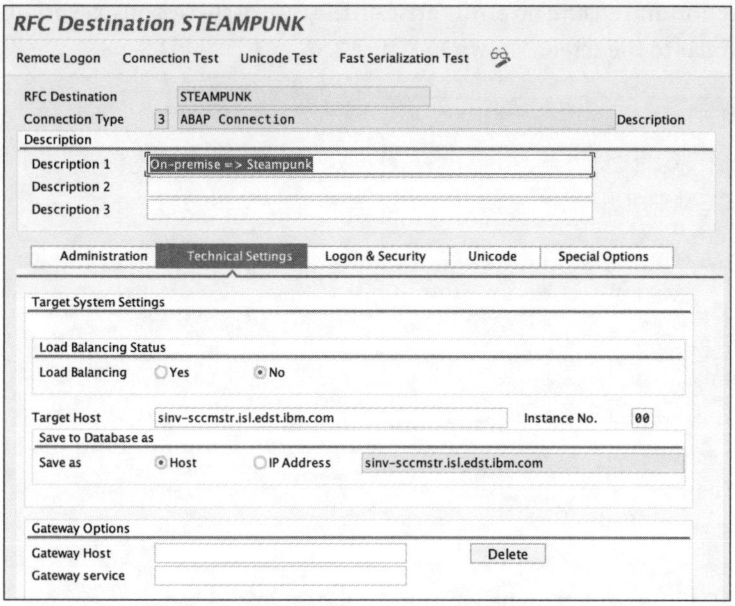

Figure 3.50 RFC Destination Details

As shown in Figure 3.51, under the **Logon & Security** tab, enter the communication user and password that was previously defined and shown earlier in Figure 3.39.

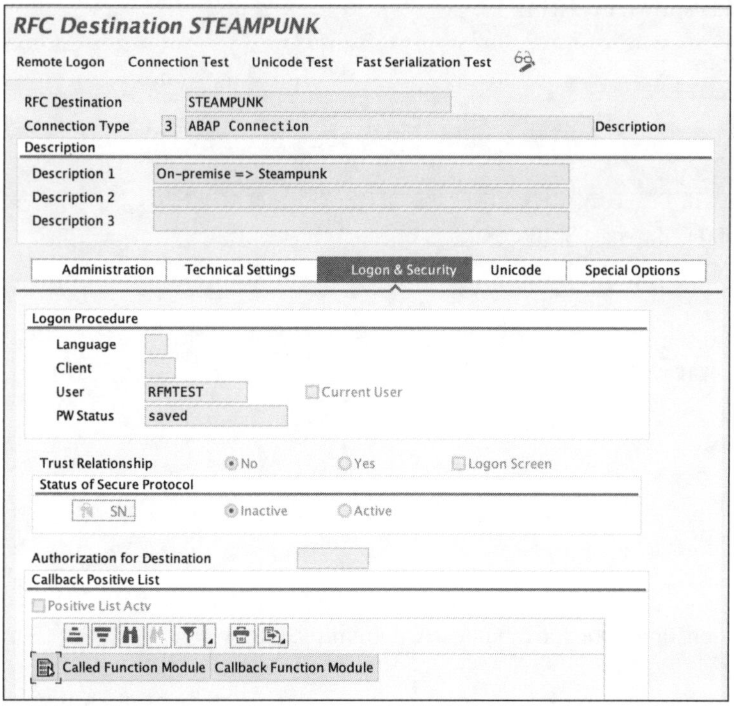

Figure 3.51 RFC Destination: Logon & Security

Now, test the connection and ensure no errors arise. In the case of successful execution, the result will be similar to the screen shown in Figure 3.52.

RFC – Connection Test

Connection Test STEAMPUNK

Connection Type SAP Connection

Action	Result
Logon	100 msec
Transfer of 0 KB	93 msec
Transfer of 10 KB	101 msec
Transfer of 20 KB	94 msec
Transfer of 30 KB	95 msec

Figure 3.52 Test Connection

Note

When calling an RFC, write its name in uppercase.

Calling the Function Module

Now, let's create a custom report to call this function module in your on-premise system by using the code shown in Listing 3.7.

```
REPORT zre_call_rfc.

DATA : lv_lb TYPE dzmeng.

SELECTION-SCREEN BEGIN OF BLOCK b1 WITH FRAME TITLE TEXT-001.
PARAMETERS : p_kg TYPE dzmeng OBLIGATORY.
SELECTION-SCREEN END OF BLOCK b1.

CALL FUNCTION 'ZFM_GET_SO'
  DESTINATION 'STEAMPUNK'
  EXPORTING
    lv_kg      = p_kg
  IMPORTING
    lv_pounds = lv_lb.

WRITE :  lv_lb.
```

Listing 3.7 Report to Consume Function Module in On-Premise System

Notice how the syntax for the function module call is standard. Provide the function module details (name, importing/exporting parameters, etc.) and RFC destination.

Thus concludes this section where you learned how to consume RFC function modules, as well as SOAP and OData services, in SAP BTP, ABAP environment from an on-premise system. You also learned how to configure communication scenarios and communication arrangements and how to consume RFC modules from an on-premise system on SAP BTP, ABAP environment.

In the next section, you'll learn how to connect cloud systems and how to consume an API on SAP S/4HANA Cloud.

3.2   Connecting to Cloud Systems

In the previous section, we covered the consumption of services that are exposed from on-premise systems.

When connecting to cloud-based systems, however, the main difference is in connectivity itself. Since the source system is available via a public IP address, you don't have to enforce additional security measures or create VPN tunnels (like cloud connector); you can simply establish a point-to-point connection.

Although SAP's software as a service (SaaS) portfolio includes multiple products, like the SAP Customer Experience product line, SAP SuccessFactors, Qualtrics, etc., from an SAP BTP, ABAP environment perspective, for building side-by-side extensions, the focus is SAP S/4HANA Cloud.

> **Note**
>
> More information about SAP S/4HANA Cloud features and capabilities can be found at *https://www.sap.com/products/s4hana-erp/features/cloud-release.html*.

Due to limited in-app extensibility options with SAP S/4HANA Cloud, side-by-side extensions are important to bridge the gaps, when business requirements exist that cannot be fulfilled otherwise. SAP BTP, ABAP environment plays an important role in building side-by-side extensions for SAP S/4HANA Cloud.

In this section, we'll focus on building extensions for the public cloud version of SAP S/4HANA Cloud, highlighting several differences from extensions for on-premise systems.

3.2.1   SAP API Business Hub

We'll start with SAP API Business Hub, which is a central location for exploring, discovering, and consuming APIs, prepackaged integrations, business services, and even sample apps created by SAP and selected partners.

To access the SAP API Business Hub, navigate to *https://api.sap.com/*. You'll see the screen shown in Figure 3.53.

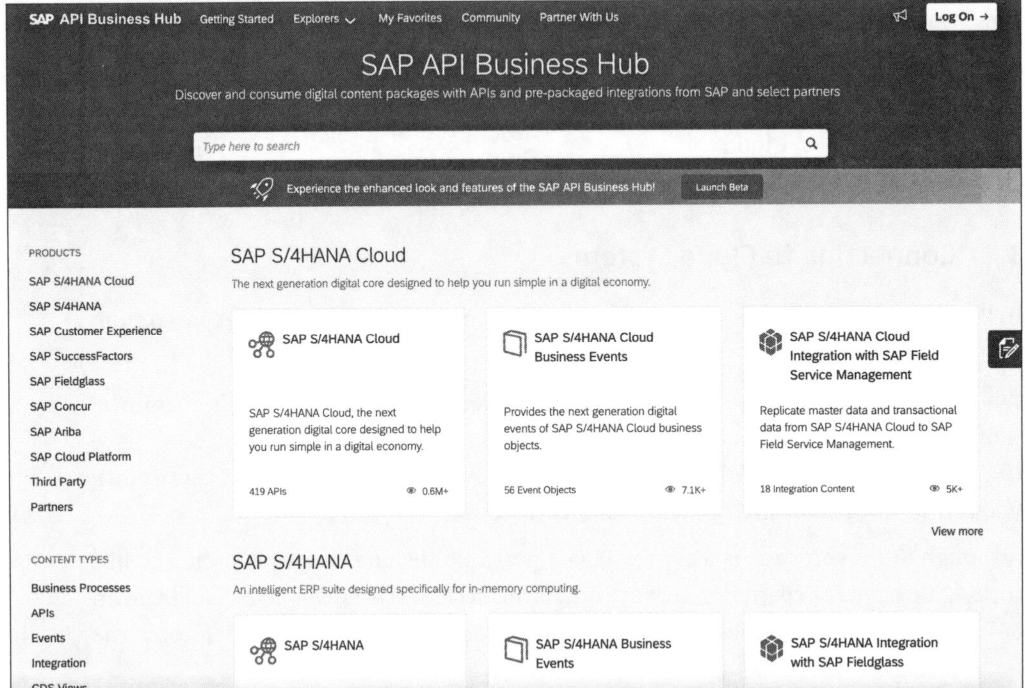

Figure 3.53 SAP API Business Hub

On the landing page, on the left, you'll see the different products, content types, and themes. From this menu, you can easily navigate to specific products/content types. The main part of the screen is the work area, which enables you to browse through different products and content.

The first product displayed in top-left corner of the work area, is the SAP S/4HANA Cloud tile. On the tile, you'll see the current number of released APIs that are available. This number increases with each new quarterly release, and at the time of this writing, almost 400 APIs are available.

Click on the **SAP S/4HANA Cloud** tile to enter the section containing the available APIs, as shown in Figure 3.54.

Note right away that two types of APIs are available: OData and SOAP. On the left, you can filter the display based on one of the two supported technologies. Also, you can filter APIs by keyword, sort APIs, and change the display (grid/list).

For example, let's say you want to know what APIs are available for sales orders. Simply type "sales order" into the filter field and submit. The result, shown in Figure 3.55, is a list of all APIs that contain both keywords ("sales" or "order") in its title or its description.

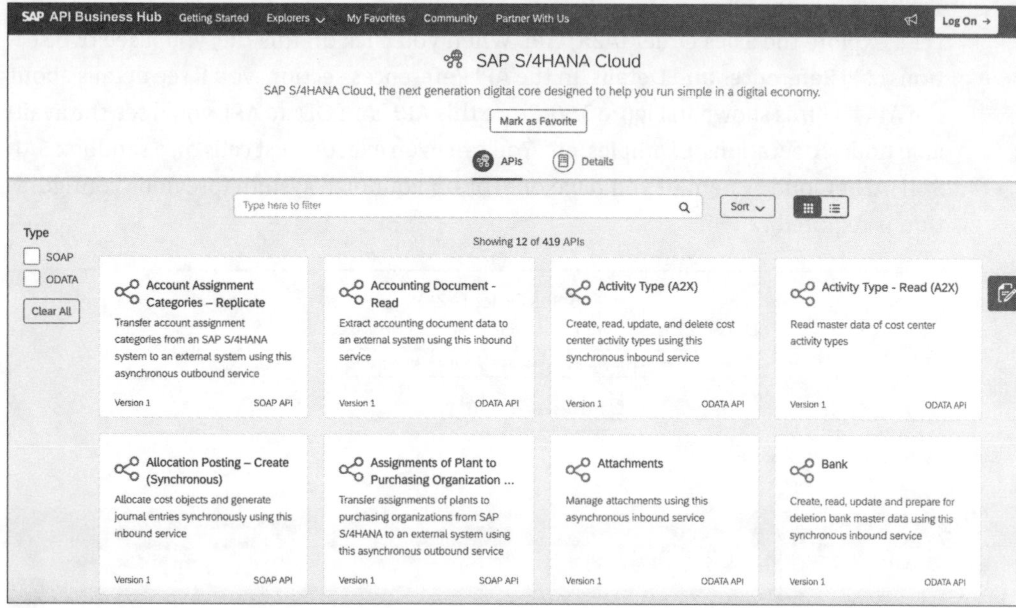

Figure 3.54 SAP S/4HANA Cloud APIs

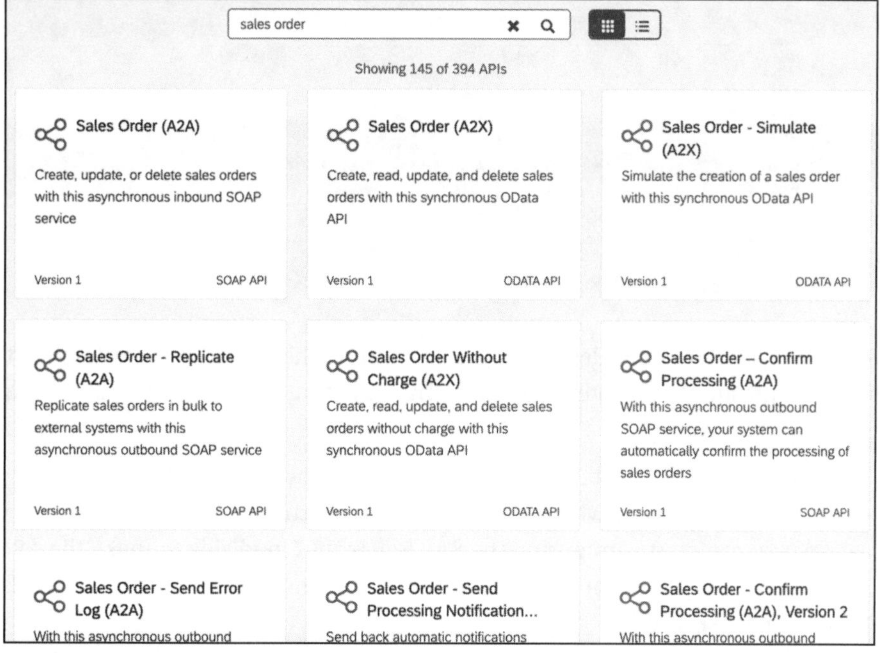

Figure 3.55 Filter Result "Sales Order"

As shown in Figure 3.55, the list contains APIs for sales orders, but also other APIs that partially match the search criteria. The most relevant APIs will be listed at the top of the result list.

Let's explore the Sales Order (A2X) API. When you click on this tile, you'll see two sections: **API References** and **Details**. In the **API References** section, you'll see details about the API itself, as shown in Figure 3.56. Since this API is an OData API, you'll see the available nodes, operations, examples, etc. You can even execute test calls on a sandbox SAP S/4HANA Cloud system (if you have one) or on your own system (previous configuration is required).

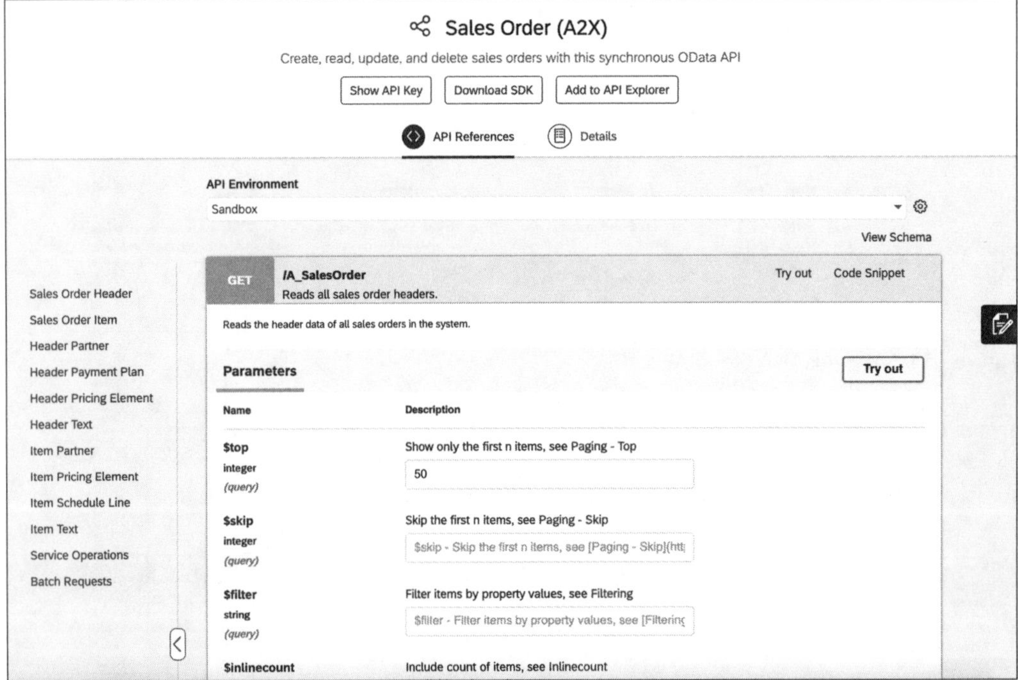

Figure 3.56 Sales Order (A2X) API References

For a SOAP API, you'll see much less information, without the capability to perform test calls. To configure an API on your source system, look at the **Details** section, shown in Figure 3.57.

Scroll down to get to the important information, such as the service URL, communication scenario, scope items in which this API is used, documentation, etc. With this information, you can start configuring the SAP S/4HANA Cloud side to make the API ready for the consumption from SAP BTP, ABAP environment, a process that we'll describe in the next section.

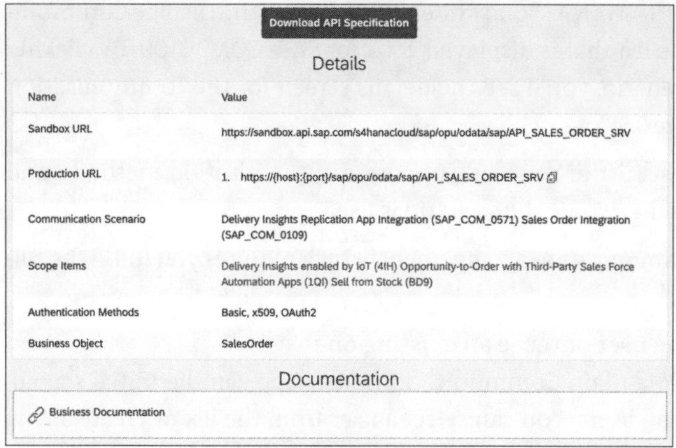

Figure 3.57 Sales Order (A2X) Details

3.2.2   Consuming an SAP S/4HANA Cloud System API

The process of exposing an API for consumption in an SAP S/4HANA Cloud system is much simpler than when working with an SAP S/4HANA on-premise system. All you need to do is configure the communication arrangement, with all the prerequisites.

Let's start with the communication scenario; in this case, we'll use SAP_COM_0109, as shown in Figure 3.58.

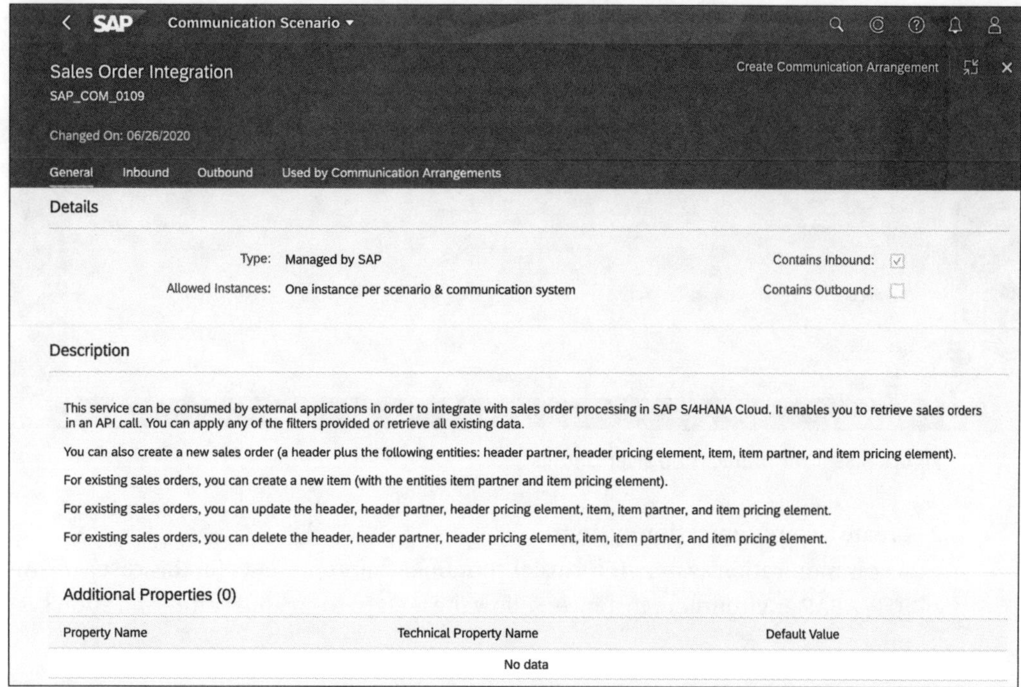

Figure 3.58 Communication Scenario SAP_COM_0109

After logging on to the SAP S/4HANA Cloud system, navigate to the Display Communication Scenarios app. In the list that is displayed, look for SAP_COM_0109. By clicking on the communication scenario, you'll see the details screen for the communication scenario, as shown in Figure 3.58.

The details screen includes a list of services, supported authentication methods, and communication arrangements where the scenario is used.

Let's now create a new communication arrangement, which requires you fulfill the following two prerequisites:

1. **Create a communication user or reuse an existing one**
   For this step, open the Maintain Communication Users app. On the initial screen, you'll see a list of existing users. You can select a user from the list or create a new one.

   To create a new user, click **New**. On the next screen, enter the user's details—user name, description, and password—and save these changes, as shown in Figure 3.59.

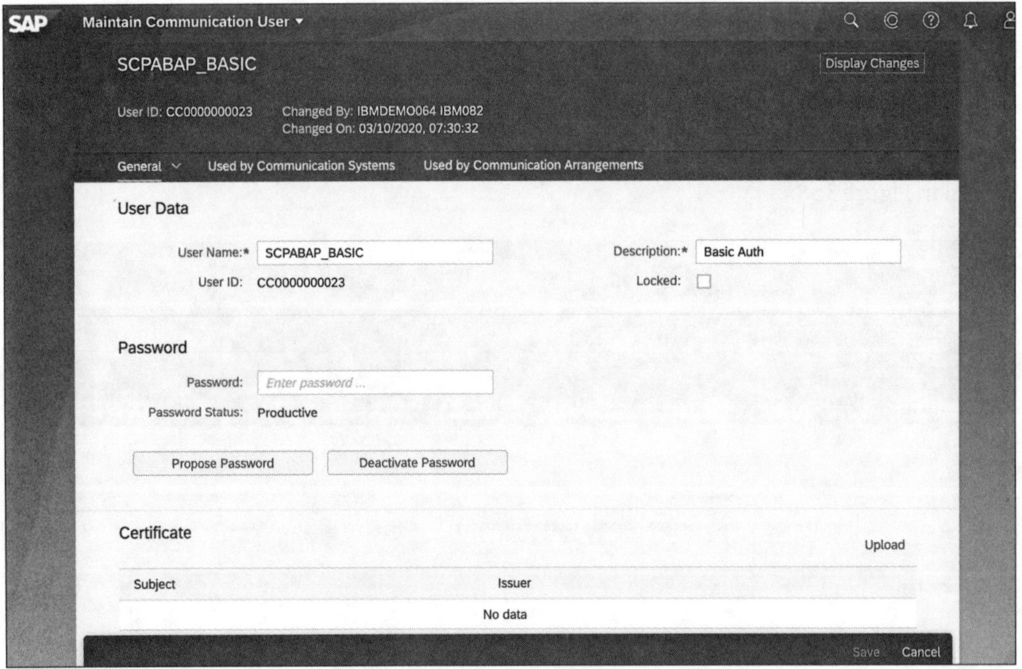

Figure 3.59 Communication User Details

2. **Create a communication system**
   A communication system is a virtual container that represents (in this case) an SAP BTP, ABAP environment instance. Follow these steps to create a communication system:

- Select **New** and, on the next screen, enter a system ID and system name in the **General Data** section. Then, in the **Technical Data** section, shown in Figure 3.60, enter details about your SAP BTP, ABAP environment system. For the host name, enter the SAP BTP, ABAP environment system host name, which is in fact the same URL as for the SAP BTP, ABAP environment instance dashboard, without the prefix *https://* and extension. The **Logical System** and **Business System** fields are free entries that must be maintained, although nothing indicates that these fields are mandatory. In our case, we entered "443" for the port number.

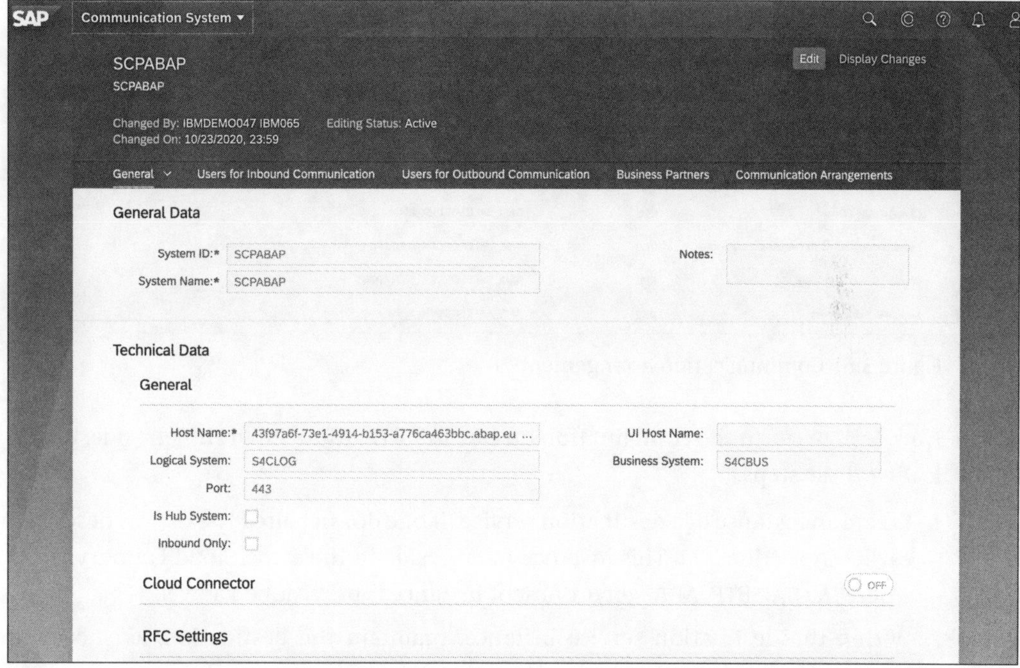

Figure 3.60 Communication System Entry Screen

- Navigate to the **Users for Inbound Communication** section and enter the communication user we created in the previous step. With the communication system is now defined, save and exit.

- Navigate to the Communication Arrangements app and then select the **New** option. In the popup window that opens, enter the arrangement name and communication scenario.

- On the next screen, enter the communication system and user for the inbound communication, as shown in Figure 3.61.

- Click on **Save**, and now, everything is ready on the SAP S/4HANA side.

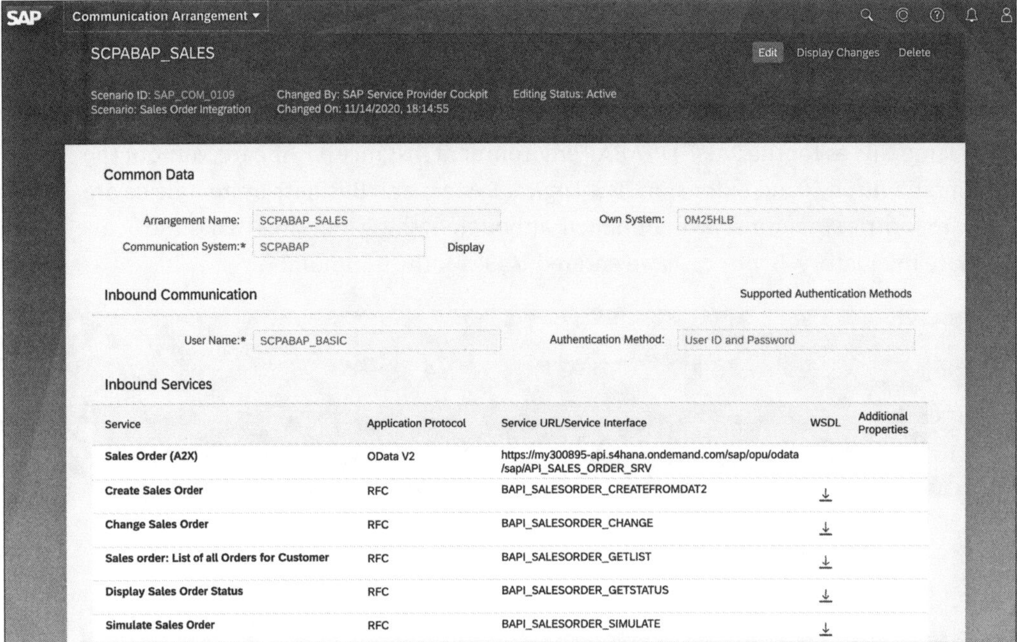

Figure 3.61 Communication Arrangement

Now, let's perform the configuration on the SAP BTP, ABAP environment side, by following these steps:

1. Create an instance of a destination service (if one doesn't already exist), as described earlier in Section 3.1.1. This instance must reside in the same Cloud Foundry space where the SAP BTP, ABAP environment instance is provisioned as well.

2. Within the destination service instance, maintain the destination, as shown in Figure 3.62. For the URL, use the SAP S/4HANA Cloud system URL. For authentication, use the communication user we created earlier.

3. Under **Additional Properties**, enter the following parameters:

```
authnContextClassRef = urn:oasis:names:tc:SAML:2.0:ac:classes:X509
TrustAll = true
```

4. Save the destination, and now, our service is ready for consumption.

Now, open Eclipse and write some code to consume the service. The code will be similar to when we consumed an OData service from an on-premise system (as described in Section 3.1.3), the only difference occurs in the line shown in Listing 3.8.

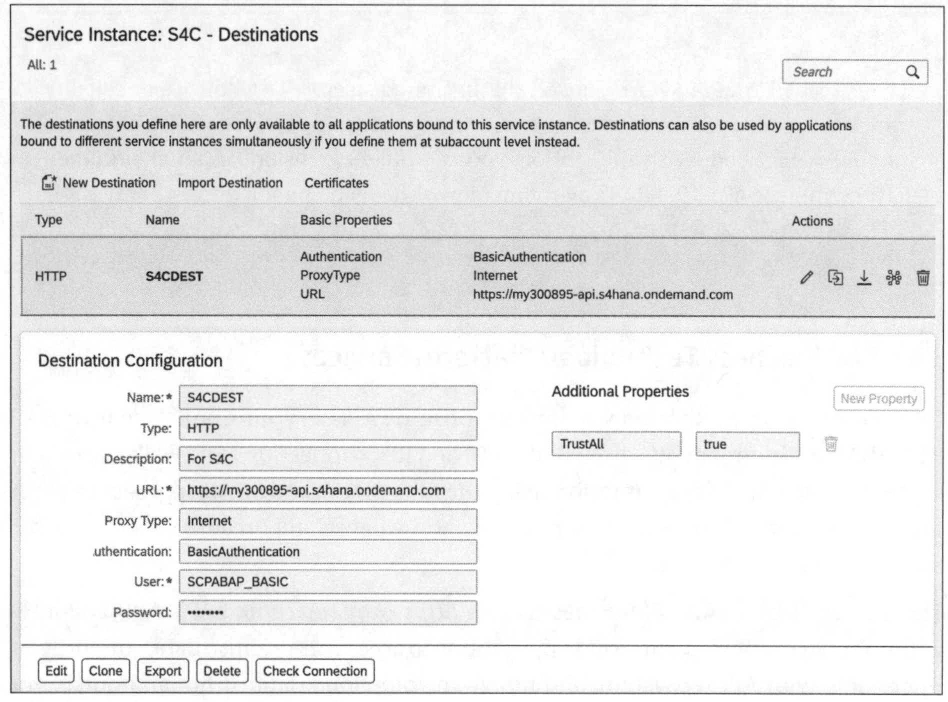

Figure 3.62 Destination Entry Screen

```
lo_http_client = cl_web_http_client_manager=>create_by_http_destination(
                cl_http_destination_provider=>create_by_cloud_destination(
                            i_name                  = 'S4CDEST'
                            i_service_instance_name = 'S4C'
                            i_authn_mode            = if_a4c_cp_
service=>service_specific ) ).
```

Listing 3.8 HTTP Client with Reference to Destination That Points to the OData Service in SAP S/4HANA Cloud

In this case, replace the following values:

```
i_name = <destination_name>
i_service_instance_name = <destination_service_instance_name>
```

With this step, you've implemented the integration with an SAP S/4HANA Cloud system and are ready to consume its services.

> **Note**
>
> At SAP TechEd 2020, SAP announced the future release of "Developer extensibility" (sometimes called "Embedded Steampunk"), another in-app extensibility option for SAP S/4HANA Cloud. With this option, you can build extensions in an environment that resembles SAP BTP, ABAP environment, which will be bundled together with the SAP S/4HANA Cloud system.

3.3   SAP Business Technology Platform Services

In earlier sections, we showed you how to consume APIs in your ABAP code in an SAP BTP, ABAP environment. We discussed HTTP and RFC connections and walked through some examples of API consumption using the RFC, OData, and SOAP approaches. With this background, let's now look more closely at the depth and breadth of the available SAP APIs for business operations.

The master URL for SAP API Business Hub is *https://api.sap.com/*. SAP's APIs and integration services enable your cloud applications to access on-premise data, support you in building your API ecosystem, and integrate your cloud data within and outside of SAP systems.

In the following sections, we'll introduce you to the APIs available in SAP API Business Hub and then provide examples of their implementation.

3.3.1   APIs from the SAP API Business Hub

Describing each API available in SAP API Business Hub is simply not possible. Therefore, in this section, we'll focus on a single example, related to SAP S/4HANA Cloud. More details about different API groups can be found at *https://api.sap.com/themes/APICONTENT*.

If you select the **SAP S/4HANA Cloud** API group, you can search within the group and also restrict the search to SOAP APIs, OData APIs, or both, as shown in Figure 3.63, using the checkboxes on the left.

Similarly, you can find content for other SAP products as well, such as SAP S/4HANA on-premise, SAP Business ByDesign, SAP Cloud for Customer, SAP Fieldglass, SAP Ariba, SAP Marketing Cloud, SAP Asset Intelligence Network, SAP SuccessFactors Employee Central, and many more.

Note that, for cloud-based solutions, like SAP S/4HANA Cloud, SAP SuccessFactors, SAP Fieldglass, etc., SAP provides a sandbox environment that you can use for testing, as shown in Figure 3.64. However, you can easily define your own testing environment to make development and testing more productive.

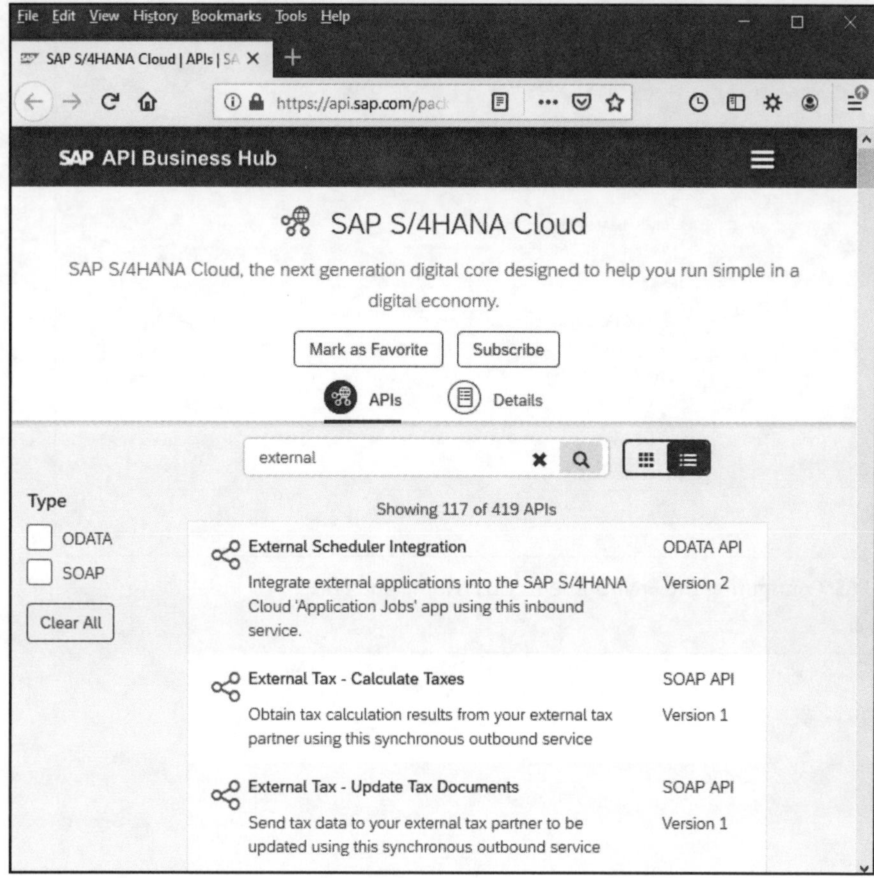

Figure 3.63 APIs for SAP S/4HANA Cloud

Figure 3.64 SAP API Business Hub: Configure Environments

In this case, we're adding our own SAP S/4HANA Cloud system to the available environments. Click on **Configure Environments**, and on the next screen, shown in Figure 3.65, enter details about your SAP S/4HANA Cloud system, including the host name and port.

Now, scroll down and enter the authentication details in the area shown in Figure 3.66. Also, specify whether this environment will be used for other APIs in the same package and/or whether you want to store the environment for future sessions as well.

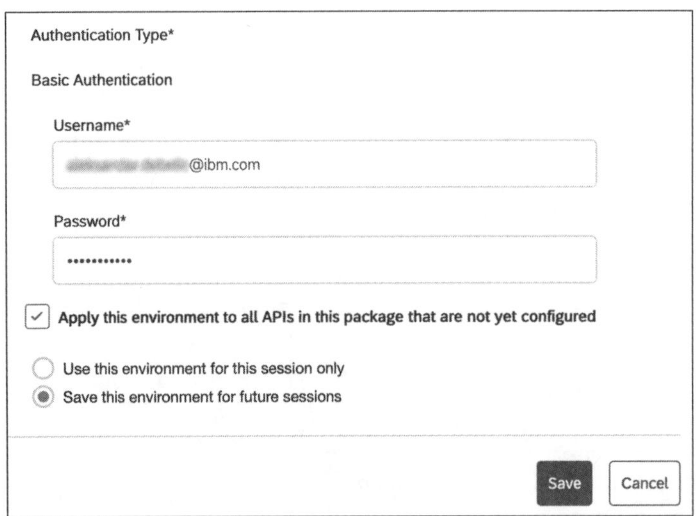

Figure 3.65 Configuring the Environment: Host Name and Port

Figure 3.66 Configuring the Environment: Authentication Details

Once you click **Save**, the environment details will be stored and are ready for use, as shown in Figure 3.67.

Figure 3.67 SAP API Business Hub: Environment List

In the following section, we'll walk you through an example of leveraging content from SAP API Business Hub to consume a service.

3.3.2   Implementation Examples

SAP API Business Hub serves as the central repository for browsing and accessing APIs created by SAP and by select partners. Let's explore how you can consume the released APIs in SAP BTP, ABAP environment.

In this section, we'll create a custom class to call whitelisted SAP APIs to consume in the cloud. You'll see in detail how to consume the whitelisted SAP APIs in the cloud by creating an HTTP service in the ABAP cloud project.

To consume this service, follow these steps:

1. Create a custom class to consume the API from SAP. To consume the API, log on to *https://api.sap.com*. In this example, the Sales Order API is used.

2. On logging on to SAP API Business Hub, an API key will be generated for each user. Copy the API key and pass it to the set_header_fields method and also pass the content_type as json, as shown in Listing 3.9.

```
class ZCL_API_MANAGE definition
  public
  final
  create public.

public section.
methods : constructor,
    get_salesorder_details RETURNING VALUE(msg) TYPE string.
protected section.
private section.
DATA: gv_url TYPE string value 'https://sandbox.api.sap.com/s4hanacloud/sap/
opu/odata/sap'.
DATA: go_http_client TYPE REF TO if_web_http_client.
ENDCLASS.

CLASS ZCL_API_MANAGE IMPLEMENTATION.
method constructor.
go_http_client = cl_web_http_client_manager=>create_by_http_destination(
  i_destination = cl_http_destination_provider=>create_by_url( gv_url ) ).
endmethod.

method get_salesorder_details.
DATA(lo_request) = go_http_client->get_http_request( ).
lo_request->set_header_fields( VALUE
#( ( name = 'Accept' value = 'application/json'   )
```

```
    ( name = 'Accept' value = 'application/json'   )
    ( name = 'APIKey' value = 'JGh41gOUhwMaYsfOaBRkt6BN6W6grAQ4' ) ) ).
  lo_request->set_uri_path(
  i_uri_path = gv_url && '/API_SALES_ORDER_WITHOUT_CHARGE_SRV/A_
SalesOrderWithoutCharge?$top=20&$format=json' ).
TRY.
DATA(lv_response) = go_http_client->execute( i_method = go_http_client-
>get )->get_text( ) .
CATCH cx_web_http_client_error.
ENDTRY.
msg = lv_response.
endmethod.
ENDCLASS.
```

Listing 3.9 API Example

As shown in Listing 3.9, you'll need to define the class ZCL_API_MANAGE. In the public
section, you should define the GET_SALESORDER_DETAILS method. Then, you'll need to
implement the class. Inside this method, use the API key and URL details and
retrieve the sales order details from the API.

3. Save and activate this class. We'll call this class in the HTTP service.

4. Create new a HTTP service in your package through ADT, as shown in Figure 3.68.

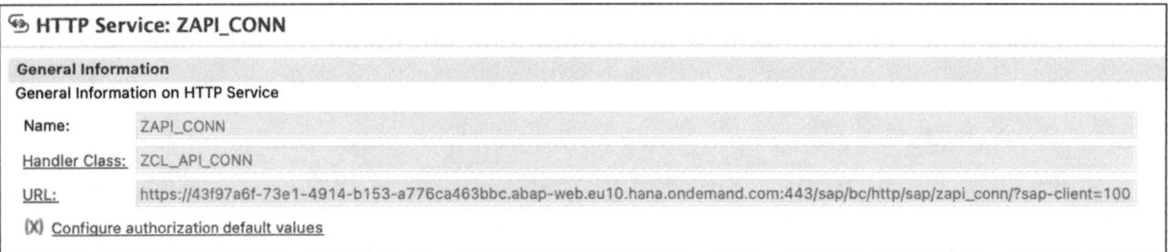

Figure 3.68 HTTP Service

5. A handler custom class will be created on creation of the service. Enter the code
   shown in Listing 3.10 in this class to call the API.

```
class ZCL_API_CONN definition
  public
  create public .
public section.
interfaces IF_HTTP_SERVICE_EXTENSION .
protected section.
private section.
ENDCLASS.


CLASS ZCL_API_CONN IMPLEMENTATION.
```

```
method IF_HTTP_SERVICE_EXTENSION~HANDLE_REQUEST.
DATA(lt_params) = request->get_form_fields( ).


READ TABLE lt_params REFERENCE INTO DATA(lr_params) WITH KEY name = 'cmd'.

IF SY-SUBRC <> 0.
 response->set_status( i_code = 400 i_reason = 'Bad Request' ).
 RETURN.
ENDIF.

CASE lr_params->value.
WHEN 'salesorder'.
response->set_text( new zcl_api_manage( )->get_salesorder_details( ) ).

WHEN OTHERS.
response->set_status( i_code = 400 i_reason = 'Bad Request').
ENDCASE.
endmethod.
ENDCLASS.
```

Listing 3.10 Consumption of the API

This example is a continuation of the example shown earlier in Listing 3.8. As shown in Listing 3.10, you'll need to define the ZCL_API_CONN class and create the class implementation as well. Then, you'll need to retrieve information using get_salesorder_details and store that information in response.

6. Save and activate the class. To get the output, click on the URL created in the HTTP service. Since the value of cmd is blank, you'll see the output shown in Figure 3.69.

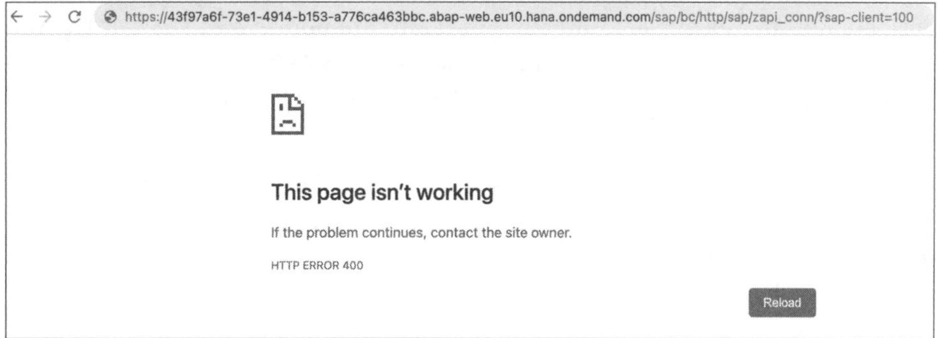

Figure 3.69 HTTP Service Output Blank cmd Value

To get the proper output, assign the value salesorder to cmd in the URL by adding &cmd=salesorder. The result is shown in Figure 3.70.

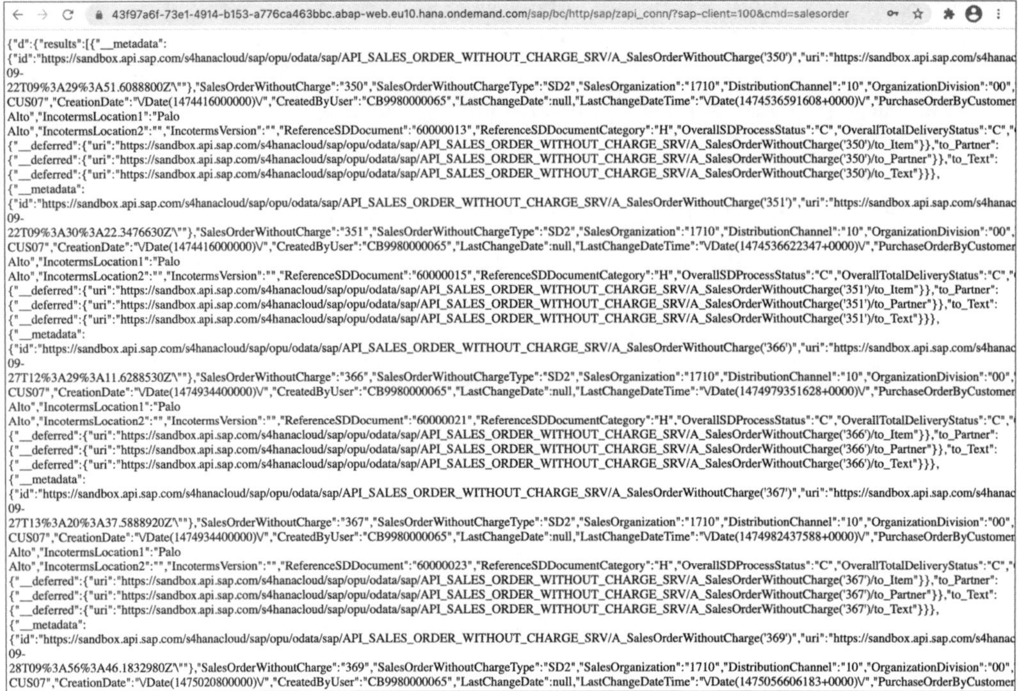

Figure 3.70 HTTP Service Output with Assigned Value

3.4   Services on Non-SAP Platforms

In the previous section, we discussed various SAP BTP services and the relevant business scenarios where you can harness true potential of these APIs. However, in some situations, you'll need to connect with other APIs as well. Specific features of some business APIs (for example, for weather forecasts) are not covered in SAP APIs. You might already have subscriptions for cloud services and APIs from other service providers. In all such cases, your SAP BTP, ABAP environment instance should be able to connect with and ingest these APIs.

In this section, we'll look more closely at some non-SAP APIs that can drastically reduce the cost and time required for software development. Moreover, since one size does not fit all, we'll discuss the industries most likely to face specific business challenges and show you how these APIs can be put to best use for such pain points.

For all the API examples that we'll discuss in this section, the objective will be to share a consulting perspective as well, so that you're not only equipped with the technical knowledge for consuming an API, but also with a business narrative to relate the API to the core business problems. In this way, you'll feel more confident recommending the most suitable set of APIs to your organization for a given business problem.

3.4.1   IBM Watson Visual Recognition API

Image recognition has gained huge momentum in all aspects of business in recent years. A large number of industries have leveraged IBM Watson Visual Recognition API for numerous use cases and projects. Its core strength is its reduction of manual labor for analyzing images and making decisions. Moreover, the fatigue from doing repetitive manual visual inspections often lead to inaccuracies and mistakes, which can be avoided with automated intelligent visual recognition.

Some common business use cases where IBM Watson Visual Recognition APIs have been deployed range from assembly line monitoring, quality inspection, asset tracking, parameter monitoring, spillage alerts, and many others, as shown in Figure 3.71.

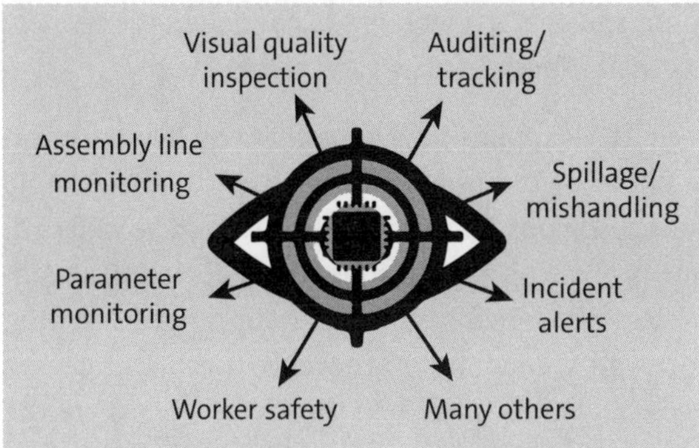

Figure 3.71 Visual Recognition: Digital Eye for Business

While building your cloud-based ABAP applications, you can leverage the following real-life examples from various industries:

- Manufacturing process may sometimes lead to rework/wastage if the individual components are not correctly placed on an assembly line. Typically, manpower must be deployed to continuously monitor and make corrections when required. With the IBM Watson Visual Recognition API, you can use images from a manufacturing setting to ensure that products are positioned correctly and raise alerts if required.

- Another example from the manufacturing industry is visual inspection and quality control. During quality inspection, the IBM Watson Visual Recognition API can identify cracks or scratches on the surface of plastic products. Heavy gears and pipes can be inspected for visible breakage. You can also utilize this API for the visual inspection of painted or polished surfaces to spot bubble marks or bare spots. All these capabilities will definitely reduce the manual effort required for quality control.

- In the utilities industry and the transportation industry, many assets may operate in open fields and require regular auditing and monitoring (windmills, fleets of trucks, planes, other assets). With the IBM Watson Visual Recognition API, not only can you establish the presence/absence of such assets, but you can also train a machine learning model to capture the rust/damage/deterioration of an asset and trigger alerts for maintenance and repair.

- Continuous monitoring of security cameras is a common pain point for almost all industries. Costs can be significantly high for some companies, particularly for industries that operate huge plants, warehouses, gas pipelines, etc. Motion sensors are not effective because these devices cannot distinguish unauthorized intruders from squirrels or birds moving around. The IBM Watson Visual Recognition API can be trained with both positive and negatives sets of images so the system can distinguish real threats from false alerts. These trained systems can then automatically monitor and trigger warning to the intruders and alerts to the security team.

- This API has also been used to enhance worker safety, for example, in large manufacturing plants to track any hot liquid spillage. If a worker falls or is lying down, the IBM Watson Visual Recognition API can trigger an alarm and call for help.

While utilizing this API in cloud-based ABAP applications, note that you'll have different sets of requirements for every phase of the implementation. In the beginning, you'll need to gather the training dataset, so your ABAP application should be well connected to image capturing and image processing applications. Then, during the training phase, you'll need to develop classification models. For a supervised learning approach, your ABAP application must provide an interface where user input can be entered and the tagging of the images can be completed. In the final stage, where the image is analyzed and inferences drawn from the continuous stream of images, the ABAP application should be able to connect with alert triggers as well so that quick responses can take care of any undesired situation. Figure 3.72 shows the data exchange flow during training and real-time usage.

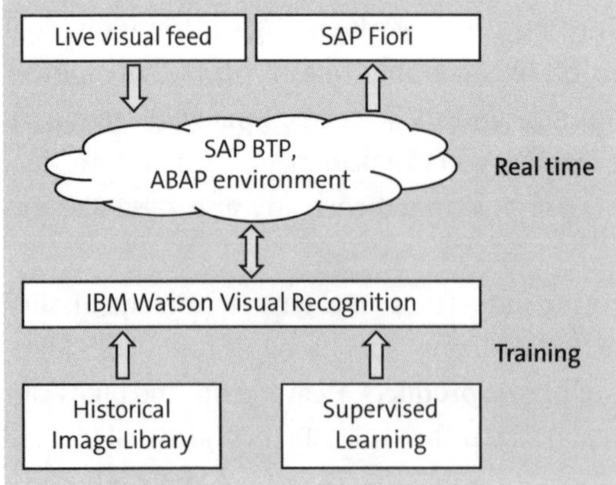

Figure 3.72 IBM Watson Visual Recognition API

In previous sections, you learned how to create connections and consume services within an ABAP application. Specific to this API, you'll need to create a class definition and class implementation to leverage the features of the IBM Watson Visual Recognition API. Some important methods that you'll need in this regard are for creating collections, to add images, training, and analysis.

To learn about using this service, and many more as well, check out the open-source project *ABAP SDK for IBM Watson*, available on GitHub at *https://github.com/watson-developer-cloud/abap-sdk-scp*.

This specific software development kit (SDK) is the first of its kind for SAP BTP, ABAP environment. As an accelerator, the ABAP SDK for IBM Watson simplifies and streamlines the consumption of IBM Watson services directly from ABAP code. This SDK is also a good resource for understanding different state-of-the-art techniques and concepts for leveraging cloud services from ABAP.

For the IBM Watson Visual Recognition API, more details on the source code are available at *https://github.com/watson-developer-cloud/abap-sdk-scp/blob/master/src/zcl_ibmc_visual_recognition_v4.clas.abap*.

Listing 3.11 shows an example class leveraging the SDK to call the IBM Watson Visual Recognition service.

```abap
CLASS zsample_classify DEFINITION
  PUBLIC
  FINAL
  CREATE PUBLIC .
  PUBLIC SECTION.
      INTERFACES if_oo_adt_classrun.
  PROTECTED SECTION.
  PRIVATE SECTION.
ENDCLASS.

CLASS ZSAMPLE_CLASSIFY IMPLEMENTATION.
  METHOD if_oo_adt_classrun~main.
   data:
     lv_image type string.
   concatenate <binary_data> into lv_image.
   data:
     lo_vr type ref to zcl_ibmc_visual_recognition_v3
   try.
       zcl_ibmc_service_ext=>get_instance(
         exporting
           i_instance_id      = 'UT_VISUAL_RECOGNITION_V3'
           "i_auth_method      = 'NONE'
           "i_host             = 'http://postman-echo.com/post'
           i_version          = '2018-03-19'
         importing
           eo_instance = lo_vr ).
```

```abap
      catch zcx_ibmc_service_exception into data(go_service_exception).
        out->write( `ERROR: get_instance`).
        return.
    endtry.

    data i_owners type zcl_ibmc_service=>tt_string.
    append 'me' to i_owners.
    append 'IBM' to i_owners.
        try.
    data(lx_imagesfile) = zcl_ibmc_service=>base64_decode( lv_image ).
    lo_vr->classify(
      exporting
        i_images_file = lx_imagesfile
        i_images_file_content_type = `image/png`
        i_owners = i_owners
      importing
        e_response = data(l_response) ).
    catch zcx_ibmc_service_exception into go_service_exception.
      out->write( `ERROR: classify: ` && go_service_exception->get_text(  ) ).
      return.
    endtry.

    loop at l_response-images into data(l_image).
      loop at l_image-classifiers into data(l_classifier).
        loop at l_classifier-classes into data(l_class).
          out->write( `Classifier: ` && l_classifier-name && `  Class: ` &&
l_class-class && `  Score: ` && l_class-score ).
        endloop.
      endloop.
    endloop.
    out->write( `DONE!` ).
endmethod.
ENDCLASS.
```

Listing 3.11 IBM Watson Visual Recognition Service

As shown in Listing 3.10, you'll need to define the `zsample_classify` class, in which an interface (`if_oo_adt_classrun`) should present in the PUBLIC SECTION of the class. Then, you'll need to create an implementation of this class. In the method named `main`, you'll need to invoke the IBM Watson Visual Recognition service to classify the input image. On successful execution of this code, the details of the classifier, the class, and the confidence score will be displayed in the output.

3.4.2   IBM's The Weather Company APIs

The Weather Company (TWC) has been part of IBM since 2016. IBM provides a set of APIs for retrieving short-term, medium-term, and long-term weather-related information, including the following details:

- Weather alerts: Headlines and details
- Daily forecast: Weather for 3, 5, 7, or 10 days
- Hourly forecast: Weather for the next 48 hours
- Intraday forecast: Weather for 3, 5, 7, or 10 days
- Almanac services: Daily and monthly weather almanac information
- Current conditions: Weather observations
- Historical data: 24-Hour historical weather observations
- Location services: Look up by address, city, or postal code

Rather than discussing each of these APIs individually, let's try to understand the importance of weather data as a whole, for different segments of business and different industries, to help you understand how to use these APIs. Figure 3.73 shows the impact of weather data on all major industries.

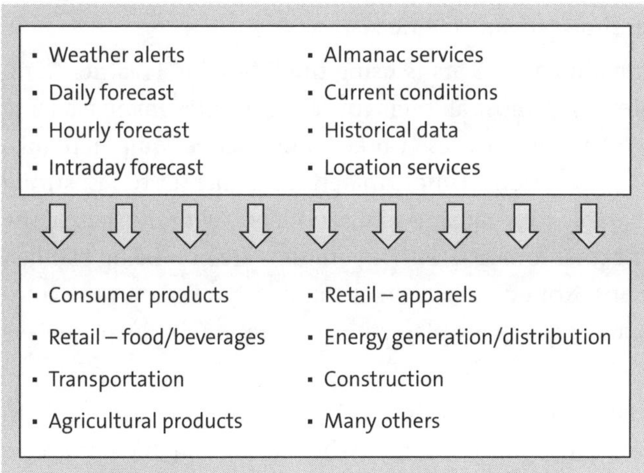

Figure 3.73 Impact of Weather Data on Major Industries

Some prominent recent success stories where these weather APIs have been extremely helpful include the following:

- A large retail food chain utilized IBM's The Weather Company APIs to forecast demand for food and beverages in its restaurants. For example, on colder days, hot beverages enjoy more demand; and on warmer days, cold drinks and beer. This business uses this information, not just for intra-day sales planning, but also for synchronizing their weekly procurement planning while considering the aggregate demand of all its stores. Direct monetary benefits include more sales volumes per week as well as fewer stock-out events, less inventory costs, and less food wastage all of which occurred because of unpredictable demand patterns.

- A big consumer products manufacturer relies on supplies of tomatoes, mint, and many other agricultural products from local farmer groups. Since weather conditions directly impact crop yield and crop quality, this company collaborated with the local government's irrigation department and with farmer supplier groups to provide mobile messages for the best time to plant seeds, to water crops, and even to protect crops through early warnings of severe weather conditions. Better crop yields (compared to same quarter in previous year), less water usage, and better backward integration with the supplier are direct benefits. Moreover, the company is more confident in its forecasts of raw material supply.

- A giant in the energy and utilities industry is heavily relying on IBM's The Weather Company APIs to forecast energy consumption. This company is also utilizing this forecast in pricing energy costs for different time slots (hourly segment pricing) considering demand situations based on temperature, humidity, and other parameters. Moreover, this company has collaborated with small local solar grids as well, from which energy can be temporarily procured during peak demand hours. As a result, both revenue and profitability have increased, and customer satisfaction has also improved because of fewer shortages and overloads.

- One of the largest construction contractors is using the IBM's The Weather Company APIs to save lives, literally. Almost all their construction sites involve a large number of construction workers. Often, these workers utilize large equipment (like cranes) during their regular operations. Unfortunately, at higher altitudes, strong winds and rains have been dangerous causing some accidents. With the help of the weekly, daily, and hourly weather forecasts, such operations are now being planned in calmer weather conditions. Not only has worker safety improved, but the company can more efficiently and proactively plan critical operations, reducing delays in the schedule.

- A top transportation company has utilized IBM's The Weather Company APIs in planning all its modes of transportation. These short-term and medium-term weather forecast help the company decide what percentage of orders meant for a destination should be sent by road versus by air. Even cargo flight schedules and road trip plans can be drawn based on the snow/rain/turbulence conditions. This capability has helped the company optimize costs as well as improve adherence to schedules by reducing weather-induced delays. The company has become preferred choice of transportation in the region which, has directly helped improve the company's bottom line.

If you want to replicate a similar success story by utilizing IBM's The Weather Company APIs in your cloud-based ABAP solution, you'll need to ensure that the right set of data exchange takes place between the SAP application and the API. Figure 3.74 shows an overview of the solution architecture for consuming IBM's The Weather Company APIs along with SAP BTP, ABAP environment. Depending on the scenario, latitude/longitude information might already be available with you, or you can extract that using the location service. You must pass these values as inputs for the forecast API, along with date and time parameters, for accurate forecasts. You can store the response received from the API in an SAP HANA database, or you can directly consume this output in SAP Analytics Cloud or in SAP Fiori, depending on the frontend application.

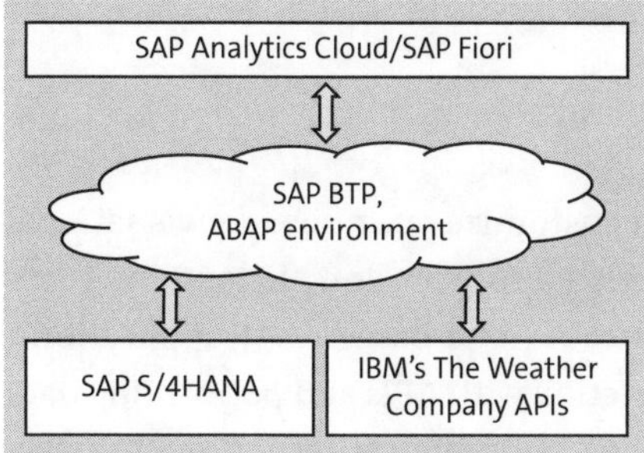

Figure 3.74 Architecture Diagram: IBM's The Weather Company APIs

3.4.3   IBM Watson Assistant

Rapid increases in queries from customers and employees has forced almost all industries to develop intelligent, conversational solutions. Banks, credit card companies, utilities, mobile services providers, and direct-to-home (DTH) services are some example industries where digital assistants have become as essential as air. These digital assistants can receive queries from customers on websites or mobile apps and then try to provide all the relevant information to customers to answer their queries. Without these digital assistants, the workforce required to answer all customer queries will be roughly 8 to 10 times larger, increasing cost pressures significantly. Moreover, these digital assistants can also reduce waiting times for customers since a large number of instances of the same assistant can work in parallel with customers. From a response quality perspective, this automation can also help ensure that responses are consistent in nature and not dependent on the limited knowledge of the customer care support staff, as in traditional customer service scenarios.

Similarly, for large corporations, even managing internal queries from an employee has become extremely easy with the help of digital assistants. Queries related to payroll, leaves of absence, intellectual capital, basic processes, and many other topics can be handled with digital assistants. Not only can you reduce the cost of support staff, you can empower your workforce with the right information, just when they need it the most. Digital assistants improve workforce efficiency as well as reduce the chance of human error while performing business operations. Figure 3.75 shows the motivation behind this approach and some business scenarios where a digital assistant can handle queries from external customers as well as from internal employees.

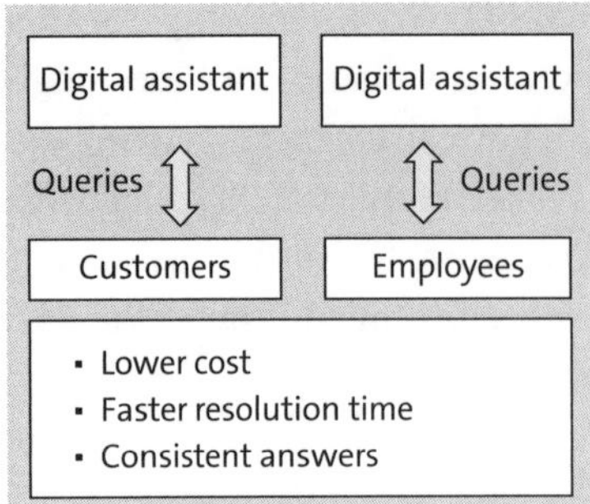

Figure 3.75 Business Scenario and Motivation for the Digital Assistant

IBM Watson Assistant enables you to create and customize your own conversational assistant, train that assistant with information related to your domain, and integrate this assistant with your applications. In the context of cloud-based ABAP applications, you'll probably find the assistant useful for connecting with APIs and performing data exchanges. You'll must use RESTful API calls based on a request-response approach. Note that the ABAP application should also be able to connect to other areas of SAP S/4HANA to access the specific information required to answer the queries.

For example, in answering questions related to a customer's bank account balance, your application should not only be able to communicate with the digital assistant but also with the SAP HANA database to fetch balance details based on customer number. For all such features, we generally recommend creating different instances of cloud-based ABAP applications dedicated to specific business modules. You can display the final output of these ABAP applications to the customer through a website or mobile app, depending on your requirements. Figure 3.76 shows how, during the training phase, historical queries can create the foundation of knowledge relation to an SAP module. This module is further connected to SAP BTP, ABAP environment in real time to display desired results with help of SAP Fiori.

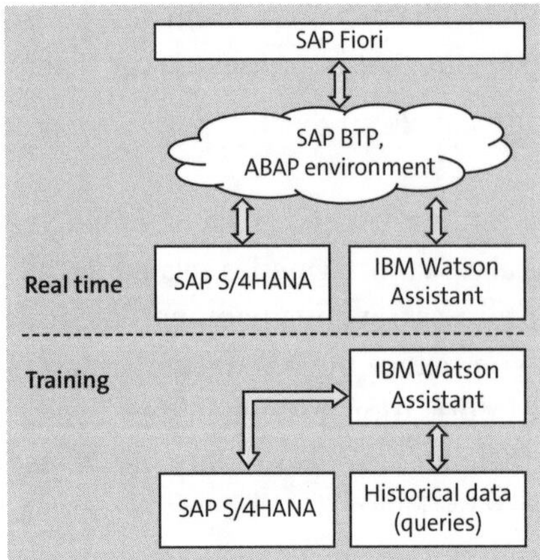

Figure 3.76 Architecture Diagram: IBM Watson Assistant

To utilize the IBM Watson Assistant API in an ABAP application, you'll need to create a class definition and class implementation to leverage the features of this API. Some important methods that you'll need in this regard are for creating workspaces, entities, synonyms, intents, and dialog nodes. You can learn more via the source code available on GitHub at *https://github.com/watson-developer-cloud/abap-sdk-scp/blob/master/src/zcl_ibmc_assistant_v2.clas.abap*.

3.4.4   IBM Watson Speech-to-Text and Text-to-Speech APIs

IBM Watson Speech-to-Text and Text-to-Speech APIs have been used across multiple industries for numerous business scenarios. As the names suggest, these APIs can convert human speech to text, and vice versa. Depending on the business scenario, you may prefer to integrate either of these APIs with your cloud-based ABAP application or perhaps even both APIs together.

Many companies combine the forces of these two APIs with the IBM Watson Assistant API (or another conversational API) to develop voice-enabled digital assistants, which may further improve the customer experience and even make conversations faster since no typing is required. Although a great use case, these two APIs have other business applications beyond just digital assistants. These APIs can been used by many industries to improve the efficiency of maintenance operations, warehouse operations, and shop floor activities, as well as for raising alerts and security warnings, as shown in Figure 3.77.

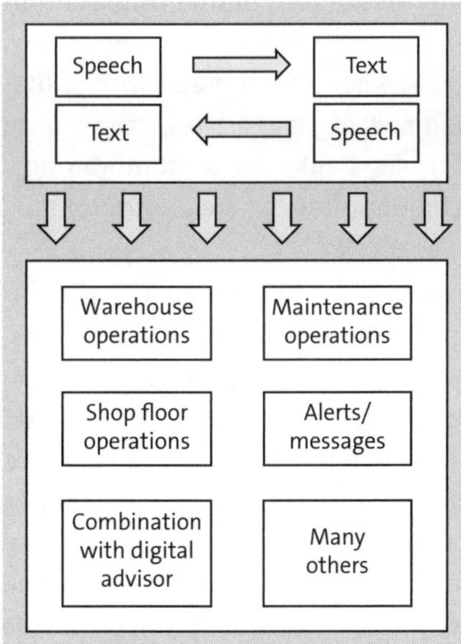

Figure 3.77 IBM Watson Speech-to-Text and Text-to-Speech APIs: Helpful for Various Domains

Let's look at some example use cases where these APIs can be used as standalone solutions addressing specific business use cases:

- A leading utilities company in North America observed that maintenance engineers typically working in the open area wore gloves to protect from the cold. Because of these gloves, these workers had difficulty updating the app on their phones and inputting maintenance operations details while in the field. Many times, they preferred to update operations all at once when winding down for the day. This behavior caused latency in the system, particularly for operations that required some additional equipment or a specialized engineer. In these cases, the maintenance planner did not have enough time to plan for additional equipment/engineer for the next day. However, with the help of the IBM Watson Speech-to-Text API, field engineers can update operational details in real time. They can use voice control on the app and update status of each operation, which automatically flows to SAP S/4HANA. This capability has helped the company get real-time visibility on the progress of maintenance operations and achieve more efficient planning of pending tasks.

- A major oil & gas company has installed pressure sensors at various chambers of a facility, which are centrally monitored. If any undue spike or drop in the pressure of any pipeline is detected, first-level alerts are sent to the operations team who can try to manage the values and normalize the pressure again. In the case of extremely high pressure variance, second-level alerts are issued to workers to evacuate that portion of the chamber and step away as a precautionary measure, until the pressure stabilizes. Previously, these alerts were only visible on a central control room screen, and someone had to shout out the warning to the workers. To avoid delay, negligence, or unusual situations, these alerts have been connected with the IBM Watson Text-to-Speech API. The new application now displays alerts on the central control room screen and immediately reads the alerts out loud, without waiting a human to initiate the verbal warning, thus further minimizing the chances of mishap.

- A large pharmaceutical company in the US manufactures baby products and feminine hygiene products in their plants, which are further stored in their warehouses. This company uses SAP Extended Warehouse Management (SAP EWM) and machine learning solutions to optimize its warehouse space, while also keeping in mind that products in higher demand are stored in such a way that these products can be more easily picked, packed, and shipped out. Workers at the warehouse had difficulties checking a pallet's details or checking location details for each package because they needed to open an app every time they wanted to enter in details and then read the storage instructions. Since most of the staff was not tech-savvy, many errors occurred, and packages were stored in the wrong bins, making the inventory reconciliation quite difficult. This company has wisely integrated the IBM Watson Text-to-Speech API with their current application to make all this process more efficient and less error-prone. Workers now just simply scan cartons using a radio frequency identification (RFID) scanner and, based on the RFID number, the app will verbally describe the location and the bin in which the packet should be placed. Even if a worker forgets the directions, a simply rescan, and the message will be replayed. This capability has significantly improved the inventory management key performance indicators (KPIs) for the company and has reduced costs as well.

These benefits of using the IBM Watson Speech-to-Text and Text-to-Speech APIs can also become instrumental for a similar success story at your organization. Note that, to integrate these APIs in your cloud-based ABAP solution, you must facilitate proper data interchange with the various SAP modules. Figure 3.78 shows the implementation scenario and data exchange between IBM Watson APIs and SAP BTP, ABAP environment. Depending on the business scenario, you may need to pass these key values as RESTful requests to provide inputs to the APIs. The response can be stored back in SAP S/4HANA (in the case of text) or can be communicated to users directly (in the case of speech).

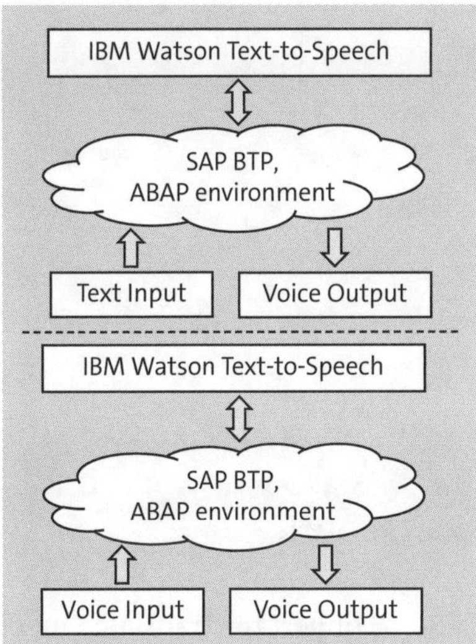

Figure 3.78 IBM Watson Text-to-Speech and Speech-to-Text APIs

3.4.5 IBM Watson Language Translator API

The IBM Watson Language Translator API can identify the language of a text and translate the text into different languages programmatically. In addition to using the provided translation models, you can build upon the existing translation models to create your own custom models for specific use cases. You can get news from across the globe and present it in your own language. You can instantly translate your webpages and apps and even your documents as well.

In addition to all these benefits, the most significant business impact of the IBM Watson Language Translator API is when used in conjunction with other useful APIs and services. Figure 3.79 shows different ways you can leverage the benefits of IBM Watson Language Translator API, ranging from document conversion, integration with digital assistants, and integration with the IBM Watson Speech-to-Text and Text-to-Speech APIs.

Let's dive more deeply into a business scenario integrating this API with a digital assistant. You can communicate with your customers in their own language and thus be able to resolve their issues much more quickly. You can also utilize this service internally and communicate with members of your workforce who may not understand your language. This capability can also be useful for broadcasting messages to a large group of diverse nationalities and backgrounds.

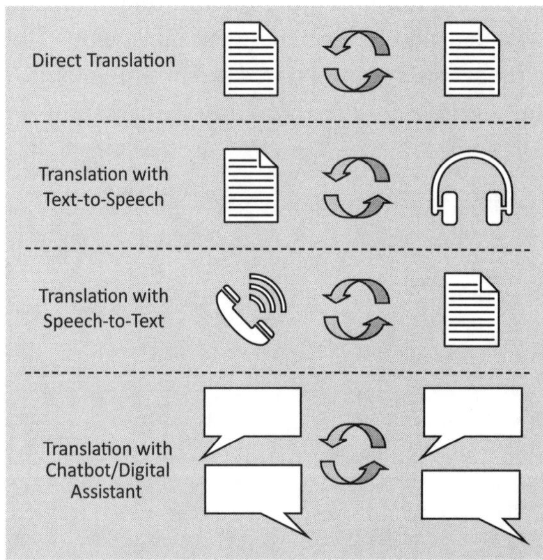

Figure 3.79 Language Translation API: Standalone and with Other APIs

This API is used by the IBM Watson Text-to-Speech API to convert the translated message into the desired language so that all the people in vicinity can quickly absorb the message. Even in the case of a one-to-one conversation with your customers, voice-enabled applications can utilize the IBM Watson Speech-to-Text API to first receive the message of the customer and convert that to text in original language. In the next step, that text can be translated to the desired language of the recipient using the IBM Watson Language Translator API and again be converted into a human voice by using the IBM Watson Text-to-Speech API.

Figure 3.80 shows the overall architecture for the best utilization of this API. You'll need proper data exchange between IBM Watson Language Translator API and SAP BTP, ABAP environment. While developing cloud-based ABAP applications for consuming this API, note that you must pass the original text message, the original language, and the desired language as input parameters to the API. The response received will be the translated text in the desired language. This capability can be further connected to a digital assistant or can be converted to speech according to the requirements of the business application.

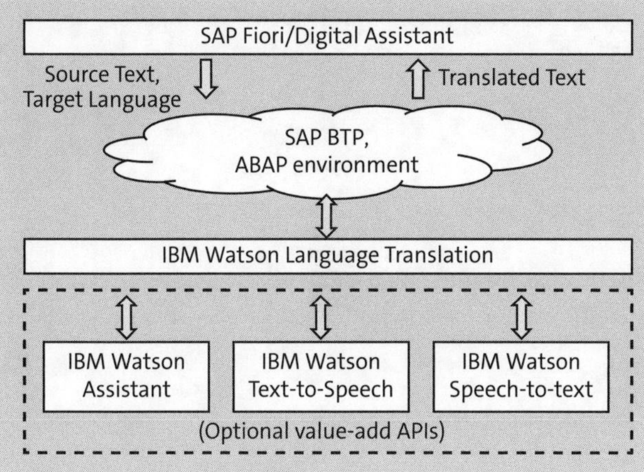

Figure 3.80 IBM Watson Language Translator

Listing 3.12 shows sample code leveraging the SDK to call the IBM Watson Language Translator service.

```
class ZCL_LANG_TRANSLATION definition
  public
  final
  create public .
public section.
INTERFACES if_oo_adt_classrun.
protected section.
private section.
ENDCLASS.


CLASS ZCL_LANG_TRANSLATION IMPLEMENTATION.
METHOD if_oo_adt_classrun~main.
  data:
    lv_apikey            type string value <api_key>,
    lo_lang_translator   type ref to zcl_ibmc_lang_translator_v3,
    lo_service_exception type ref to zcx_ibmc_service_exception,
    ls_request           type zcl_ibmc_lang_translator_v3=>t_translate_request,
    lv_text              type string,
    ls_trans             type zcl_ibmc_lang_translator_v3=>t_translation_result,
    lv_response          type string.
```

```abap
" get Watson Language Translator service instance
zcl_ibmc_service_ext=>get_instance(
  exporting
    i_url =
'https://api.au-syd.language-translator.watson.cloud.ibm.com/instances/
a8d0ad7c-1ef8-481e-acc0-f93257933dc0'
    i_apikey    = lv_apikey
    i_version   = '2018-05-01'
  importing
    eo_instance = lo_lang_translator ).

" store text to be translated into ls_request and set the languages
lv_text = 'Welcome to ABAP Cloud '.
append lv_text to ls_request-text.
ls_request-model_id = 'en-de'.
ls_request-source = 'EN'.
ls_request-target = 'DE'.

try.
    lo_lang_translator->translate(
      exporting
        i_request     =   ls_request
        i_contenttype = 'application/json'
        i_accept      = 'application/json'
      importing
        e_response = ls_trans ).
  catch zcx_ibmc_service_exception into lo_service_exception.
*     message lo_service_exception type 'E'.
  endtry.
        out->write( ls_trans ).
endmethod.
ENDCLASS.
```

Listing 3.12 IBM Watson Language Translator

As shown in Listing 3.12, you'll need to define a class ZCL_LANG_TRANSLATION, in which an interface (if_oo_adt_classrun) is present in the PUBLIC SECTION of the class. Then, you'll need to create an implementation of this class. In the main method, you'll invoke the IBM Watson Language Translator service to translate the input text, in source language English, to the target language German. Upon successful execution of the code, the translated text will be shown in the output. This scenario is an example of the translation service in action, and any pair of languages can be used in your application, depending on business scenario.

3.4.6   IBM Watson Discovery

We're living in an information age. Every second, organizations generate such a vast amount of data that can't be manually processed and contextualized. When you really need the information, you might feel the literal meaning of the phrase "finding a needle in a haystack." Thus, in the information era, your business growth depends on your cognitive search capabilities, which includes search engines that can derive information from an entire range of internal and external sources or an intelligent artificial intelligence (AI) that can filter out and identify insights that are the most relevant and actionable in the context of a search.

As shown in Figure 3.81, cognitive search engines can crawl, index, and mine both structured and unstructured data. These information models are then further refined and optimized by context models to enrich enterprise data and to build correlations so that the information is easily discoverable. Cognitive search engines significantly focus on utilizing machine learning and AI technologies for on-demand knowledge discovery and can proactively predict a user's intent to make the most context-relevant and personalized results.

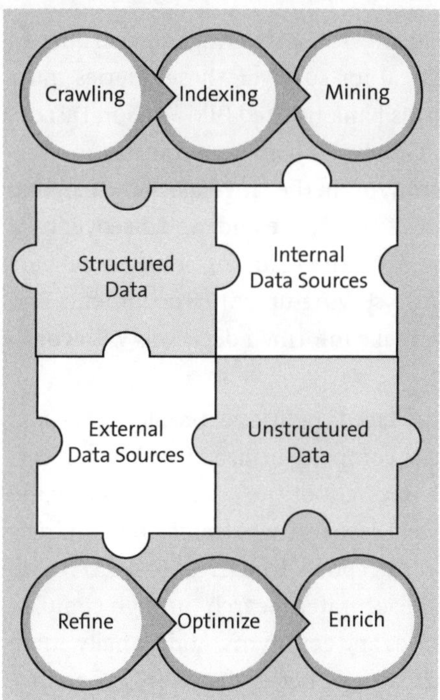

Figure 3.81 Cognitive Search Engine

IBM Watson Discovery is a leader in the cognitive search engine space. This solution utilizes a mixture of supervised and unsupervised relevancy training combined with simple graphical tools to engage subject matter experts (SMEs) in shaping capability.

SMEs can utilize Smart Document Understanding (SDU) to visually label industry-specific content within complex business documents, allowing IBM Watson Discovery to surface the exact answers from that content. Moreover, out-of-the-box integration with IBM Watson Assistant enables natural language question answering (NLQA), delivered through conversational interfaces.

To completely appreciate the business benefits of IBM Watson Discovery, let's look at some success stories of organizations that have integrated this API in their business applications:

- A leading insurance company harnessed power of IBM Watson Discovery to augment their financial consultants' intelligence to help them make smarter decisions and complete tasks more quickly. The system was trained on insurance products, policies, and business processes. Data ingestion intelligence techniques were used to boost the relevancy and completeness of results. IBM Watson Discovery was integrated with a digital assistant to automate the answering of frequently asked questions as well. With this combination of man and machine, this insurance company reduced pressure on its workforce and delighted customers with consistent and accurate contextual information.

- A leading banking organization was facing challenges in answering internal queries to the central team from branch offices. Critical for survival, these queries must answered correctly and in a timely manner. This bank utilized IBM Watson Discovery and worked with IBM to train the system about its banking products and services, including commonly used terms and acronyms in the language of the banking domain. Machine learning techniques were able to analyze trends and discover relationships buried in the company's enterprise data. The resulting solution delivers specific answers to employee queries while also serving up entire documents and supporting links, thus allowing employees to make informed decisions with confidence.

- A major player in the energy/oil & gas industry faced traditional problems like loss of expert knowledge as their SMEs retired. This company utilized IBM Watson Discovery to make existing knowledge available throughout the company and to preserve decades of collective wisdom. IBM Watson Discovery was trained by hundreds of thousands of pages of documentation, from reports to correspondence. With embedded machine learning, the system could accurately search through complex business documents for industry-specific phrases and acronyms. Additionally, intelligent intent interpretation resulted in an increasingly better discovery experience by identifying the most relevant context of the queries based on its text, the business role of the user, and the overall business scenario.

Figure 3.82 shows various components of the solution and how data is exchanged between multiple sources and IBM Watson Discovery. Considering the different phases of implementing cognitive search engine APIs—training, testing, and deployment—

you should create three variants of cloud-based ABAP applications for these scenarios. For training the system, your ABAP applications should focus on uploading documents and conversational details. In this context, RESTful API calls will have larger input (request) sections and short output (response) sections. When moving to the testing phase, the request-response sections of API calls might both be well populated. Finally, when moving to the deployment stage, your ABAP applications must have connections and data exchange set up with user-friendly interfaces, either through a digital assistant or through an SAP Fiori UX.

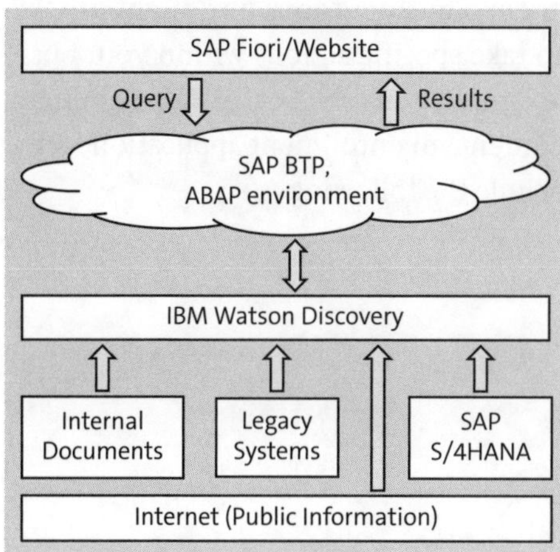

Figure 3.82 IBM Watson Discovery

To utilize the IBM Watson Discovery API in an ABAP application, you'll must create a class definition and class implementation to leverage the features of this API. Some important methods that you'll need are for creating environments, collections, tokenization dictionaries, stopword lists, queries, and query notices. Sample code is available on GitHub at *https://github.com/watson-developer-cloud/abap-sdk-scp/blob/master/src/zcl_ibmc_discovery_v1.clas.abap*.

3.4.7   Amazon Textract API

A common problem for many large SAP clients is that when clearing customer invoices, the information needs to be extracted from remittance advices, and the two (invoices and remittances) then need to be matched up. Many large corporations, where the volume of documents might be in the millions of pages (every year), usually hire a business process outsourcing company or a banking/financial companies to take care of these activities for them. This kind of outsourcing does come with a significantly high cost when considering large volumes of documents.

With advancements in AI, this kind of repetitive manual activity can be easily automated. The Amazon Textract service is one such service in this area of opportunity.

With Amazon Textract, you can quickly automate manual document activities, enabling you to process millions of document pages in hours. You can extract structured and unstructured data quickly and accurately, while maintaining security, confidentiality, and compliance. In addition to the core optical character recognition (OCR) conversion, a few advanced features are available with Amazon Textract. For example, you can capture tabular data for ease of processing. You can also capture form style data from a document and store this data in the form of key-value pairs.

Just one example of business problems related to extracting information from scanned documents, depending on the industry and business area, you might use the information gleaned from Amazon Textract to take specific actions within your business applications, as shown in Figure 3.83. For example, you can use this data to initiate the next steps for a loan application, a tax document, an enrollment application form, or a medical expense claim. Additional uses include the following:

- Custom invoice processing
- Vendor invoice processing
- Enrollment application processing
- Expense report processing

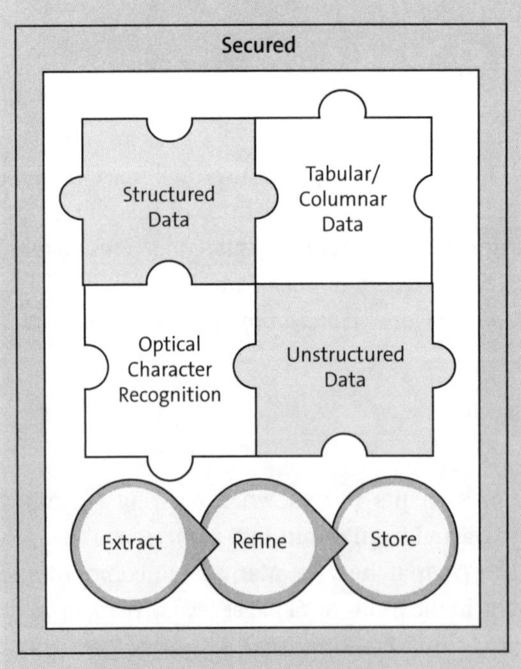

Figure 3.83 Amazon Textract API

In your cloud-based ABAP application, you can incorporate the Amazon Textract API in various flavors. The simplest approach is shown in Figure 3.84. In this context, you'll create a frontend where a user can upload a document, and then, with help of Amazon Textract, your ABAP application can perform the clearing of financial documents in SAP S/4HANA. Another variant might store all documents received from every customer every day in an application server folder, then process all of these using Amazon Textract at the end of the day to perform financial clearing. Depending on your business requirements, other variants of the ABAP application can also be created. The results of clearing can be displayed using SAP Analytics Cloud or SAP Fiori or can simply be sent via email for further review and corrections.

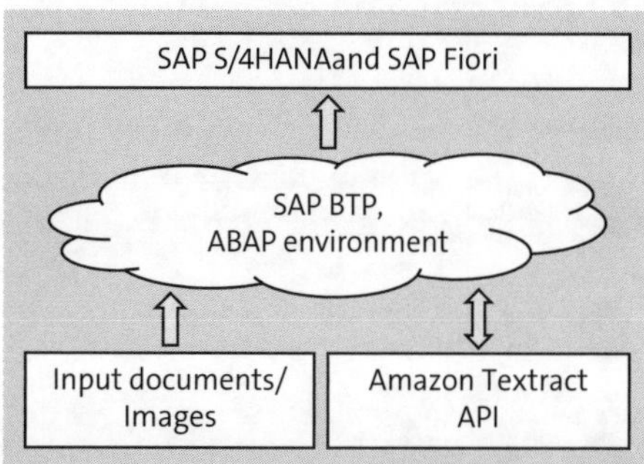

Figure 3.84 Amazon Textract API

3.4.8   APIs from Other Providers

In this section so far, we've covered only some example services that are available and that can be leveraged in SAP BTP, ABAP environment. Of course, many more other APIs exist, from different providers, which can help solve business challenges that would be difficult to address otherwise.

Let's take Google Cloud Platform as an example. Navigate to *https://cloud.google.com/ products* where you'll see the currently available offerings. For example, let's say you want to use some services from the **AI and Machine Learning** section. Navigating to **AI Building Blocks** will provide an overview of the different services available, grouped by their coverage area, as shown in Figure 3.85.

By digging deeper into the documentation for each service, you'll find more useful information, for example, on how to leverage services using REST APIs, including examples.

Another example is Microsoft Azure. Navigate from the Microsoft Azure home page (*https://azure.microsoft.com/*) to **Products**, as shown in Figure 3.86, for an overview of the whole offering. From this page, you can navigate to a specific section to learn more about specific services you might want to leverage.

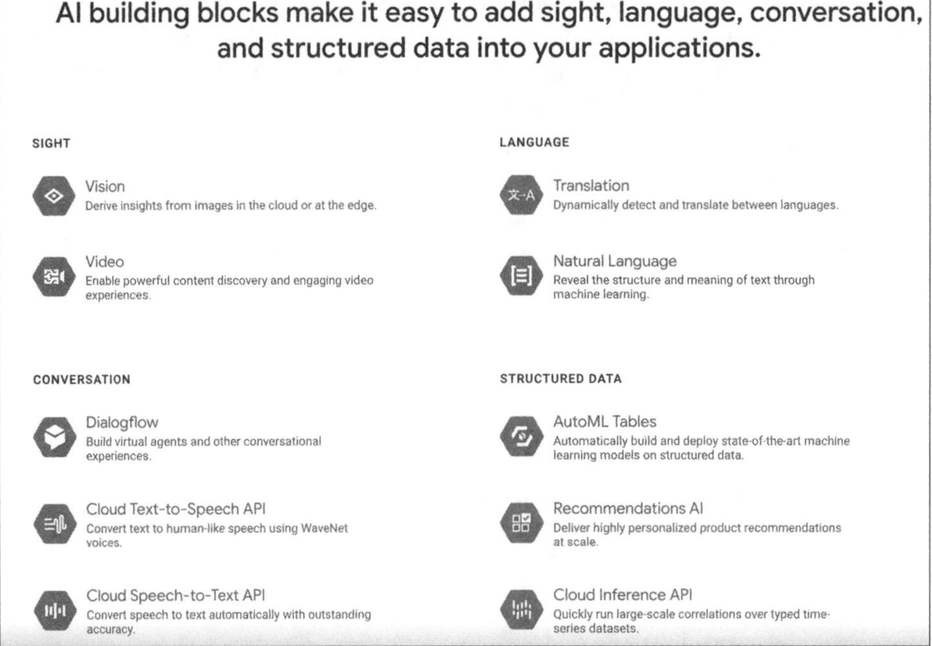

Figure 3.85 Google Cloud Platform: AI Building Blocks

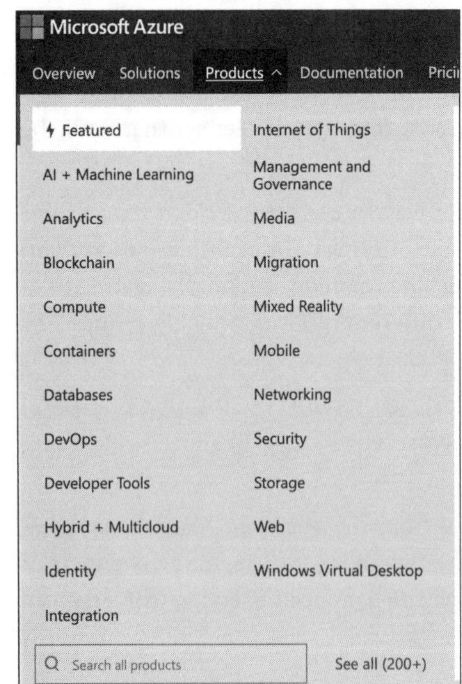

Figure 3.86 Microsoft Azure Product Offerings

What is also interesting about Microsoft Azure is the available open-source ABAP SDK for Azure. More information about this SDK is available at *https://github.com/microsoft/ABAP-SDK-for-Azure*. This SDK supports currently eight services and is built for on-premise SAP systems. Although the ABAP SDK for Azure doesn't support SAP BTP, ABAP environment, you can use this solution as a starting point, and some parts can be reused, to simplify the integration with services available on Microsoft Azure.

To call RESTful APIs from external providers, use the template shown in Listing 3.13.

```
TRY.
"we create http destination by url to API endpoint
DATA(lo_http_destination) =
      cl_http_destination_provider=>create_by_url( '<API_Endpoint>' ).
"create HTTP client by destination
DATA(lo_web_http_client) = cl_web_http_client_manager=>create_by_http_
destination( lo_http_destination ).
"add header
DATA(lo_web_http_request) = lo_web_http_client->get_http_request( ).
lo_web_http_request->set_header_fields(VALUE #(
( name = 'Content-Type' value = 'application/json' )
( name = 'Accept' value = 'application/json' )
( name = '<...>' value = '<...>' )
 ) ).
"add body
lo_web_http_request->set_text( '<...>' ).
"Available Security Schemes for productive API Endpoints
"Bearer and Basic Authentication
"lo_web_http_request->set_authorization_bearer( i_bearer = '<...>' ).
"lo_web_http_request->set_authorization_basic( i_username = '<...>' i_password =
'<...>' ).
"set request method and execute request
DATA(lo_web_http_response) = lo_web_http_client->execute( if_web_http_client
=><METHOD, for example GET or POST> ).
DATA(lv_response) = lo_web_http_response->get_text( ).

CATCH cx_http_dest_provider_error cx_web_http_client_error cx_web_message_error.
    "error handling
ENDTRY.
```

Listing 3.13 Call External REST API: Code Sample

As shown in Listing 3.13, you can create a local reference for the HTTP reference, which can be further used to create the HTTP client. You can then follow the usual request-response approach to submit the input to the API and receive the processed output.

This code template is generic in nature and can be utilized with multiple API service providers.

3.5   Summary

In this chapter, you've learned how to set up connections and explored some examples of how services can be consumed, including real-life examples of APIs in action. You can now leverage the code samples and business scenarios to rapidly develop use cases and reduce your time to market.

Since business scenarios are constantly changing, we highly recommend conducting design workshops. You can propose solutions, perhaps using the success stories mentioned in this chapter, and explore the technical feasibility of the solution by carefully selecting a connection type and implementation approach.

Also, you learned how to leverage open-source projects to improve your solutions and improve turnaround times by leveraging accelerators made available by the open-source community.

The list of SAP and non-SAP cloud-based APIs is always growing and cannot be entirely covered in any book. However, since a significant number of cloud-based APIs were covered in this chapter, you'll be able to leverage the hands-on exposure of these APIs and enable any other API as well, since you're equipped with right tools and resources.

Chapter 4
Operating Applications

In this chapter, we'll show you how to manage, maintain, and monitor ABAP cloud applications. We'll cover several options for application life-cycle management and describe numerous monitoring tools and the best practices for their use. You'll also learn to manage roles and security and learn how to trace, troubleshoot, and debug your applications.

This chapter may not sound interesting if your attention is centered on application development only. However, when you begin working on a real project, questions may arise beyond the scope of application development itself, for instance, regarding the transport of ABAP objects, debugging options, the running of traces, how to monitor your applications, assigning business roles, and so on. In anticipation of those questions, we've dedicated this full chapter on operating ABAP applications on SAP Business Technology Platform (SAP BTP). Keep in mind that SAP BTP is continuously evolving, introducing new and enhanced features every quarter. Hopefully, this chapter will provide the latest and greatest to help you overcome many practical problems.

We'll start this chapter with an overview of Git and how it relates to SAP BTP, ABAP environment development in Section 4.1 and then continue our journey by explaining debugging, monitoring, and tracing options in Section 4.2. In Section 4.3, we'll describe role management and other security enablement for building enterprise-grade business applications. Finally, in Section 4.4, we'll conclude this chapter by discussing application jobs. In this chapter, you'll learn everything you need to know to operate applications, so let's get started.

4.1 Application Lifecycle Management

Application lifecycle management is primarily the process of managing an application's lifecycle from the development phase to its deployment and also includes the continuous management of its changes in an efficient way. In a traditional ABAP world, we would use a transport management solution for managing application lifecycles. In this section, we'll how you how to reuse traditional concepts in a modern way while retaining the benefits of both using abapGit and Git-enabled Change and Transport System (gCTS). This section includes transforming ABAP objects from on-premise systems to SAP BTP, ABAP environment and transporting the objects from one SAP BTP,

ABAP environment system to another. We'll start this section with an overview of Git in case you're not familiar with the concept.

4.1.1   Git

Git is a free and open-source distributed version control system designed to handle everything from small to very large projects with speed and efficiency.

Thus, as a version control system, you can store your development artifacts in Git with automatic version control. Git allows multiple people to work on the same software and has the option of integrating the work undertaken by different developers. Git is lightweight, and you can easily install Git for your local project. For corporate projects, various Git hosting services are available. GitHub is one such a web-based hosting service based on Git. Open to all, GitHub is quite popular among full-stack developers. Other popular hosted Git services include Gerrit, GitLab, Bitbucket (Atlassian), etc.

If you're a classic ABAP developer, you may not be familiar with Git, since usually, you're developing your code in a central ABAP editor (e.g., Transaction SE80). The use of Git is also not mandatory when working in ABAP Development Tools (ADT). However, Git exists in various forms for both SAP S/4HANA and SAP BTP, ABAP environment. So, let's briefly look at the terminology used with Git and GitHub to help you to better understand the subsequent sections:

- **Repository**
  A repository is equivalent to a project folder and contains many files and their version numbers.

- **Commit**
  A commit is a revision or a set of revisions to a file. Developers actually save the file to a local Git repository. Each save after a change can be considered a commit.

- **Master**
  The main branch with the codebase currently running for the user.

- **Branch**
  A parallel version of the repository, which means you can work freely without impacting the master or primary branch.

- **Merge**
  A merge is the act of combining the changes of one branch with another branch, for instance, the master.

- **Fetch**
  A fetch is the act of getting the latest changes without merging branches.

- **Pull**
  A pull is the act of fetching and merging with the current branch you're requesting from.

- **Push**
  A push is the act of sending your locally committed changes to a centralized or remote repository, e.g., GitHub.

- **Pull request**
  A pull request contains the proposed changes to a repository requested by a developer. A collaborator can accept or reject them.

- **Clone**
  Cloning is the act of copying a remote repository to your local repository, in a newly created directory.

Now that you understand the basics of Git and are familiar with its terminology, let's shift our focus to how to use Git in your system landscape.

4.1.2   abapGit and gCTS

abapGit was introduced by the SAP community in 2014 for the easy exchange and installation of code across different systems that may exist in different organizational cloud or on-premise environments. Even though traditional copy-paste may work for some scenarios, ABAP community members saw the need for a consistent way of exchanging ABAP objects to avoid repetitive installation effort and manual errors. The result is abapGit, which has become quite popular since. The high-level architecture for using abapGit is shown in Figure 4.1.

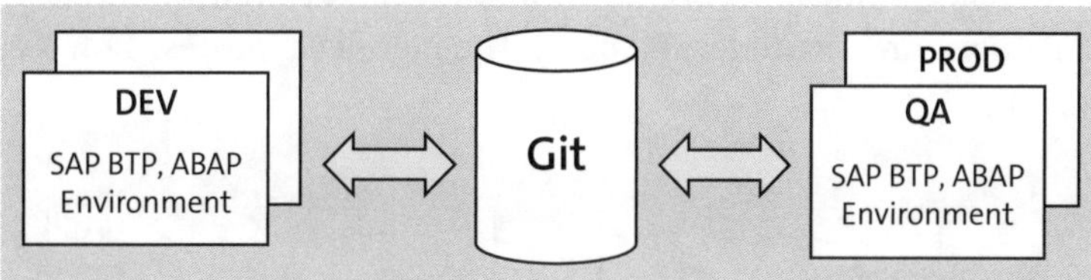

Figure 4.1 abapGit Architecture

As shown in Figure 4.1, the basic idea behind abapGit is to exchange source code between any ABAP systems using a Git-based repository (i.e., GitHub). However, you can also use this architecture and store your project-specific code in a GitHub repository, later exchanging this code with other projects, which may work in different environments.

Even though abapGit is an open-source project started by SAP community members, SAP has been contributing to abapGit since 2018. However, SAP does not officially support this project as a part of their product portfolio or support strategy. SAP's position is that any issues that may arise while using abapGit must be addressed by SAP community members without any liability or warranty.

After this brief look into abapGit, we'll now switch our focus to Git-enabled Change and Transport System (gCTS). For now, don't worry if you don't understand when and how to use abapGit or gCTS, both of which will be revisited in following sections.

Before explaining gCTS, let's try to understand the benefits and limitations of lifecycle management in traditional ABAP development. ABAP developers generally write code to their development systems, assign this code to packages and transport requests, and then release the transport requests after finishing development tasks and performing unit testing. The SAP Basis team monitors the import queue and imports this code manually or using a batch job. One advantage of using ABAP is its centralized development environment. ABAP developers log on to an SAP application server and write the code directly on that server. Once development objects are created, these objects are visible to other developers in the system and can be reused after activation. This concept is completely different compared to other traditional, non-SAP development environments, where individual developers might write code in their local environments and then deploy this code to servers at specified frequencies.

The advantage provided by central development is tremendous, since fellow developers can quickly check what's happening in related developments and raise issues earlier. However, central development prevents multiple developers from working on the same objects simultaneously. By introducing gCTS, SAP is aiming for the best of both worlds, that is, repository-based development where concurrent work on the same object is possible, while retaining all the benefits of central development.

gCTS was formally introduced as a part of SAP S/4HANA 1909. However, gCTS is also used in SAP BTP, ABAP environment, under the hood. The high-level process flow for gCTS is shown in Figure 4.2.

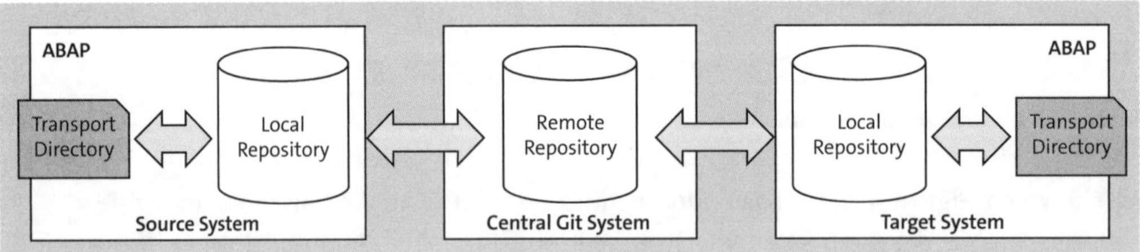

Figure 4.2 gCTS Process Flow

Using the gCTS process flow shown in Figure 4.2, any ABAP object can be enabled in gCTS and can be stored in a Git repository. Objects can be attached to transport requests/tasks, just as in traditional development. Changes attached to a task will be committed to a local repository and then finally be pushed to a remote repository. Finally, objects in the remote repository can be transferred to the target system's local repository can then be imported by using traditional tools.

The following use cases can be addressed using gCTS:

- **Distribution of custom development using Git**
  You can use gCTS to distribute your custom code through a central repository, which will enable possible rollbacks before importing to the target system, if required.

- **Maintenance and feature development simultaneously without using an N+1 landscape**
  Currently, many SAP customers use an N+1 landscape to support production issues and new projects in parallel. As shown in Figure 4.3, F* stands for new feature landscape, M* for maintenance landscape, D for development, Q for quality, and P for production. At some point, custom code and custom configurations are merged in the landscape manually via a tool, e.g., an SAP Solution Manager retrofit. With the full functionality of gCTS, including branching and complex merging, we believe that an N+1 landscape can eventually be eliminated. By using gCTS, your maintenance team can work on a maintenance branch while the project team works on a project branch. In the end, they can merge all their code with the master branch using the merging functionality.

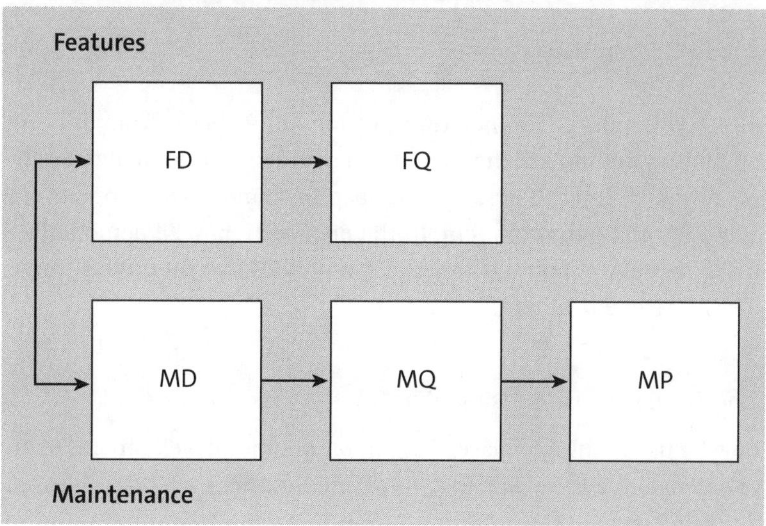

Figure 4.3 N+1 Landscape

- **Collaborative development**
  gCTS allows multiple developers from the same project or from different projects to work on development objects simultaneously without any issues. This feature speeds up the development timeline. Developers can merge their code later in the integration environment if merging is supported by the environment.

- **Continuous integration in ABAP**
  With the introduction of gCTS, SAP allows you to set up a complete continuous integration pipeline. With continuous integration, as soon as a change is made, it is pushed to the remote repository.

You should now have a fair understanding of both abapGit and gCTS. Figure 4.4 shows some possible use cases that involve both abapGit and gCTS in your landscape.

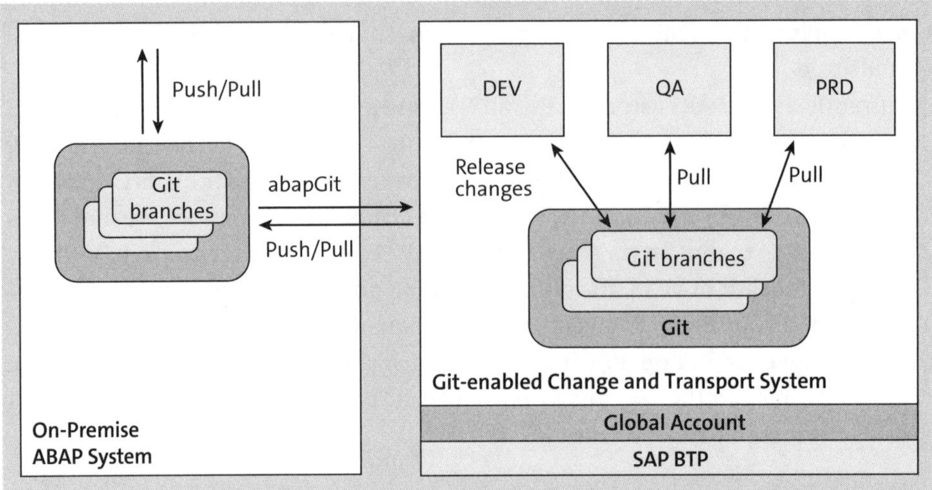

Figure 4.4 abapGit and gCTS: Together in Action

As shown in Figure 4.4, you must gCTS when transporting ABAP objects from one SAP BTP, ABAP environment system to another (e.g., from development to quality assurance to production systems). You can possibly use abapGit to move your on-premise ABAP objects to SAP BTP, ABAP environment. In the next section, we'll demonstrate how to use abapGit to move ABAP objects from your SAP S/4HANA on-premise environment to SAP BTP, ABAP environment.

4.1.3   Custom Code Transformation Using abapGit

Since ABAP can now be used multiple places (e.g., in on-premise development, in an app, or a key-user extension) and in SAP BTP, ABAP environment, SAP has released multiple versions of ABAP languages:

1. Standard ABAP (Unicode)
2. ABAP for key users
3. ABAP for SAP BTP (option **ABAP for SAP Cloud Platform** in Figure 4.5)
4. Other obsolete versions (e.g., non-Unicode ABAP, etc.)

When creating a new ABAP program in your on-premise SAP S/4HANA environment, or under the **Properties** tab of a class, you can select the desired ABAP language version, as shown in Figure 4.5.

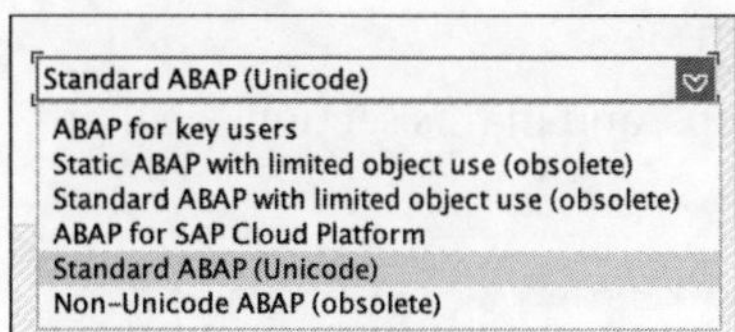

Figure 4.5 ABAP Language Versions

While developing an object, you'll select the desired language version to make the object ready for a target environment. For example, if you're creating a class, we recommend that you select, as the language version, the **ABAP for SAP Cloud Platform** so that the class can be later moved to SAP BTP, ABAP environment if needed. Nevertheless, transforming on-premise ABAP code to SAP BTP requires three primary steps: evaluating your current code, installing abapGit, and finally transforming your code. We'll look at each of these steps next.

Evaluating Your On-Premise Code for Compatibility

Before you can move your code to the cloud, make sure that your code is using the ABAP for SAP BTP language version (again, **ABAP for SAP Cloud Platform** on screen). You can perform this check either by changing the language version under the **Attributes/ Properties** tab of the development objects or running the Code Inspector (Transaction SCI) or the ABAP Test Cockpit (ATC) check. Both checks are available through SAP Note 2830799, which you can apply to your system if using SAP S/4HANA 1709 or above (SAP_BASIS 7.52 and above).

Once you implement SAP Note 2830799, you should see a Code Inspector/ATC check variant, **SAP_CP_READINESS_REMOTE**. Under this check variant, two primary items are selected, as shown in Figure 4.6. The first item—**Test for Restricted Language Scope (ABAP Language Version)**—will check your ABAP code with respect to ABAP version for SAP BTP, and the second item—**Whitelist Check (on-premise)**—will make sure that you're using standard, whitelisted objects only.

Selection	Doc...	Attri...	RFC...	Extr...	Checks
⌄ 📁 ☑	ⓘ				List of Checks
⟩ 📁 ☐	ⓘ				General Checks
⌄ 📁 ☑					Cloud Readiness
• ☐					SSCUI ATC Healthcheck
• ☑	ⓘ	📑	z→		Test for Restricted Language Scope (ABAP Language Versi
• ☐	ⓘ		z→	🖥	Whitelist Check
• ☑	ⓘ	📑	z→	🖥	Whitelist Check (on-premise)

Figure 4.6 List of Checks under the Transaction SCI/ATC Check Variant

Once you execute the cloud readiness check and make adjustments, if any, you're now ready to execute the next step—installing abapGit.

Installing abapGit On-Premise and as an Eclipse Plugin

Installing and configuring abapGit requires several steps, and in this section, we'll provide granular details of each step.

Prerequisites

As a prerequisite, you'll need to work with your SAP Basis administrator to install a root certificate from *https://github.com* or from *https://digicert.com*. This certificate must be installed in your on-premise SAP S/4HANA environment using Transaction STRUST. This requirement is mandatory because your on-premise environment will be interacting with the cloud-based GitHub repository. More details can be found at *https://docs.abapgit.org/guide-ssl-setup.html*.

To install abapGit on-premise and as an Eclipse plugin, follow these steps:

1. Create an account at *https://github.com* and then create a new repository, as shown in Figure 4.7.

Figure 4.7 Creating a New Repository

2. Enter a meaningful name for the repository and then click the **Create repository** button, as shown in Figure 4.8. The name of our repository in this case is **SCP_ABAP**.

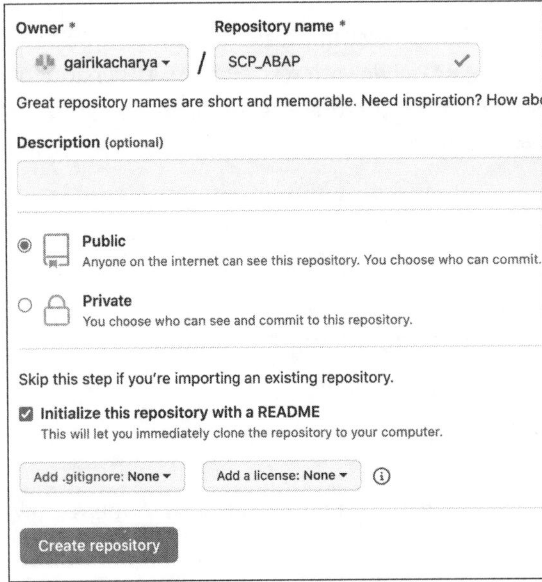

Figure 4.8 Create Repository Button

3. To install abapGit, go to *https://github.com/larshp/abapGit* and download the latest build. The **Latest build** link for the ABAP program will appear somewhere down the page, as shown in Figure 4.9.

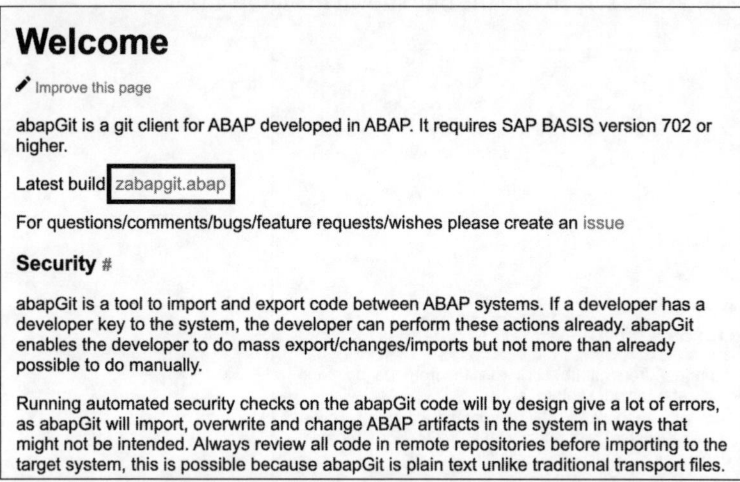

Figure 4.9 Identifying the Latest Build Link

Clicking on the link should show you the source code, as shown in Figure 4.10.

```
REPORT zabapgit_full LINE-SIZE 100.

* See http://www.abapgit.org

*****************************************************************************
* The MIT License (MIT)
*
* Copyright (c) 2014 abapGit Contributors
*
* Permission is hereby granted, free of charge, to any person obtaining a copy
* of this software and associated documentation files (the "Software"), to deal
* in the Software without restriction, including without limitation the rights
* to use, copy, modify, merge, publish, distribute, sublicense, and/or sell
* copies of the Software, and to permit persons to whom the Software is
* furnished to do so, subject to the following conditions:
*
* The above copyright notice and this permission notice shall be included in all
* copies or substantial portions of the Software.
*
* THE SOFTWARE IS PROVIDED "AS IS", WITHOUT WARRANTY OF ANY KIND, EXPRESS OR
* IMPLIED, INCLUDING BUT NOT LIMITED TO THE WARRANTIES OF MERCHANTABILITY,
* FITNESS FOR A PARTICULAR PURPOSE AND NONINFRINGEMENT. IN NO EVENT SHALL THE
* AUTHORS OR COPYRIGHT HOLDERS BE LIABLE FOR ANY CLAIM, DAMAGES OR OTHER
* LIABILITY, WHETHER IN AN ACTION OF CONTRACT, TORT OR OTHERWISE, ARISING FROM,
* OUT OF OR IN CONNECTION WITH THE SOFTWARE OR THE USE OR OTHER DEALINGS IN THE
* SOFTWARE.
*****************************************************************************

CLASS zcx_abapgit_2fa_error DEFINITION
  inheriting from CX_STATIC_CHECK
  create public .

public section.

  data MV_TEXT type STRING read-only .

  methods CONSTRUCTOR
    importing
      !TEXTID like TEXTID optional
```

Figure 4.10 Latest Build Code of abapGit

4. Copy the latest build code and create a program, called ZABAPGIT, in your on-premise SAP S/4HANA environment using Transaction SE38.

5. If you now execute your ZABAPGIT program in your on-premise SAP S/4HANA environment, you should see a screen like the one shown in Figure 4.11.

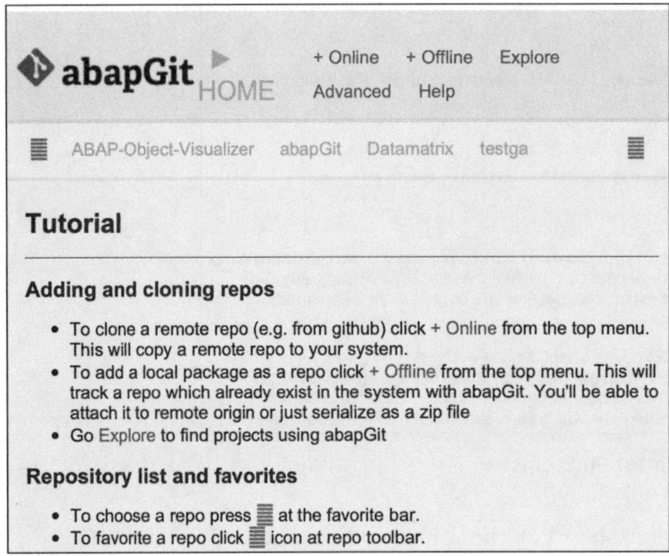

Figure 4.11 abapGit Open

6. abapGit has now been installed to your on-premise environment. Let's install the abapGit Eclipse plugin by opening ADT and selecting **Help · Install New Software**, as shown in Figure 4.12.

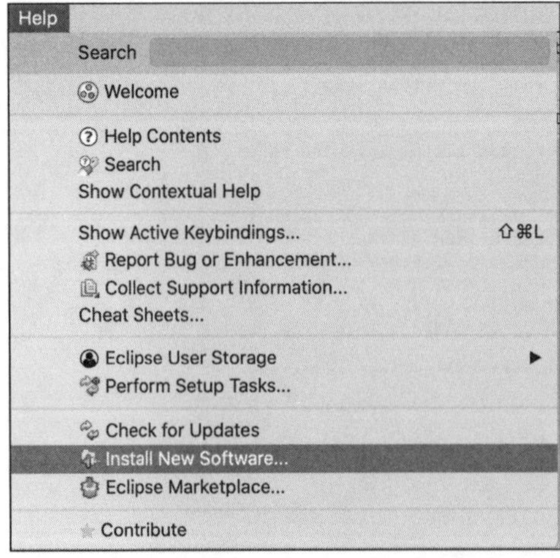

Figure 4.12 Install New Software... Option

Note

The abapGit plugin requires that the ADT plugin be installed first. We're assuming that you have the latest version of Eclipse and that ADT has been installed using the link *https://tools.hana.ondemand.com/latest*.

7. To install the abapGit plugin for Eclipse, go to *http://eclipse.abapgit.org/updatesite/* and select **abapGit for ABAP Development Tools (ADT)**, as shown in Figure 4.13. Complete the installation by clicking **Next**.

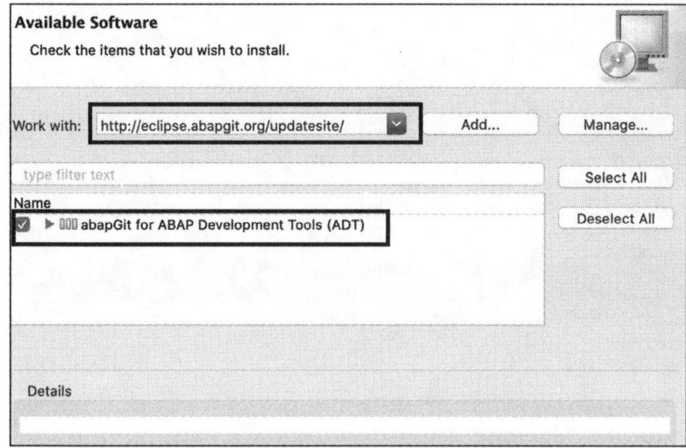

Figure 4.13 abapGit Eclipse Plugin Installation

You've now installed abapGit for your on-premise environment and as an Eclipse plugin. Yes, it's that easy! Let's now move on to the final step to transform your code.

Transforming Your Code from On-Premise to the Cloud

Transforming your code from your on-premise ABAP system to SAP BTP, ABAP environment requires the following steps:

1. Copy the URL for your GitHub repository, as shown in Figure 4.14.

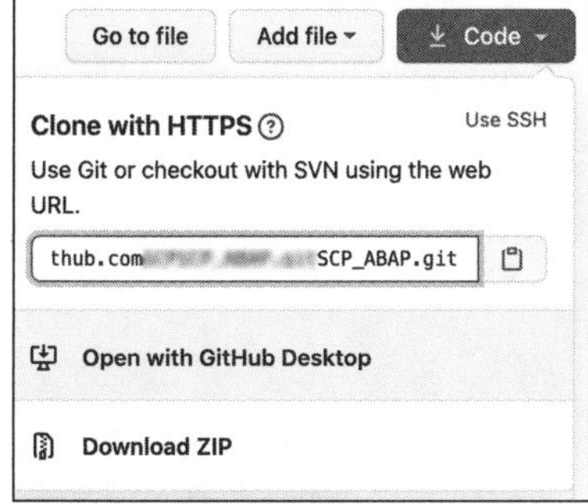

Figure 4.14 Copying the GitHub Repository URL

2. Once you're done installing abapGit in Eclipse, you can go back to the on-premise system where you previously installed abapGit program and execute the program ZABAPGIT (or whatever name you used while creating the program). In the output of the ZABAPGIT program, click the **+ Online** link, as shown in Figure 4.15.

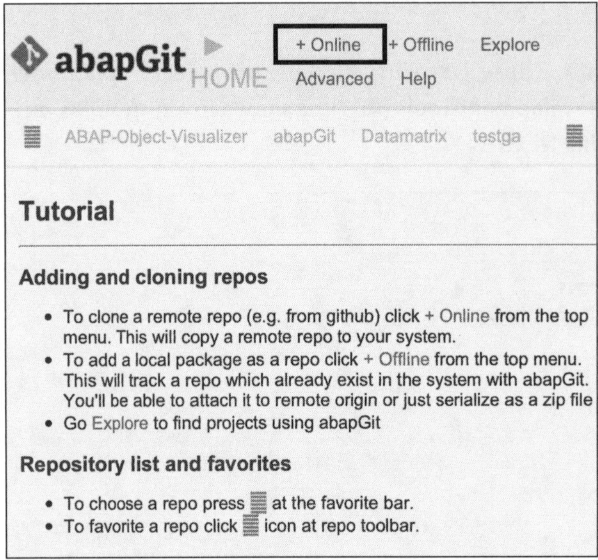

Figure 4.15 + Online Option in abapGit On-Premise

3. The **+ Online** link will display a popup window where you'll need to enter your previously copied GitHub repository URL. As shown in Figure 4.16, click on **Create Package** to create a new package for this exercise. On the next screen, shown in Figure 4.17, enter some details for that package.

4. After executing the previous step, in abapGit, you should see the cloned Git repository, **SCP_ABAP**, as shown in Figure 4.18.

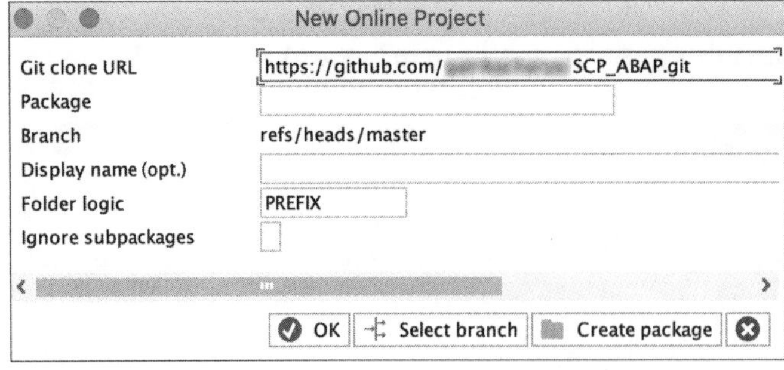

Figure 4.16 Entering the GitHub Repository URL

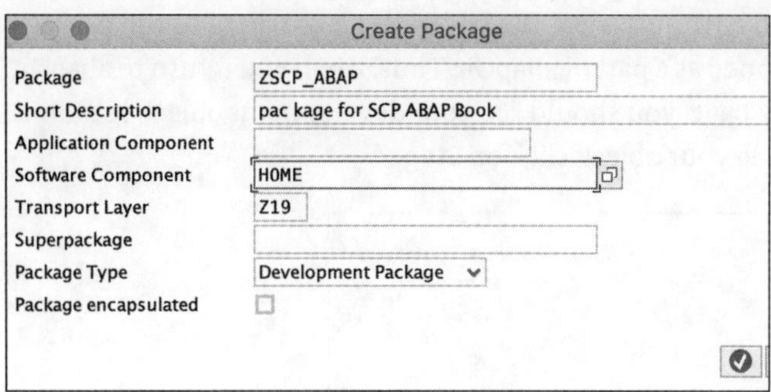

Figure 4.17 New Package Creation for the GitHub Repository

Figure 4.18 Cloned Git Repository in an On-Premise ABAP System

5. Now, you can create any compatible ABAP object and can assign the package ZSCP_ ABAP. In this case, our class is ZCL_TEST_ABAPGIT, as shown in Figure 4.19.

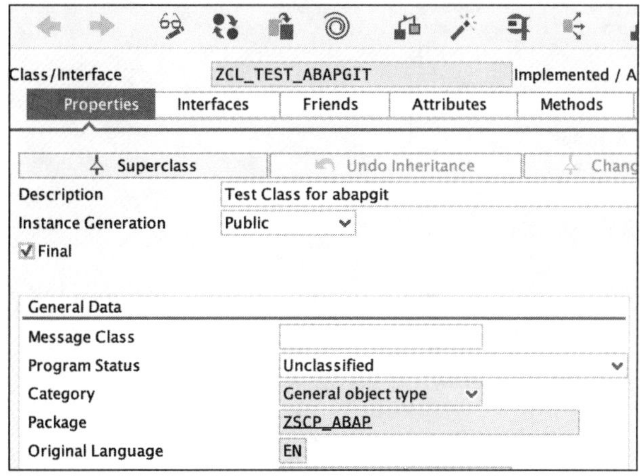

Figure 4.19 ABAP Class to Be Transformed

6. Our development object (i.e., the class `ZCL_TEST_ABAPGIT`) is assigned to the package ZSCP_ABAP, which was cloned as a part of abapGit. Thus, when you return to abapGit and refresh the repository page, you should see your development object, as shown in Figure 4.20. Once you see your object, click on **Stage**.

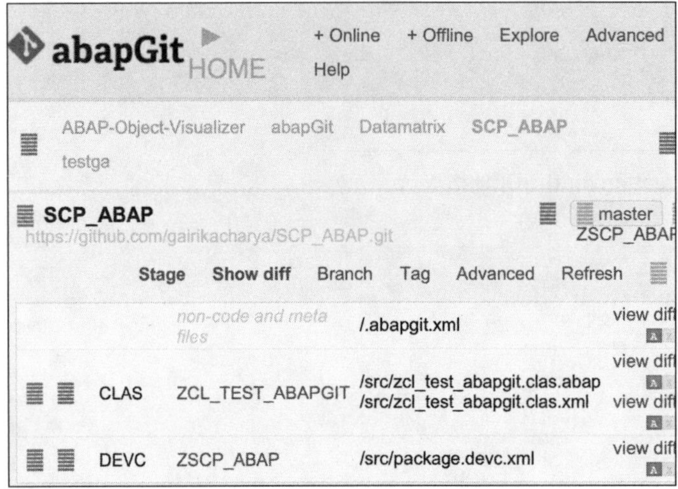

Figure 4.20 Objects Added to the Local Repository

7. After staging your object, you can now commit this object to the repository, as shown in Figure 4.21.

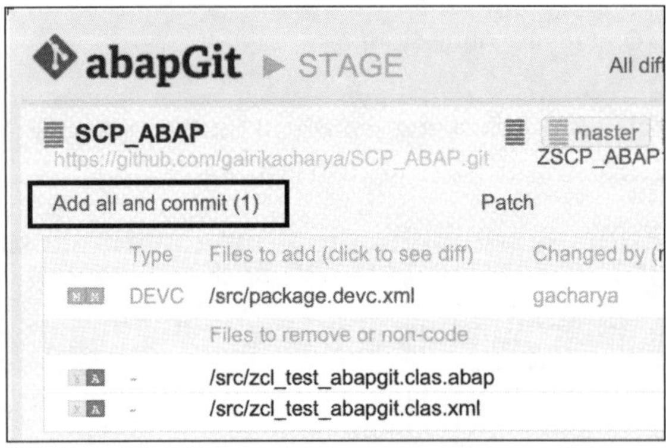

Figure 4.21 Committing to the Remote Repository

As shown in Figure 4.22, triggering a commit will prompt you for your email address and for a comment for each commit. Additionally, you'll also need to enter your GitHub credentials on the next screen, as shown in Figure 4.23.

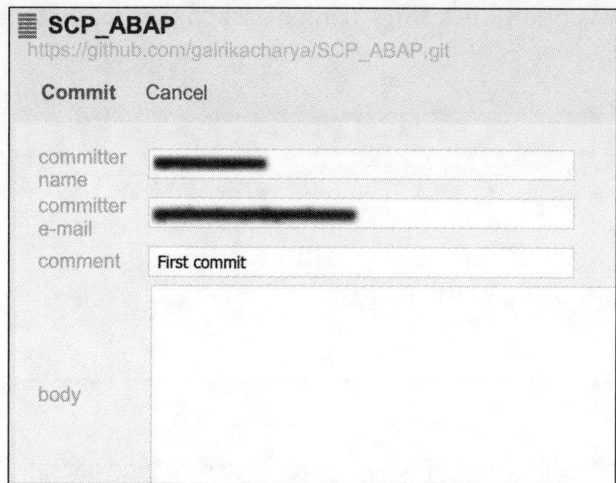

Figure 4.22 Email and Comment for Each Commit

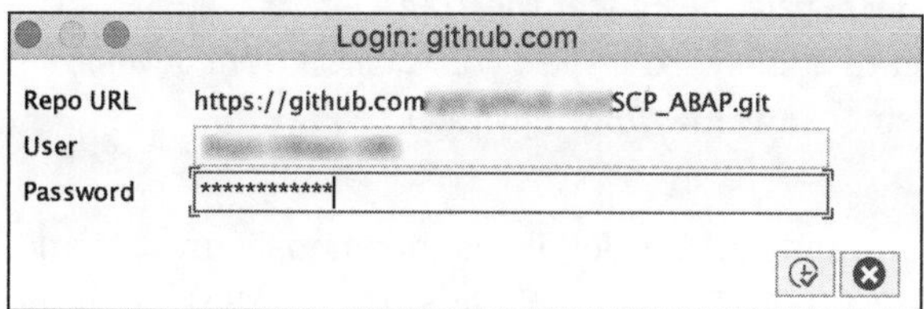

Figure 4.23 GitHub Credentials

After a successful commit, when you log on to your GitHub account, you should see your object(s), as shown in Figure 4.24.

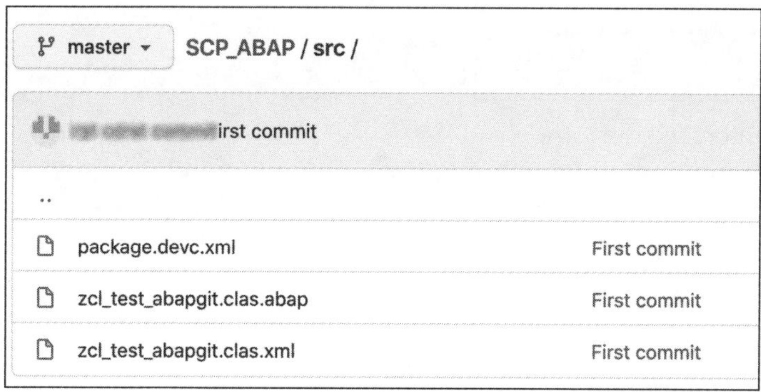

Figure 4.24 On-Premise Objects Available in the GitHub Repository

8. Our next step is to pull the objects from the repository. Open ADT and open the **abapGit Repositories** view using the Eclipse menu path **Window · Show View · Other**. Make sure you select your existing SAP BTP ABAP project before opening this view.

9. Once the **abapGit Repositories** view is open, click the **+** (**Link new abapGit repositories**) button, as shown in Figure 4.25.

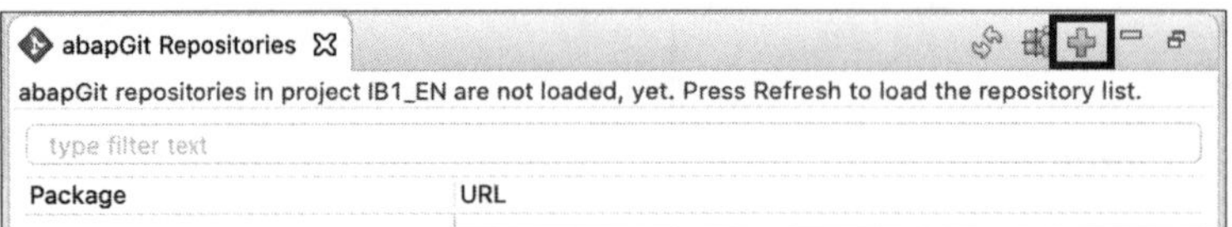

Figure 4.25 Adding a New Repository in Eclipse for SAP BTP ABAP

Note

You must create a *software component*, which is the highest-level structure type package, before creating any package in SAP BTP, ABAP environment. Software components can be created using the Manage Software Components app, which we'll describe in Section 4.1.4. You can only create local packages under the ZLOCAL package without creating software components and thus cannot transport your objects from development to quality to production systems in SAP BTP, ABAP environment.

10. Enter your GitHub repository URL and package details, as shown in Figure 4.26 and Figure 4.27. Note that this package should be available in SAP BTP, ABAP environment (i.e., target package). You can also optionally select the **Pull after link** checkbox to pull it automatically.

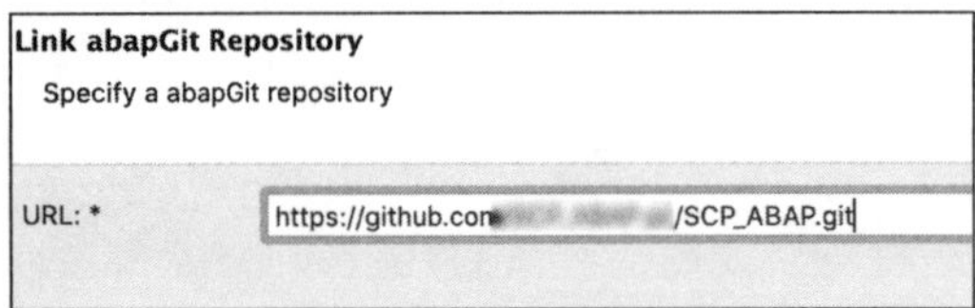

Figure 4.26 GitHub Repository URL

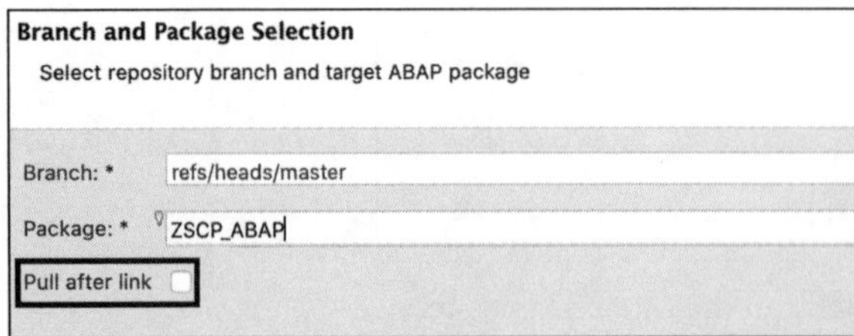

Figure 4.27 Branch and Package Selection and Optional Pull

11. If you don't select the **Pull after link** checkbox, you can pull the object using the **Pull** option in the context menu, as shown in Figure 4.28.

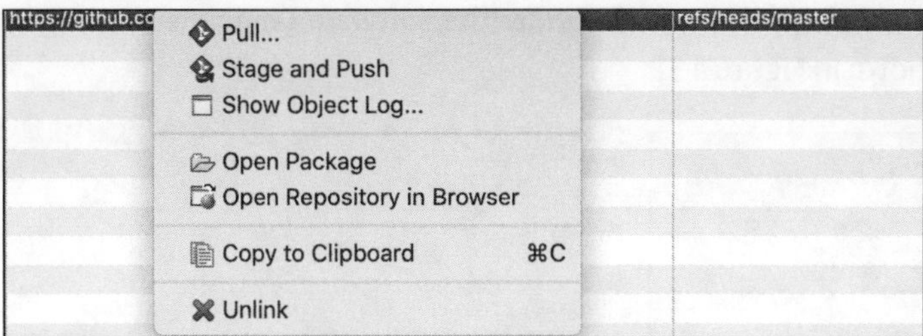

Figure 4.28 Pulling Object(s) from the Repository

12. Your object will now be displayed in the target SAP BTP, ABAP environment, as shown in Figure 4.29.

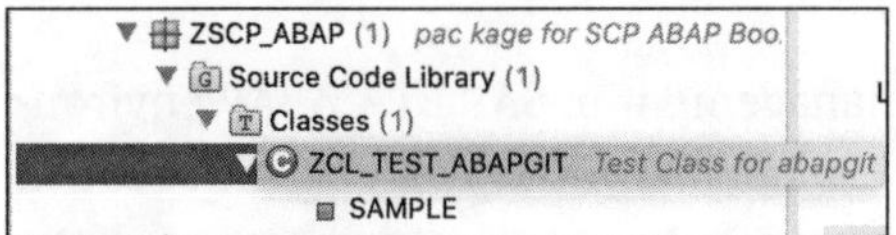

Figure 4.29 ABAP Class Available in SAP BTP ABAP

Congratulations! You've now successfully created a class in SAP S/4HANA on-premise, validated the code with respect to the language version (ABAP for SAP BTP) and transformed the code to SAP BTP, ABAP environment using abapGit. In the next step, we'll demonstrate how to transport objects from one SAP BTP ABAP system to another using gCTS.

4.1.4   Manage Software Components App

The Manage Software Components app is an SAP Fiori app available for administrators to configure, execute, and monitor the lifecycle management of the ABAP objects in SAP BTP, ABAP environment. This app can be displayed in the administrator dashboard of the service instance. For example, Figure 4.30 shows two ABAP services running in our Cloud Foundry development space. In this section, we'll also show you how to move objects from one instance to another.

Name	Service	Plan	Referencing Applications
1	abap	standard	None
2	abap	standard	None

Figure 4.30 Two ABAP Instances

Clicking on any instance name will display the **Open Dashboard** option for that instance. This dashboard is simply an SAP Fiori launchpad primarily used by administrators. The

Manage Software Components app is available under the **Software Component Lifecycle Management** group, as shown in Figure 4.31.

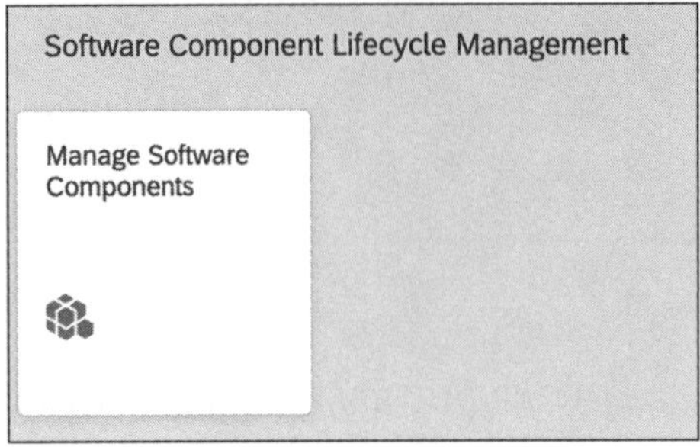

Figure 4.31 Manage Software Components App

This app is the bread and butter of lifecycle management in SAP BTP, ABAP environment and uses some gCTS components under the hood. If you're working with gCTS in SAP S/4HANA, then some configuration is required along with the creation of the GitHub repository. This configuration and repository creation steps are not required for SAP BTP, ABAP environment, which takes care of all this complexity behind the scenes. As shown in Figure 4.30, let's say you have two ABAP instances in your Cloud Foundry organization/subaccount. To transport objects from one instance to the other, follow these steps:

1. **Create a software component**
   Open the Manage Software Components app and click on **Create.** Enter a name and description for the software component and select an option from the **Type** dropdown list (**Development** or **Configuration**), as shown in Figure 4.32. Then, click **Save**.

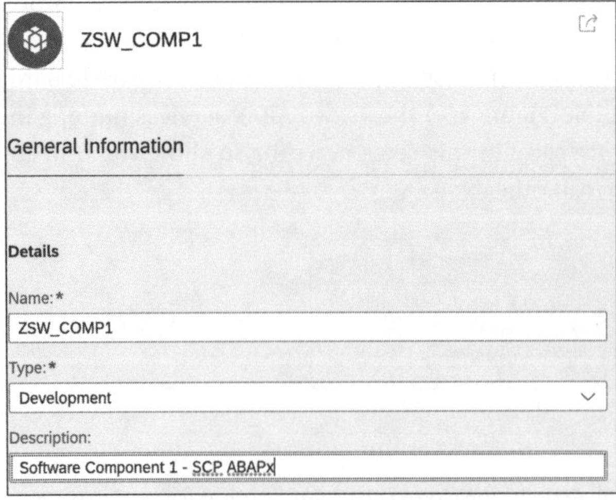

Figure 4.32 Creating a Software Component

The creation of a software component automatically creates a GitHub repository behind the scenes. Thus, you should **Clone** and **Pull** the repository, as shown in Figure 4.33, before starting your development.

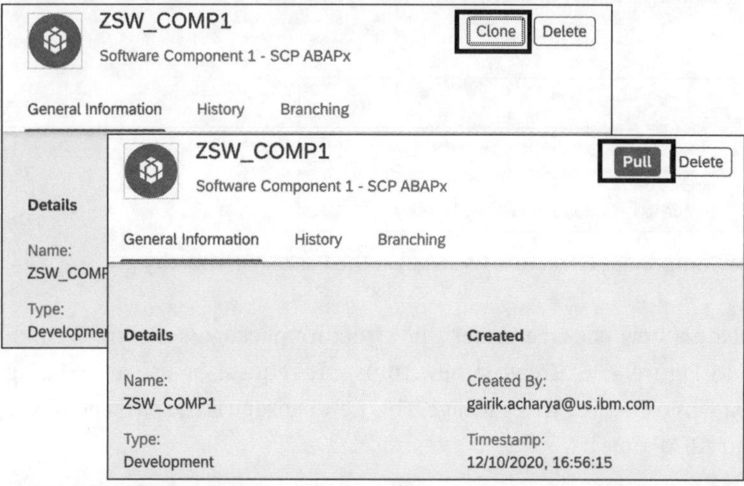

Figure 4.33 Cloning and Pulling from an Automatically Created Repository

After saving the software component, you should see its details under its **History** and **Branching** tabs, as shown in Figure 4.34. You won't be able to create any branches until a development object is assigned to this software component. We'll describe this process later in this section.

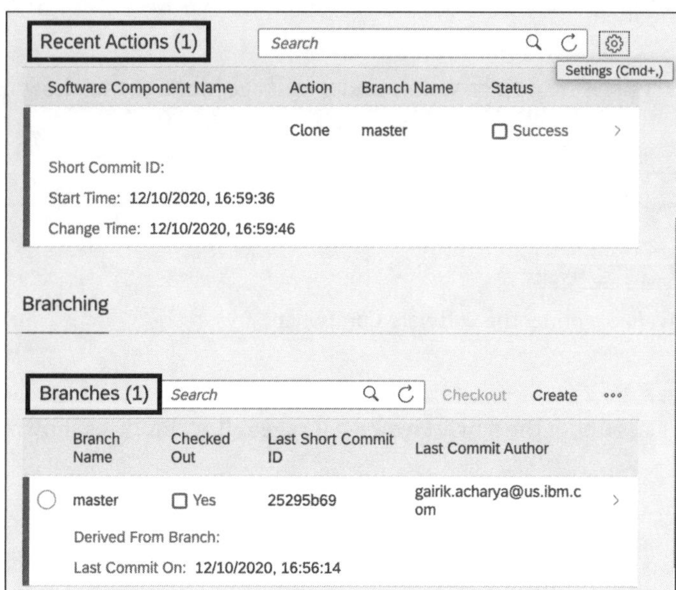

Figure 4.34 Newly Created Software Component's History and Branching

2. **Create an ABAP package**

   In this next step, you'll create a package under the software component we just created. If you now open ADT and select **Favorite Package · Add Package** under your SAP BTP ABAP system, you can search for your software component, as shown in Figure 4.35.

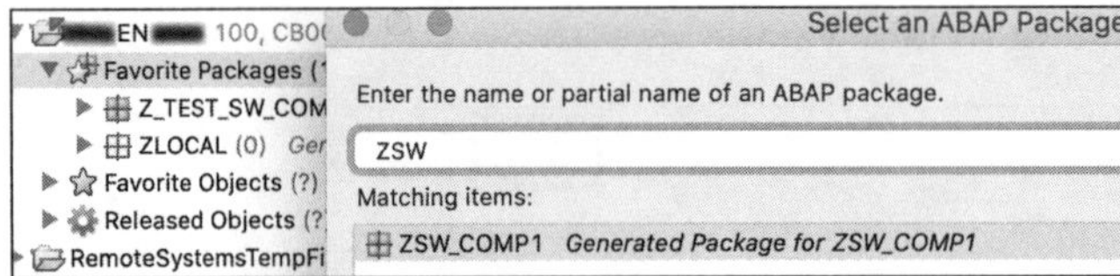

Figure 4.35 Software Component (Structure Package) in the Package Hierarchy

You can now create a new package under this structure package/software component, as shown in Figure 4.36. Create a new transport request or use an existing transport request while creating this package. This new package is the package where you'll assign your ABAP objects.

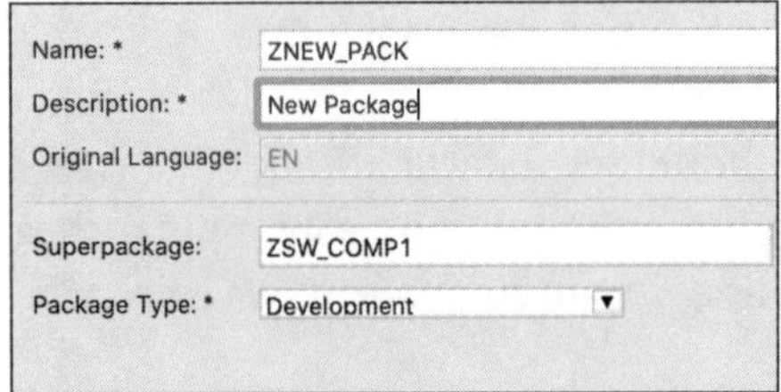

Figure 4.36 New Package Creation

The newly created package should now be visible under package hierarchy, as shown in Figure 4.37.

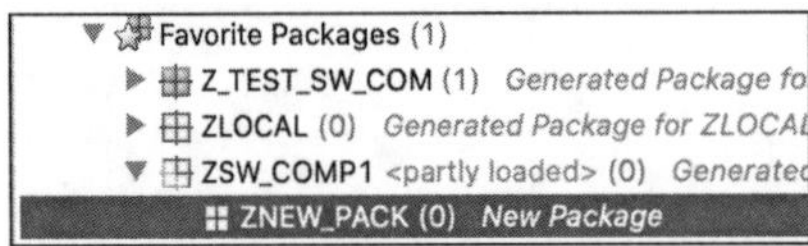

Figure 4.37 New Package Available under the Software Component

3. **Create a new ABAP class**

   Now, let's create a new class under the newly created package ZNEW_PACK, as shown in Figure 4.38.

Figure 4.38 New ABAP Class

For illustration purposes, we've provided an example of the logic of a simple ABAP class in Listing 4.1.

```abap
CLASS z_new_test_class DEFINITION
  PUBLIC
  FINAL
  CREATE PUBLIC .

  PUBLIC SECTION.
    METHODS test_meth.
  PROTECTED SECTION.
  PRIVATE SECTION.
    data v_var1 type c.
ENDCLASS.
CLASS z_new_test_class IMPLEMENTATION.
  METHOD test_meth.
    v_var1 = 'A'.
  ENDMETHOD.
ENDCLASS.
```

Listing 4.1 ABAP Class Logic

4. **Releasing the transport request**

   You can now release your task(s) and the main transport in the **Transport Organizer** view in ADT by right-clicking on your transport number and selecting **Release**, as shown in Figure 4.39.

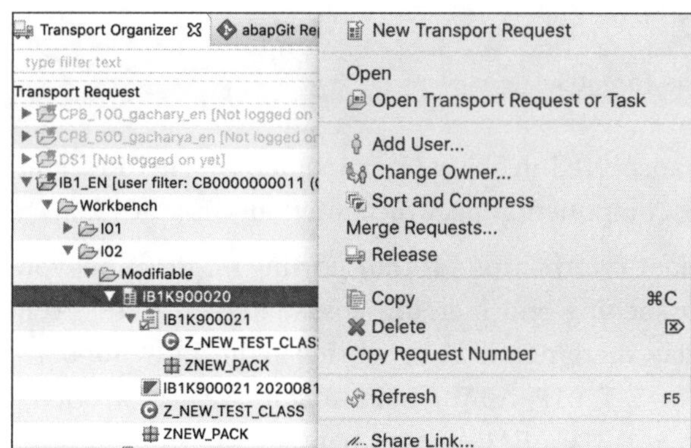

Figure 4.39 Releasing Tasks and the Main Transport

5. **Pull from target system**

   Once the transport is released, you can open the admin dashboard of your target (e.g., test) system and open the Manage Software Components app. You should see your software component with an indicator that the software component has not been created in this system (**In System: No**), as shown in Figure 4.40.

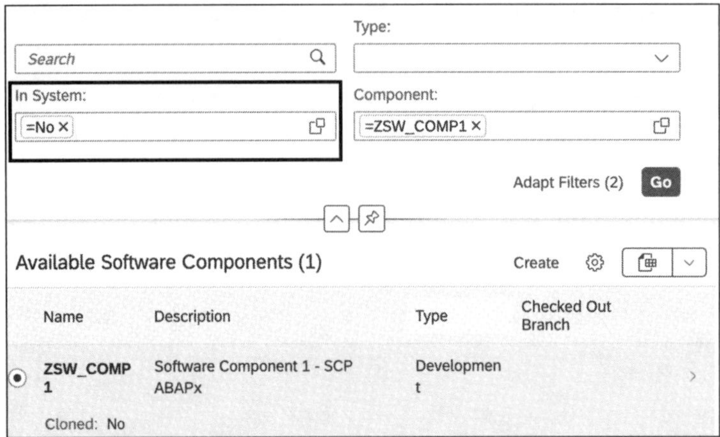

Figure 4.40 Software Component in the Target System

You can now pull objects from the repository by selecting your software component and clicking the **Pull** button, as shown in Figure 4.40.

6. **Check your object(s) in target system**

    After pulling the repository, your transported ABAP object(s) should be visible under the target system in ADT, as shown in Figure 4.41.

Figure 4.41 ABAP Class Available in the Target System

At this point, you've successfully transported an object from a source system to a target system using the Manage Software Components app, which internally uses gCTS.

Even though you now have sufficient information for transporting information, you probably have questions about branching and merging possibilities in gCTS. Even though branching is supported today, merging will be supported in the future for both gCTS in SAP S/4HANA and also for SAP BTP, ABAP environment. We'll now briefly explain the branching capability in SAP BTP, ABAP environment.

To work with multiple versions of the code simultaneously, you'll need to create multiple branches. When you create a software component, its master branch is created automatically. To create other branches, go to the software component and pull it once. After the pull is successful, you can create further branches by clicking the **Create** option. Figure 4.42 shows two branches: the **master** branch and the **Maintenance** branch.

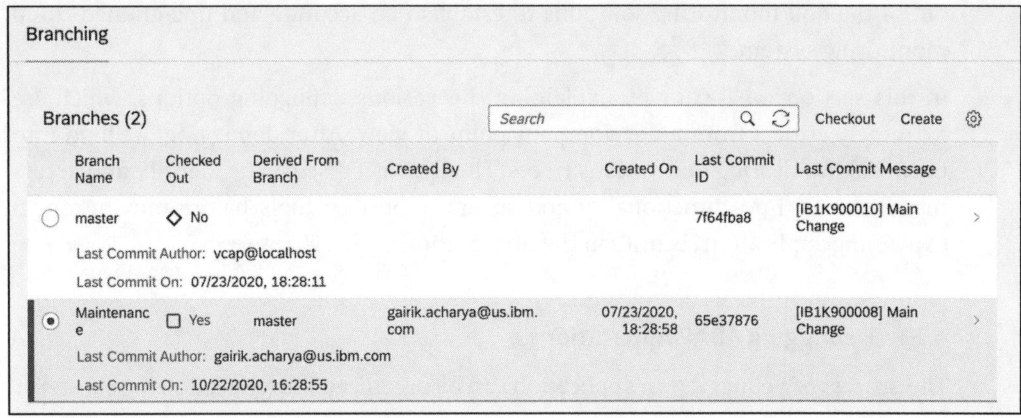

Figure 4.42 Master and Maintenance Branches

You can now check out individual branches by clicking the **Checkout** button. Once you check out any branch, that system will have the version available in that branch. Let's test that. Make sure your master branch code is released and transported and then create a maintenance branch. You can check out the maintenance branch and add some additional code using ADT. Since you've checked out the maintenance branch, these additional changes will be kept in that branch only. You can now release the new transport request created for the maintenance branch. After releasing the transport, you can now selectively check out either the master or the maintenance branch and then try to display the source code in ADT. You'll now see different versions of code in each branch.

Currently, the gCTS used in SAP BTP, ABAP environment (i.e., the Manage Software Components app) does not support the merging of the branches. Merging is possible with the gCTS available in SAP S/4HANA 2020, and we expect a similar capability in SAP BTP, ABAP environment in the future. Keep an eye on the product roadmap to understand future directions for this feature.

Let's now talk about the available monitoring, debugging, and tracing options in SAP BTP, ABAP environment.

4.2   Monitoring

Application monitoring plays a critical role in the success of a hybrid cloud environment. A successful monitoring environment provides continuous monitoring of applications, triggering alerts for any exceptions, logging all the critical actions, and finally analyzing the content in a meaningful way for better decision-making. Over time, your organization may build tons of cloud applications on SAP BTP, ABAP environment. Without a proper strategy for monitoring, most applications will degrade and become inefficient. SAP BTP, ABAP environment provides a proper infrastructure and many out-of-the-box monitoring solutions to establish an accurate and fine-grained cloud monitoring system.

In this section, we'll start by explaining the various debugging options, which are extremely critical from a development point of view. After debugging, we'll go into technical monitoring and various traces. Finally, we'll discuss the identity and access management (IAM) functionality and several reporting tools before moving on to explaining application security in the final section of this chapter.

4.2.1   Debugging ABAP Applications

The process of debugging an application can be organized into the following six steps:

1. **Setting up static breakpoints**
   Setting up a static breakpoint is as simple as selecting the correct line of code, right-clicking on it, and setting a breakpoint by selecting the **Toggle Breakpoint** option from the context menu, as shown in Figure 4.43.

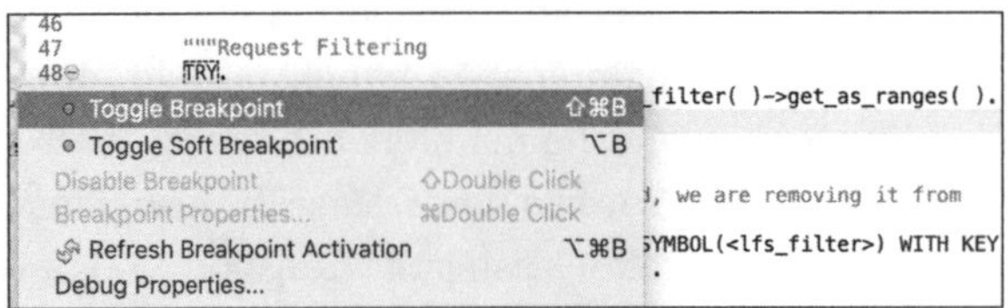

Figure 4.43 Setting Up a Breakpoint

   You can also set this breakpoint by double-clicking on the ruler (i.e., column with the line number or the first column) of the Eclipse editor.

2. **Setting up and managing dynamic breakpoints (optional)**
   You can set up a dynamic breakpoint by using the Eclipse menu **Run · ABAP Breakpoints · Add Statement Breakpoint** or **Run · ABAP Breakpoints · Add Exception Breakpoint**. Make sure you select the **Soft Breakpoint** checkbox, as shown in Figure 4.44. Selecting this option will ensure that debugger stops here only when debugging has been initiated by some other means (e.g., using a static breakpoint).

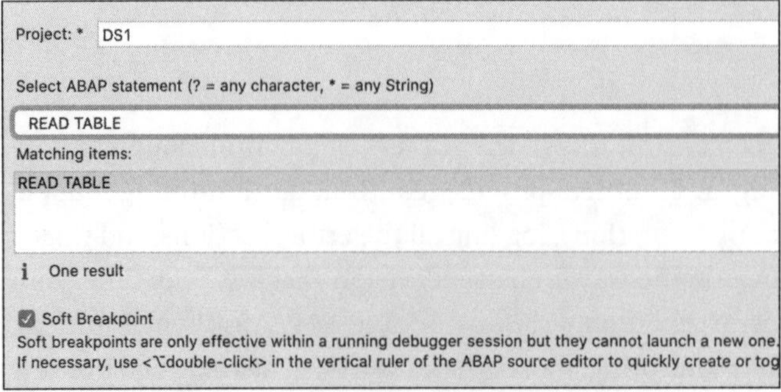

Figure 4.44 Dynamic Soft Breakpoint Option

3. **Adding conditions to a static breakpoint**

   You can set some conditions for a static breakpoint to trigger that breakpoint conditionally. Right-click on a static breakpoint and select **Breakpoint Properties...** to enter any condition. The syntax of a condition is validated during runtime, and you can use variables, literals, built-in functions, and more in a condition. Some valid conditions are `sy-tabix = 10`, `lines( itab ) <> 0`, `strlen( <chars/string>) >= sy-tabix`, etc. For more details about the syntax, refer to the SAP Help at *http://s-prs.co/v523600*.

4. **Executing the application from Eclipse or SAP Fiori launchpad**

   Once you set a breakpoint, you can execute your ABAP objects in various ways to trigger debugging and allow the program to stop at the breakpoint. For example, test any class by following the menu path **Run As · ABAP Application (Console)**, as long as the class uses the interface `if_oo_adt_classrun` in the main method of the interface, as shown in Listing 4.2. In our example, we have a method (`get_http_client`) to get the instance of an HTTP destination. To test this class directly, we added the interface `if_oo_adt_classrun` and implemented its main method accordingly. As a result, this class can be tested using the context menu options **Run As · ABAP Application (Console)**.

```
CLASS zrap_cl_vendor_aux_c_cp DEFINITION
  PUBLIC
  FINAL
  CREATE PUBLIC .
  PUBLIC SECTION.
    interfaces:
       if_oo_adt_classrun.
    CLASS-METHODS:
     get_http_client  RETURNING VALUE(http_client) TYPE REF TO if_web_http_
client
                      RAISING  cx_web_http_client_error
                               cx_http_dest_provider_error .

  PROTECTED SECTION.
  PRIVATE SECTION.
ENDCLASS.

CLASS ZRAP_CL_VENDOR_AUX_C_CP IMPLEMENTATION.
  METHOD get_http_client.

DATA(http_destination) = cl_http_destination_provider=>create_by_cloud_
destination(
 i_name = 'S19_HTTP_ABAP'
 i_authn_mode = if_a4c_cp_service=>service_specific
 ).
```

```
    http_client = cl_web_http_client_manager=>create_by_http_
destination( http_destination ).
  ENDMETHOD.
  METHOD IF_OO_ADT_CLASSRUN~MAIN.
    try.
        DATA(http_test_client) = zrap_cl_vendor_aux_c_cp=>get_http_client( ).
      catch cx_web_http_client_error cx_http_dest_provider_error.
        "handle exception
    endtry.
  ENDMETHOD.
ENDCLASS.
```

Listing 4.2 Interface for Direct Class Debugging

An Eclipse debugger session will open, as shown in Figure 4.45.

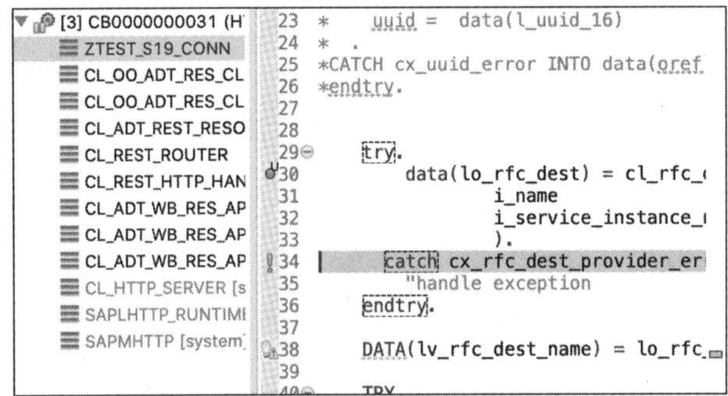

Figure 4.45 Debugging Session in Eclipse

> **Tips**
>
> By default, a breakpoint will be set for the current ADT user. You can change the user ID by right-clicking on the breakpoint and selecting **Debugging Properties…**. This approach is particularly helpful when you plan to debug using a frontend SAP Fiori app and the app is using a communication user for your SAP BTP, ABAP environment services. If you're unsure about the communication user ID, you'll need to work with your system administrator, who can provide that information from the launchpad/dashboard of the SAP BTP, ABAP environment.

5. **Managing and analyzing variables and internal tables**

   Just as in an SAP GUI ABAP debugger session, you can analyze the content of variables and internal tables in the **Variables** view of the ADT debugging screen, as shown in Figure 4.46.

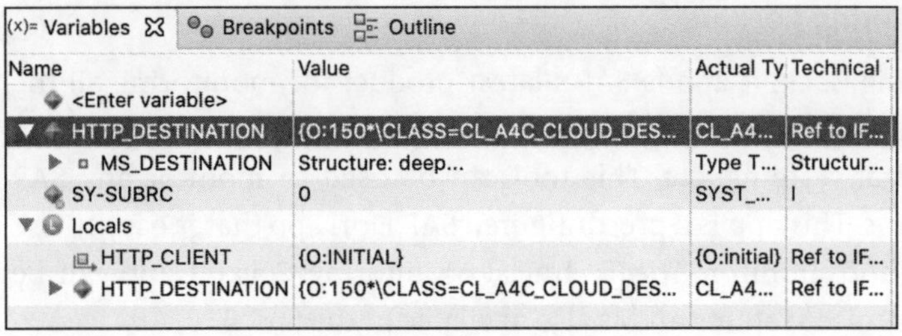

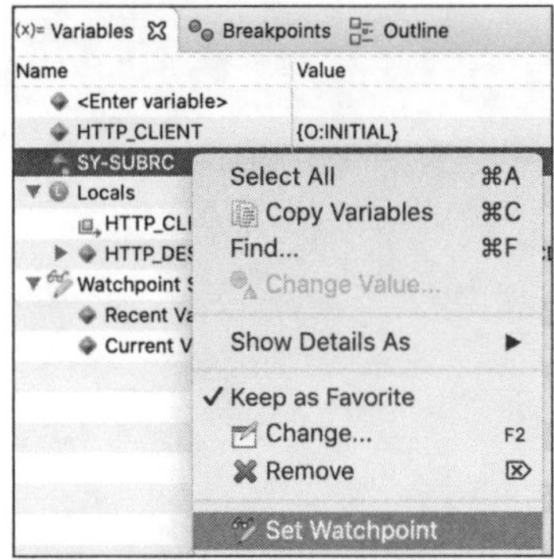

Figure 4.46 Variable Display in Eclipse Debugging

6. **Using watchpoints while debugging**

   You can create watchpoints from the **Variables** view of the debugging session. Right-click on any variable and select the **Set Watchpoints** option, as shown in Figure 4.47. Once a watchpoint is set, you can enter a condition by right-clicking a watchpoint in the **Breakpoints** view and selecting **Breakpoint properties…** from the context menu, as shown in Figure 4.48.

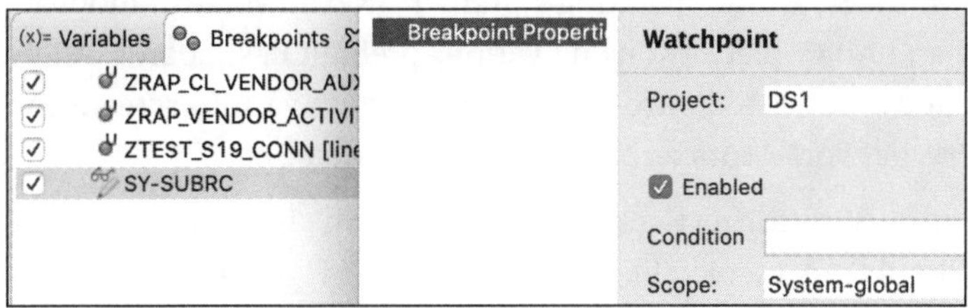

Figure 4.47 Setting Up a Watchpoint in a Debugging Session

Figure 4.48 Watchpoint Condition

We hope you now have a fair understanding of the basics of ABAP application debugging in ADT. Let's now change our focus to the available monitoring and tracing options.

4.2.2   Technical Monitoring Cockpit

The Technical Monitoring Cockpit app is SAP's latest product for monitoring all the technology stacks of an SAP system, including the database and host, the application server, physical or virtual systems, etc. This tool can be used to monitor any SAP NetWeaver-based systems. This app is a preconfigured SAP Fiori app that is easy to use and simple to operate and monitor. The Technical Monitoring Cockpit app can be installed for your on-premise systems and is also available in the SAP BTP, ABAP environment, which is our primary focus in this book.

Once you configure your SAP BTP, ABAP environment system, the Technical Monitoring Cockpit app is available in the admin dashboard SAP Fiori launchpad, as shown in Figure 4.49.

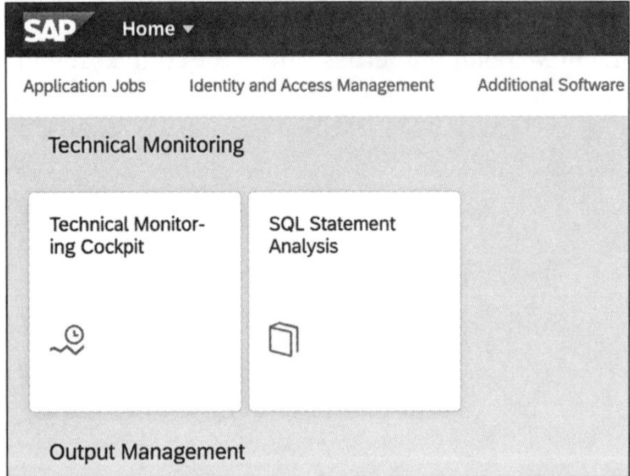

Figure 4.49 Technical Monitoring Cockpit App in the Admin Dashboard/Launchpad

The Technical Monitoring Cockpit app of the SAP BTP, ABAP environment does not support the monitoring of the host and operating system. However, you can monitor the SAP HANA database and ABAP platform using various prebuilt dashboards. Additionally, multiple traces and analysis tools are available, along with various screens for monitoring CPU, memory, network, storage, performance, workload, etc.

Once you open the Technical Monitoring Cockpit app and click the hamburger icon in the top-left corner, you should see a screen like the one shown in Figure 4.50.

Under the **Dashboards** menu item, you can navigate to application server- or database system-related dashboards. These dashboards will help you monitor and analyze memory utilization, work process utilization, CPU utilization, SQL workload, CPU utilization peaks, column tables by size, etc.

Under the **Analysis** menu item, you'll see the following three options:

- **SQL Statement Analysis**
  To analyze the performance of SQL statements for a given time range.

- **SQL Trace Analysis**
  You can view any results of SQL trace you previously executed through ADT. More details on this topic will be described in Section 4.2.3.

- **Table Analysis**
  This option will provide the characteristics, resource consumption, and workload of a table so you can better optimize your tables and indexes.

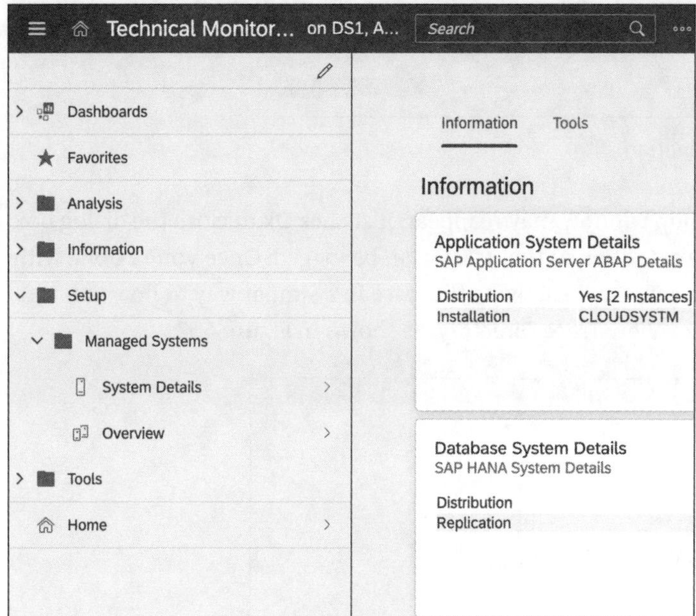

Figure 4.50 Tree Menu of Technical Monitoring Cockpit

Under the **Information** menu item, you'll see various options for monitoring CPU, configuration, memory, network, storage, etc. These individual analytics are integrated to the dashboards mentioned earlier.

As an ABAP developer, you'll rarely use the Technical Monitoring Cockpit app, which is primarily used by the system administrators. However, understanding the Technical Monitoring Cockpit app will always help you when troubleshooting and monitoring your app. Now that we've briefly elucidated the basic features of the Technical Monitoring Cockpit app now, let's turn to a key element of this app, the SQL trace, which ABAP developers will use frequently.

4.2.3   SQL Trace

Using SQL trace, you can analyze and trace any SQL statements executed directly or indirectly through an ABAP application. This feature is similar to the classic SQL trace option available in Transaction ST05. You can activate and deactivate traces from ADT and then traces can be evaluated or analyzed using the SQL trace app.

To activate the SQL trace, right-click on the ABAP cloud project in ADT and select **SQL Trace**. A popup window will appear, as shown in Figure 4.51, to confirm the activation.

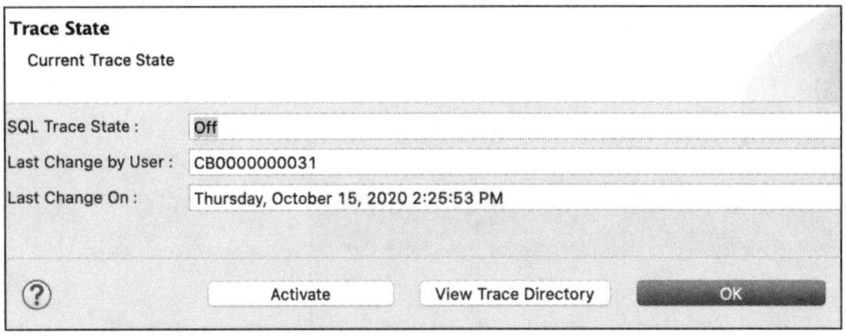

Figure 4.51 SQL Trace Activation

After activation, the popup window may reappear; just click **Ok** to close the dialog box and then perform some activities so that traces can be logged. Once you're done with testing your application, you can deactivate the trace in a similar way to how you activated the trace. Now, click **View Trace Directory,** as shown in Figure 4.52.

Figure 4.52 Opening a Trace Directory from ADT/Eclipse

The **Trace Directory** view will now be opened as an SAP Fiori app, as shown in Figure 4.53.

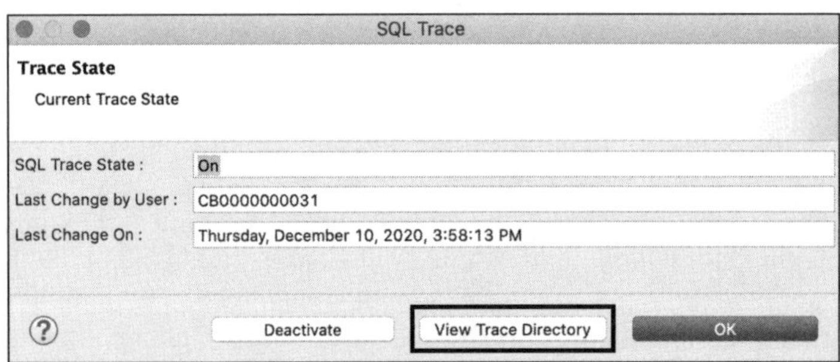

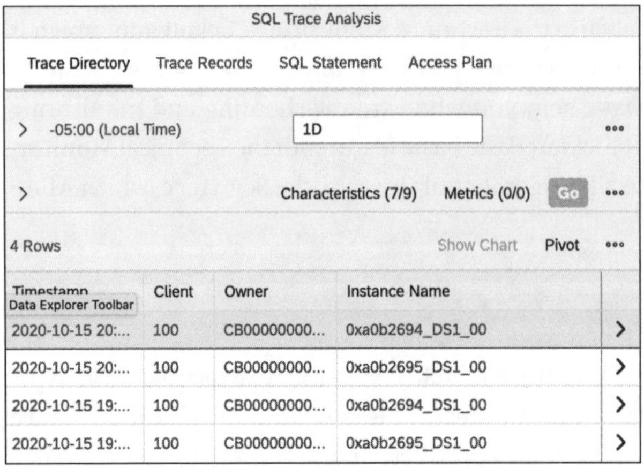

Figure 4.53 SQL Trace App/Trace Directory

The SQL Trace app can only be opened from the admin dashboard of the SAP BTP, ABAP environment system. This app has the following four views:

- **Trace Directory**

  This default view, shown in Figure 4.53, displays a list of traces that have run previously for the user. If you want to view the traces of another user or for all users, you can adjust the filter, as shown in Figure 4.54.

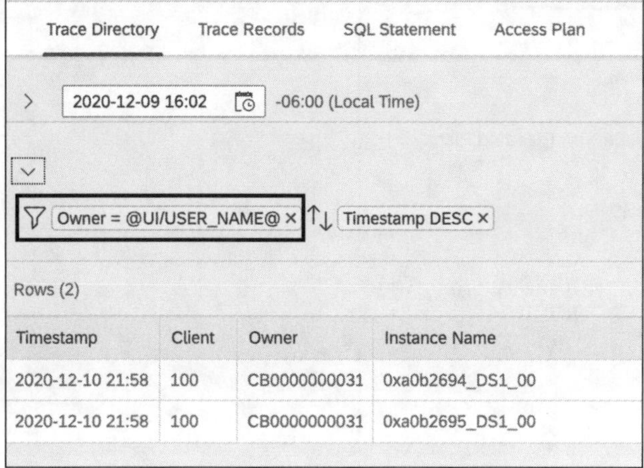

Figure 4.54 Changing/Removing User IDs from the Filter

- **Trace Records**

  A list of SQL statements for each trace execution, as shown in Figure 4.55.

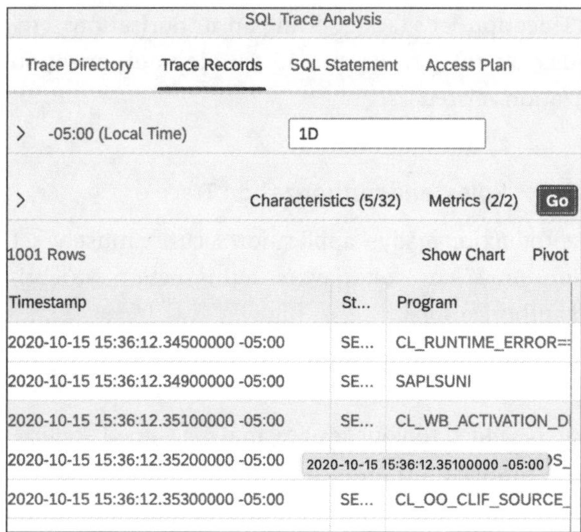

Figure 4.55 Trace Records

- **SQL Statement**

  By selecting any row in the **Trace Records** view and then navigating to the **SQL statement** view, you'll see the actual SQL statement that was executed, as shown in Figure 4.56.

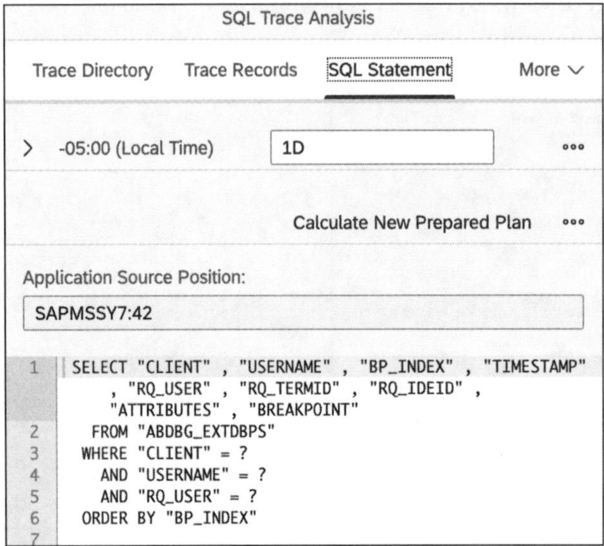

Figure 4.56 SQL Statement

- **Access Plan**

  This view displays the SQL access plan and the average cost for an SQL statement.

Now that we've discussed the SQL Trace app, let's tackle setting up authorizations, creating business roles, assigning business roles to users, and running authorization traces to troubleshoot any authorization-related issues.

4.2.4   Authorization Model, Business Roles, and Authorization Trace

Creating a service does not make it production ready—application security must be set up. Security for any cloud application comes in various flavors. In this section, we'll primarily focus on business roles and authorizations.

Authorization Object and Fields

Let's look at how authorizations can be added to your service, making a good application into great application. Follow these steps:

1. **Creating authorization fields**

   Authorization fields are dynamic authorization components whose values can be passed from the application. You can create an authorization field by right-clicking your desired package and selecting **New · Other Repository Objects · Authorizations · Authorization Field**, which opens the screen shown in shown in Figure 4.57.

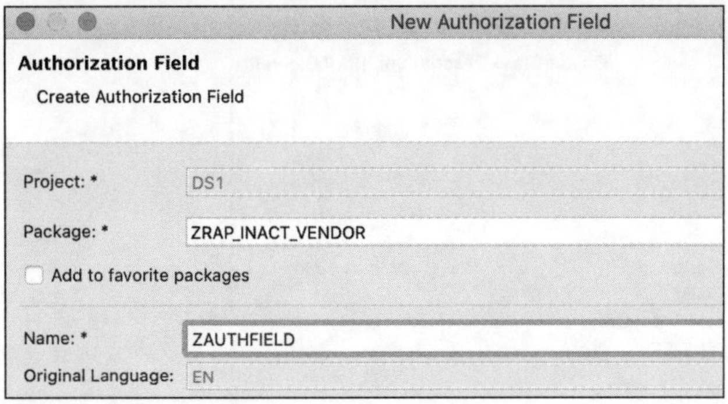

Figure 4.57 Authorization Field

Activate the field after maintaining the **Data Element** field and (optionally) the **Check Table** field, as shown in Figure 4.58.

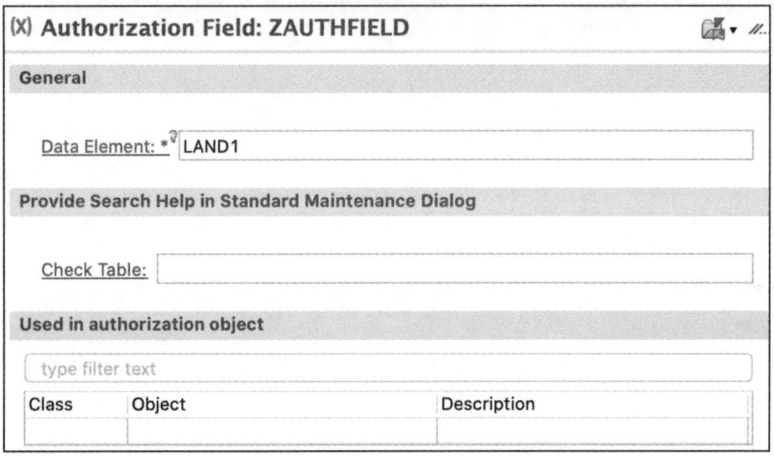

Figure 4.58 Data Element and Check Table for the Authorization Field

2. **Creating authorization objects**
   Our next step is to create an authorization object in a similar way to how you created the authorization field. Field **ACTVT** (activity) will be automatically added, and you can maintain the list of activity values (e.g., **01: Create**, **02: Change**, etc.). Now, add your previously-created authorization fields, as shown in Figure 4.59.

3. **Maintaining the default values of the authorization objects**
   Now, you'll add the previously created authorization object to your service, which can be performed in the service bindings in the specified link, as shown in Figure 4.60.

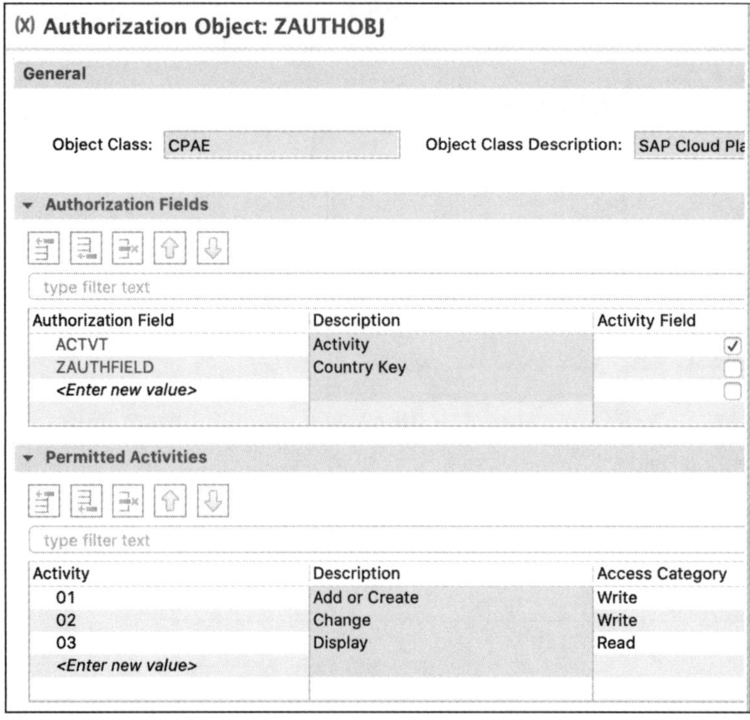

Figure 4.59 Activity and Other Authorization Fields

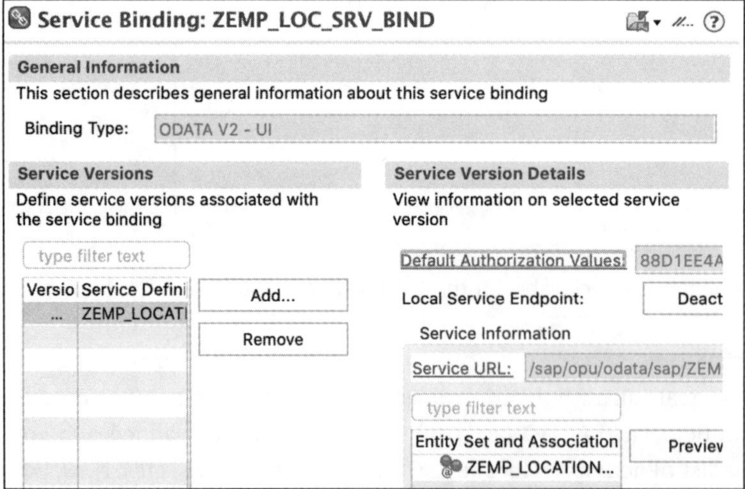

Figure 4.60 Authorization Object in Service Bindings

Figure 4.61 shows the added custom authorization object and default value of **03** (**Display**). This default value indicates that, if no values are transferred from the application, then this value should be considered the default activity.

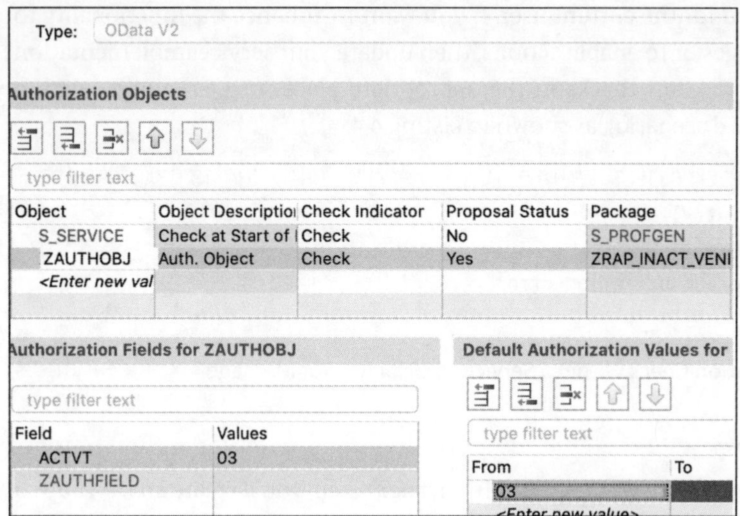

Figure 4.61 Adding an Authorization Object and Maintaining Default Values

Access Control and Roles

Next, you'll create an access control and enhance your service behavior. You can create an access control from the context menu of your package by selecting **New · Other Repository Objects · Code Data Services · Access Control**. You should select the **Define Role with PFCG Aspect** template. Depending on the template, some default source code for role template will be generated, as shown in Listing 4.3.

```
@EndUserText.label: 'ACCESS CONTROL'
@MappingRole: true
define role ZACCESS {
    grant
        select
            on
                ZEMP_LOCATION_CDS
                    where
                        (country) = aspect pfcg_
auth(ZAUTHOBJ, ZAUTHFIELD,  ACTVT = '03');
}
```

Listing 4.3 Source Code for Role Definition

As shown in Listing 4.3, this role is for querying (**READ** access) the core data services (CDS) view ZEMP_LOCATION_CDS with the activity as **03 (Display)**. The country value will come dynamically from the application or the service.

By default, any select query to your CDS view, in this case, ZEMP_LOCATION_CDS, will apply the Data Control Language (DCL) ZACCESS and will fetch only records of the countries authorized for the current user. If you want to enhance the functionality to display error messages or to enable actions, then update your service implementation with manual authorization checks in the appropriate places (e.g., in any validation method of a managed scenario), as shown in Listing 4.4.

```
     AUTHORITY-CHECK OBJECT 'ZAUTHOBJ' ID 'ACTVT' FIELD '03' ID 'ZAUTHFIELD'
FIELD wa_result-country.
         IF sy-subrc <> 0.
           "Handle authorization error
         ENDIF.
```

Listing 4.4 Authorization Check within a Service to Display Error Message

Creating the IAM App

Now, your basic preparation work is done. In this next step, you'll create an IAM app so that your service can be consumed by an SAP Fiori app, as described in detail in Chapter 2. Make sure you select the **External App** option while creating the IAM app for this SAP Fiori scenario, as shown in Figure 4.62.

Name: *	ZIAM_APP
Description: *	IAM App
Original Language:	EN
Application Type	EXT - External App
Application ID Suffix	_EXT

Figure 4.62 Property Selection while Creating an IAM App

The IAM app has two tabs: **Services** and **Authorization Objects**. You'll need to add your service (e.g., OData V2 or V4) under the **Services** tab, which should automatically add the authorization object, if maintained in the service level, as shown in Figure 4.63 and Figure 4.64, respectively.

(X) Services: ZIAM_APP_EXT

Service Type	Service Name
OData V2	ZEMP_LOC_SRV_BIND_0001

Figure 4.63 Services Tab with an OData V2 Service

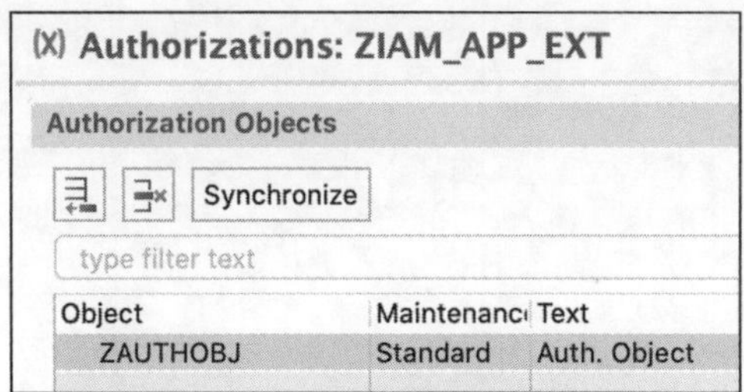

Figure 4.64 Authorization Object Added Automatically

Business Catalogs

You'll learn how to create business roles in a later section, but while creating business roles, you'll have an option of adding business catalogs. Within a business catalog, you can add restriction types and restriction fields for control authorization at a granular level. Using restriction fields and restriction types, you'll make certain objects not visible or not editable for certain users.

To create restriction fields, right-click an ADT project and select **New** · **Other Objects** · **Cloud Identity and Access Management** · **Restriction Field**. Then, enter the authorization field in the maintenance screen of the restriction field, as shown in Figure 4.65.

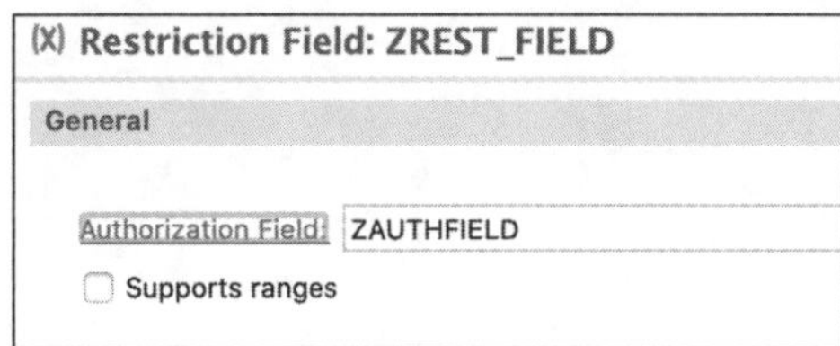

Figure 4.65 Authorization Field Assigned to a Restricted Field

After creating restriction field in a similar way from the project's context menu. Restriction fields and one more authorization object should be assigned to a restriction type, as shown in Figure 4.66.

After creating a restriction type and a restriction field, you can now create a business catalog, which will be assigned to a business role later. To create a business catalog, right-click a project and select **New** · **Other Repository Objects** · **Cloud Identity and Access Management** · **Business Catalog**. After creating a business catalog, you can add the IAM app we created earlier under the **Apps** tab of the newly created business catalog, as shown in Figure 4.67.

You should now add the created restriction type to this business catalog along with **Read**, **Write**, or **F4** access to decide how you would like to restrict the field for the user, as shown in Figure 4.68.

Figure 4.66 Restriction Type

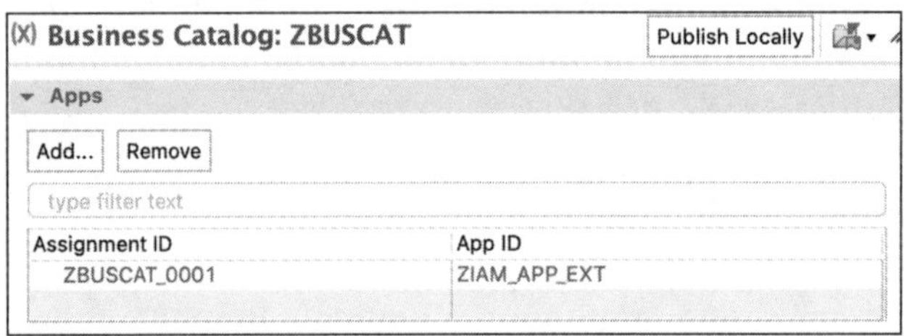

Figure 4.67 Business Catalog and IAM App Assigned

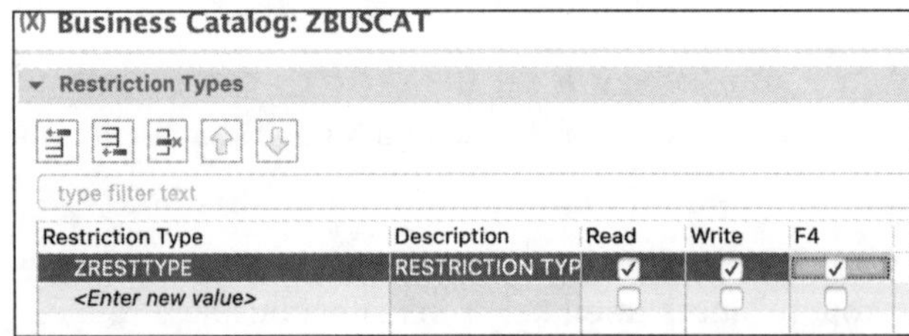

Figure 4.68 Restriction Type Assigned to a Business Catalog

Once you assign the IAM app and restriction type to your business catalog, you can publish the business catalog by clicking the **Publish Locally** button so that this catalog can be assigned to a business role in the next step.

Business Roles

Your business role can now be used by a system admin, who can add then to a business catalog by logging on to the SAP BTP, ABAP environment dashboard/launchpad and opening the Maintain Business Roles app. In this app, you can create a new business role and assign the business catalog, as shown in Figure 4.69.

New Role for ABAP Cloud Book
ZNEW_ROLE

Write Access: Restricted   Draft Last Changed By: Gairik Acharya
Read Access: Unrestricted   Draft Last Changed On: 10/20/2020, 17:03:48
Value Help Access: Unrestricted

	1	1
General Role Details	Assigned Business Catalogs	Assigned Business Users

Assigned Business Catalogs (1)

	Business Catalog	Business Catalog ID
☐	**Business catalog**	ZBUSCAT

Figure 4.69 Adding Business Catalog to a Business Role

You can now assign one or more users to the business role under the **Assigned Business Users** tab. Select one or more business users and then click **Maintain Restrictions** to maintain the restricted values, as shown in Figure 4.70.

Values	Ranges

	Value
☐	CA
☐	US

Figure 4.70 Restricted Values

Display Authorization Trace App

At this point, you've now created and used multiple variations of authorization-related objects and may have a fair understanding about using these concepts in your real project. However, over time, you may end up with too many services and apps, built with many authorization objects and fields, and finding out which authorizations are missing can be extremely difficult. To troubleshoot authorization-related issues, you can use the Display Authorization Trace app. This app can activate or deactivate traces. Activation is possible only for a single user. The trace list displays the information, as shown in Figure 4.71.

From the **Authorization Check Status** field, you can determine whether the check was successful and take the necessary actions. Additionally, you can find more details about each check by navigating to the object page of a particular row.

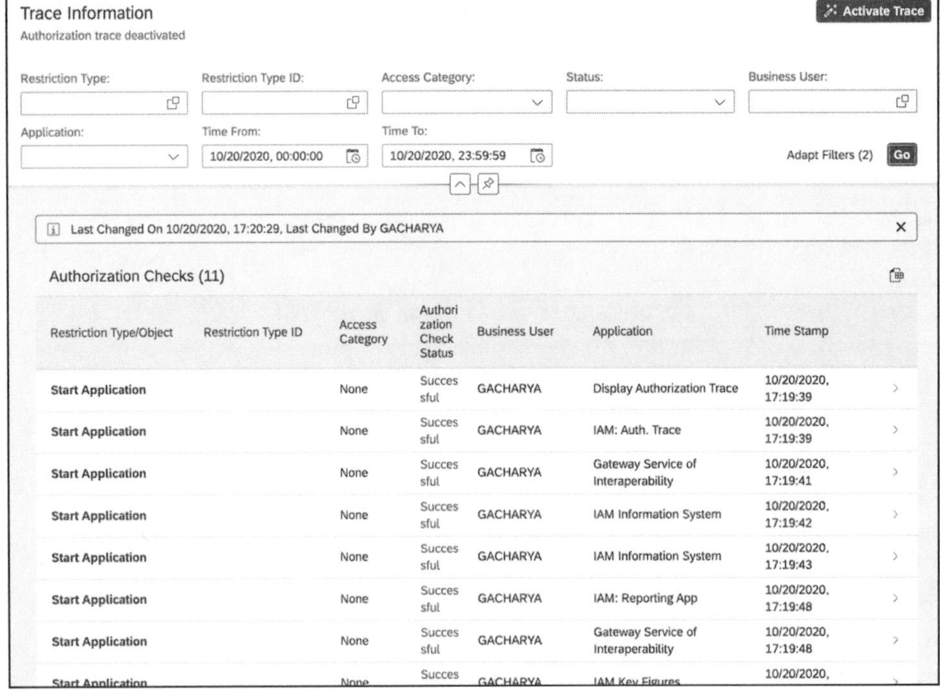

Figure 4.71 Authorization Trace

After a detailed walkthrough of authorization, business role, and authorization traces, let's now look at the possible analytical and reporting options in the IAM area.

4.2.5   IAM Reporting

SAP provides various out-of-the-box reporting and analytical apps to help you understand your SAP BTP, ABAP environment's IAM usage. The two primary apps are the IAM Information System app and the IAM Key Figures app.

Using the IAM Information System app, you can learn the various usages of business roles, business catalogs, restrictions, applications, and business role templates. Under the **Business Role - Business User** tab, you'll see which roles are assigned to which users, as shown in Figure 4.72. You can navigate to the role or user details screen from this tab.

The **Business Role - Business Catalog** tab represents the various business catalogs assigned to business roles. Similarly, you can see the restrictions assigned to each role in the **Business Role - Restrictions** tab, as shown in Figure 4.73.

Additionally, you'll also get a list of applications and their business roles and catalogs under the **Business Role - Application** tab.

Various selection criteria are available to filter these lists, and you can download the list as a comma-separated values (CSV) file.

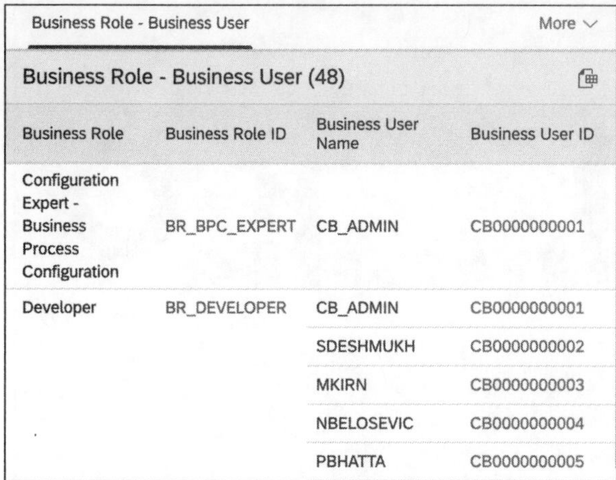

Figure 4.72 Business Role - Business Users Tab

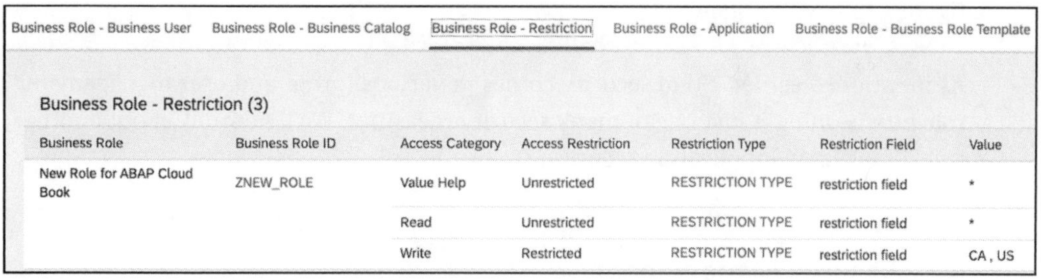

Figure 4.73 Business Role - Restrictions Tab

The IAM Key Figures app is a dashboard type (overview page) app where you can access the following information:

- Number of business users assigned to business roles
- Month of a business user's last logon
- Number of locked and unlocked business users
- Validity of business users

This app is great for reviewing whether any users are locked and the validity of the lock, whether any activity from any user occurred for a specific time period, whether some users have too many roles, and more. A typical output from the app will look like the screen shown in Figure 4.74.

You can also navigate to further details, for example, if you want a list of users who have more than three business roles. Other configuration options can change the look and feel of the charts.

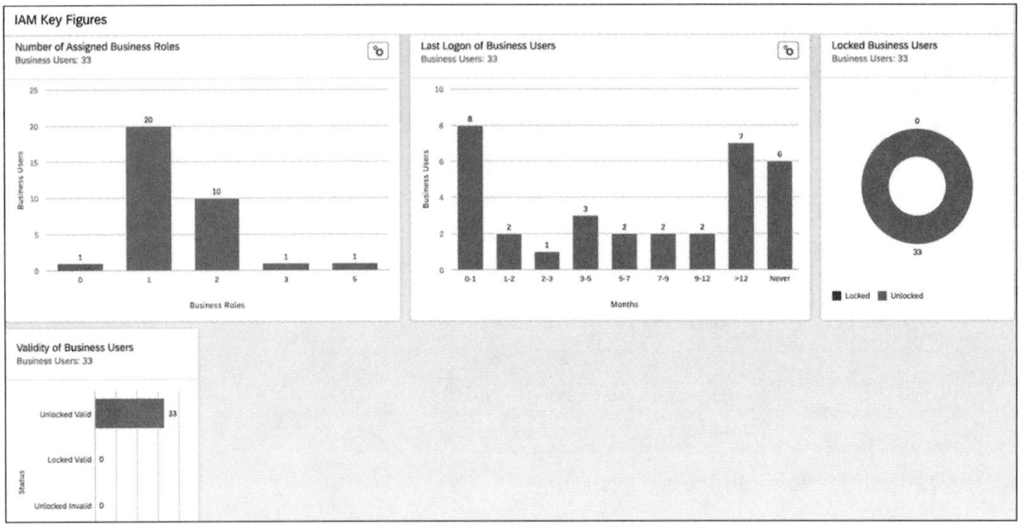

Figure 4.74 IAM Key Figures

As mentioned earlier, cloud security comes in various forms, and user management/ role provisioning is just one of many security measures. We'll now talk about another level of security—application security.

4.3   Security Management

As mentioned earlier in this chapter, cloud security comes in various flavors: data center security, identity and access management (IAM), identity authentication, application security, data security, data transmission security, and so on. In Section 4.2.4 and Section 4.2.5, we described IAM security in detail. Although we can't cover all types of security in this book, we'd like to highlight some key features of application security in this section.

The phrase *application security* describes security measures at the application level to prevent application data or code from being stolen or hijacked. This task involves necessary approaches to protect apps in the process of development and as well as after deployment. As a part of application security, we'll describe three primary elements provided out of the box in this section: maintaining certificates for secured communication, clickjacking protection, and managing content security.

4.3.1   Maintaining the Certificate Trust List

The Maintain Certificate Trust List app is where you'll maintain the list of certificates for secured communication. This app is equivalent to Transaction STRUST in an on-premise ABAP system. SAP provides some common certificates out of the box, as shown in Figure 4.75. Additional certificates can be installed based your requirements. Installed certificates are considered trustworthy, and thus, you can set up secure communications with the other system. You can also review additional details of the certificate and also delete certificates that are not required, although SAP-provided certificates cannot be deleted.

Issued To	Issued By	Type	Valid To	
DST Root CA X3	DST Root CA X3	Managed by SAP	09/30/2021	>
GeoTrust Global CA	GeoTrust Global CA	Managed by SAP	05/21/2022	>
Baltimore CyberTrust Root	Baltimore CyberTrust Root	Managed by SAP	05/12/2025	>
DigiCert Global Root G2	DigiCert Global Root G2	Managed by SAP	01/15/2038	>
DigiCert Global Root CA	DigiCert Global Root CA	Managed by SAP	11/09/2031	>
DigiCert High Assurance EV Root CA	DigiCert High Assurance EV Root CA	Managed by SAP	11/09/2031	>
Entrust Root Certification Authority - G2	Entrust Root Certification Authority - G2	Managed by SAP	12/07/2030	>
Entrust Root Certification Authority	Entrust Root Certification Authority	Managed by SAP	11/27/2026	>
VeriSign Universal Root Certification Authority	VeriSign Universal Root Certification Authority	Managed by SAP	12/01/2037	>
VeriSign Class 3 Public Primary Certification Authority - G5	VeriSign Class 3 Public Primary Certification Authority - G5	Managed by SAP	07/16/2036	>
SAP Cloud Root CA 01	SAP Cloud Root CA 01	Managed by SAP	08/29/2036	>
COMODO RSA Certification Authority	COMODO RSA Certification Authority	Managed by SAP	01/18/2038	>
SAP Global Root CA	SAP Global Root CA	Managed by SAP	04/26/2032	>

Figure 4.75 List of Installed Certificates

Let's now talk about clickjacking and what SAP has provided to overcome the clickjacking issue.

4.3.2   Maintaining Clickjacking Protection

The word "clickjack" came from the word "hijack" and signifies that someone is hijacking your clicks on a webpage. Before explaining SAP's solution on clickjacking, let's look at clickjacking a little more closely.

Clickjacking allows attackers to hide malware and other threats under the content of legitimate webpages. When visitors click on links on pages that have been clickjacked, they can unknowingly become victims of malware or other unintended actions. One webpage can completely overlaid with another webpage, with all the possible transparency options using HTML's Iframe. Iframe allows multiple layers on a single webpage, and sometimes, identifying the individual layers can be difficult. This feature is good for web developers building cool stuff, but this capability also allows hackers to trick you into performing harmful activities inadvertently.

Several variations of clickjacking exist:

- Likejacking: Manipulating the **Like** button on a webpage
- Cursorjacking: Manipulating your cursor position, forcing you to perceive that cursor is somewhere else
- Cookie jacking: Stealing your browser cookies
- File jacking: Tricking you to enable hackers to access files on your device

The typical flow of a clickjacking event is shown in Figure 4.76.

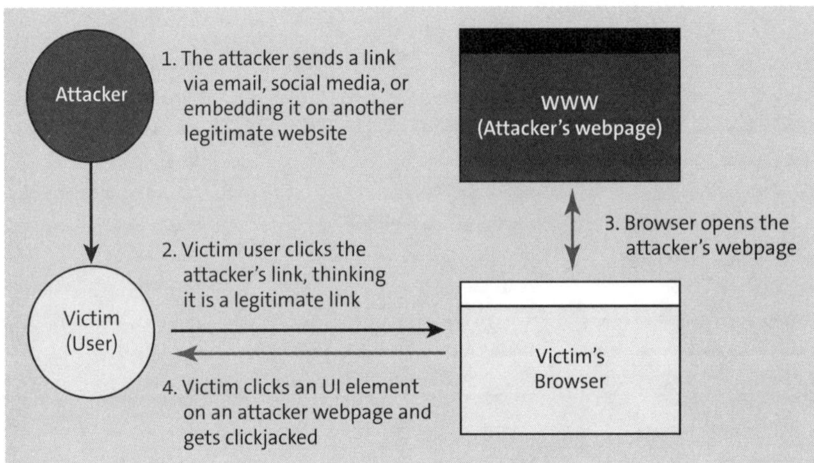

Figure 4.76 Clickjacking Process Flow

SAP provides the Maintain Protection Whitelists app to add trusted hosts to the clickjacking protection whitelist. If your SAP BTP, ABAP environment system is embedded into another application, your launchpad or SAP Fiori application is running under another application. Maintaining this clickjacking protection whitelist will ensure that your parent application is secure. Figure 4.77 shows a sample entry for maintaining the whitelist so that real malicious hosts can be identified.

Clickjacking Protection	Trusted Network Zones		
Trusted Hosts (1)			
Trusted Host Name	Schema	Port	Type
trustedhost.com	HTTPS	443	Managed by Customer

Figure 4.77 Trusted Host Whitelisting

In the same app, as shown in Figure 4.78, you can also maintain trusted network zones and list URL patterns for which redirections are allowed or blocked.

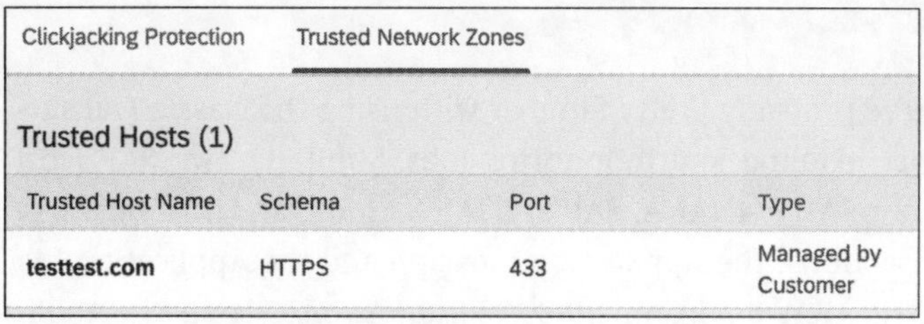

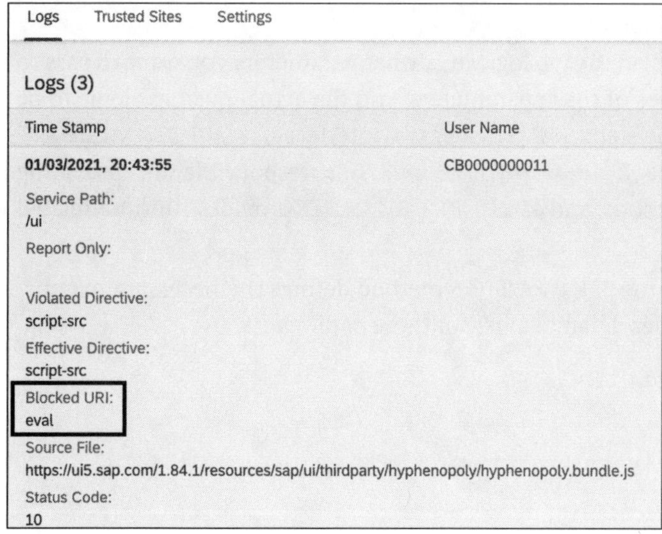

Figure 4.78 Trusted Network Zones

4.3.3   Managing Content Security

A *content security policy* is a standard that allows you to disable certain HTML/JavaScript features to reduce the attack surface of applications running in a browser.

With the Manage Content Security app, you can perform the following tasks:

- Display a whitelist of allowed sources
- Display logs of violations

As shown in Figure 4.79, multiple entries exist, which are logged as violations. However, some of these violations can be ignored, for example, the **eval** violation can occur in Google Chrome, as shown in Figure 4.79.

Figure 4.79 Example Log with EVAL Violation

As described so far, some key application security elements are already available in SAP BTP, ABAP environment. Let's now shift our focus to the next and final section of this chapter, on application jobs.

4.4   Application Jobs

As an ABAP developer, you're probably pretty familiar with using the classic Transactions SM36 and SM37 for scheduling and monitoring jobs. Similarly, two SAP Fiori applications are available for SAP S/4HANA, SAP S/4HANA Cloud, and SAP Business Suite to replace these transactions: the Application Jobs app and the Application Job Templates app. As of SAP BTP, ABAP environment 2008 (August 2020), only the Application Jobs app is available for SAP BTP, ABAP environment to define jobs based on predefined or custom templates. However, you can still use ADT to create custom job templates.

Some standard job templates available in SAP BTP, ABAP environment include the following:

- The Build Starter job template
- The Delete Global Lock History for SEPA Mandates job template
- The Delete IBANs job template
- The Lock Unused Business Users job template

In general, application jobs are created in four steps. The first three steps must be executed in ADT since no apps are available as of now. Only the last step (i.e., scheduling a batch job) can be performed using an app. Let's look at these steps in more detail next:

1. **Implementing business logic**
   In this step, you'll decide on the job logic (e.g., what parameters you want to pass to the job, the default values of these parameters, and the actual business logic to be executed as a part of a job step). SAP provides two interfaces: `IF_APJ_DT_EXEC_OBJECT` and `IF_APJ_RT_EXEC_OBJECT`. `IF_APJ_DT_EXEC_OBJECT` is responsible for providing design-time-related methods, and `IF_APJ_RT_EXEC_OBJECT` provides runtime-related methods.

   As shown in Listing 4.5, the `GET_PARAMETERS` method defines the necessary parameters for the job and defines default values for these parameters.

```
class ZCL_JOB_LOGIC definition
  public
  final
  create public.

public section.
  interfaces: IF_APJ_DT_EXEC_OBJECT,
  IF_APJ_RT_EXEC_OBJECT.
protected section.
private section.
ENDCLASS.
```

```abap
CLASS ZCL_JOB_LOGIC IMPLEMENTATION.

  METHOD IF_APJ_DT_EXEC_OBJECT~GET_PARAMETERS.
    et_parameter_def = VALUE #(
        ( selname = 'S_CCODE'    kind = if_apj_dt_exec_object=>select_option
datatype = 'C' length = 4 param_text = 'Company Code'
changeable_ind = abap_true )


    ).


    " Return the default parameters values here
    et_parameter_val = VALUE #(
        ( selname = 'S_CCODE'    kind = if_apj_dt_exec_object=>select_option
sign = 'I' option = 'EQ' low = 'US99' )
    ).

  ENDMETHOD.

  METHOD IF_APJ_RT_EXEC_OBJECT~EXECUTE.
    types ty_ccode type c length 4.
    data s_ccode type range of ty_ccode.
    LOOP AT it_parameters INTO DATA(ls_parameter).
      CASE ls_parameter-selname.
        WHEN 'S_CCODE'.
          APPEND VALUE #( sign   = ls_parameter-sign
                          option = ls_parameter-option
                          low    = ls_parameter-low
                          high   = ls_parameter-high ) TO s_ccode.
      ENDCASE.
    ENDLOOP.
" Implement business logic
"e.g., call company code related API using the value s_ccode

  ENDMETHOD.
```

Listing 4.5 Implementing the Business Logic for the Application Job App

2. **Defining a job catalog entry and creating a job template (if not already available)**
   To create a job catalog entry and a job template, you'll need to create and implement
   a class and then run the class in console mode. As described earlier, to execute a class
   in console mode, you must implement the interface `if_oo_adt_classrun`. Listing 4.6
   shows an example implementation where we're first creating a job catalog entry by
   instantiating the singleton class `cl_apj_dt_create_content` and calling its method

`create_job_cat_entry`. Then, we're instantiating the business logic class, created in the previous step.

Finally, we'll create the job using the method `create_job_template_entry`. Make sure you use the main transport number and not the task while maintaining value for `lc_transport_request` and `lc_class_name` is the business logic class name created in the previous step. This entire class needs to be run using the ADT context menu **Run As · ABAP Application (Console)**. After running the class successfully, you should see success messages in your **Console** view. Running the class is a mandatory step because this step creates the job catalog and the job template.

```abap
class ZCL_JOB_CATALOG_TEMPLATE definition
  public
  final
  create public .

public section.
  interfaces if_oo_adt_classrun.
protected section.
private section.
ENDCLASS.

CLASS ZCL_JOB_CATALOG_TEMPLATE IMPLEMENTATION.
  METHOD IF_OO_ADT_CLASSRUN~MAIN.
    CONSTANTS lc_catalog_name TYPE cl_apj_dt_create_content=>ty_catalog_name
VALUE 'ZCAT_DEMO'.
    CONSTANTS lc_catalog_text    TYPE cl_apj_dt_create_content=>ty_text
VALUE 'Application job'.
    CONSTANTS lc_class_name    TYPE cl_apj_dt_create_content=>ty_class_name
VALUE 'ZCL_JOB_LOGIC'.

    CONSTANTS lc_template_name    TYPE cl_apj_dt_create_content=>ty_
template_name VALUE 'ZCL_JOBDEMO_TEMPL'.
    CONSTANTS lc_template_text    TYPE cl_apj_dt_create_content=>ty_text
VALUE 'Steampunk - Demo job template'.

    CONSTANTS lc_transport_request TYPE cl_apj_dt_create_content=>ty_
transport_request VALUE 'DS1K900196'.
    CONSTANTS lc_package        TYPE cl_apj_dt_create_content=>ty_package
VALUE 'ZSCP_ABAP'.

    DATA(lo_dt) = cl_apj_dt_create_content=>get_instance( ).
    TRY.
    "Create Job catalog entry
        lo_dt->create_job_cat_entry(
```

```abap
            iv_catalog_name      = lc_catalog_name
            iv_class_name        = lc_class_name
            iv_text              = lc_catalog_text
            iv_catalog_entry_type = cl_apj_dt_create_content=>class_based
            iv_transport_request = lc_transport_request
            iv_package           = lc_package
        ).
        out->write( |Job catalog entry created successfully| ).

      CATCH cx_apj_dt_content INTO DATA(lx_apj_dt_content).
        out->write( |Creation of job catalog entry failed: { lx_apj_dt_
content->get_text( ) }| ).
    ENDTRY.


    " Create a job template
    DATA lt_parameters TYPE if_apj_dt_exec_object=>tt_templ_val.
    " Create an instance of the business logic and get the parameter details
    NEW zcl_job_logic( )->if_apj_dt_exec_object~get_parameters(
      IMPORTING
        et_parameter_val = lt_parameters
    ).


    TRY.
        lo_dt->create_job_template_entry(
            iv_template_name     = lc_template_name
            iv_catalog_name      = lc_catalog_name
            iv_text              = lc_template_text
            it_parameters        = lt_parameters
            iv_transport_request = lc_transport_request
            iv_package           = lc_package
        ).
        out->write( |Job template created successfully| ).

      CATCH cx_apj_dt_content INTO lx_apj_dt_content.
        out->write( |Creation of job template failed: {
lx_apj_dt_content->get_text( ) }| ).
        RETURN.
    ENDTRY.
  ENDMETHOD.

ENDCLASS.
```

Listing 4.6 Creating a Job Catalog Entry and Job Template

Once you run your class in the console, an IAM app is automatically created, as shown in Figure 4.80. This IAM app will be used while setting up the authorizations, which we'll do next.

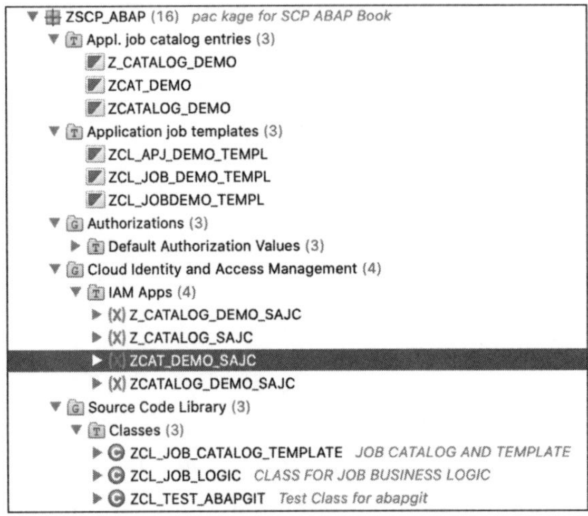

Figure 4.80 IAM App Created Automatically

3. **Setting up authorizations**

   The process of setting up authorizations was described in detail in Section 4.2.4. Since our IAM app is already created, you'll only now need to create a business catalog in ADT by referring to the IAM app name. Then, you'll create a business role and assign users to this role using the Maintain Business Roles app.

4. **Scheduling the job**

   Our final step is scheduling the job using the Application Jobs app. Figure 4.81 shows the initial screen of the app, where a list of already scheduled or executed jobs, and their statuses, will be displayed.

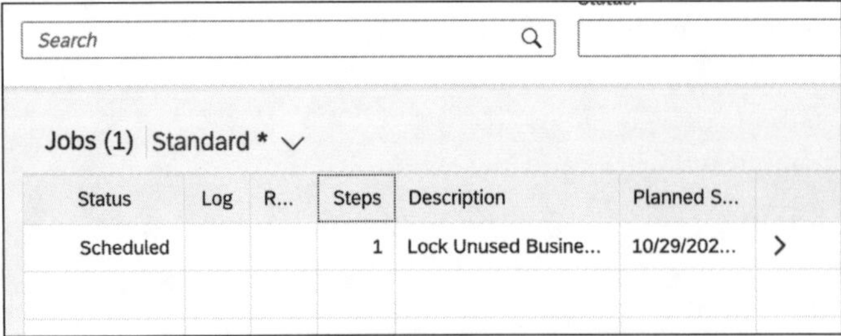

Figure 4.81 Initial Screen of the Application Job App

Click **Create** to create a new job based on an existing template. Your custom template should now appear in the list of available templates, as shown in Figure 4.82.

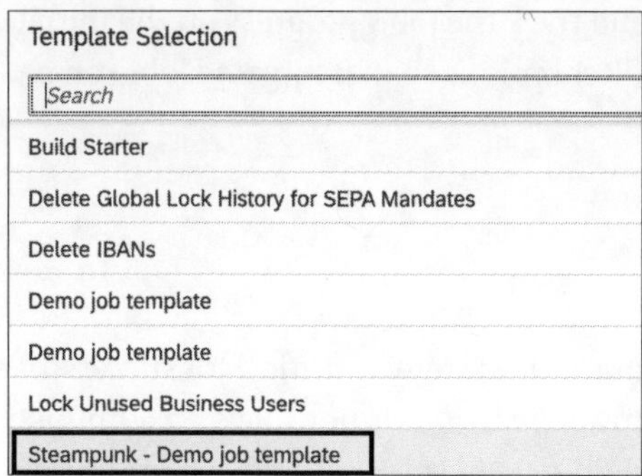

Figure 4.82 Custom Job Template

You also have the option of scheduling the job as recurrent, as shown in Figure 4.83.

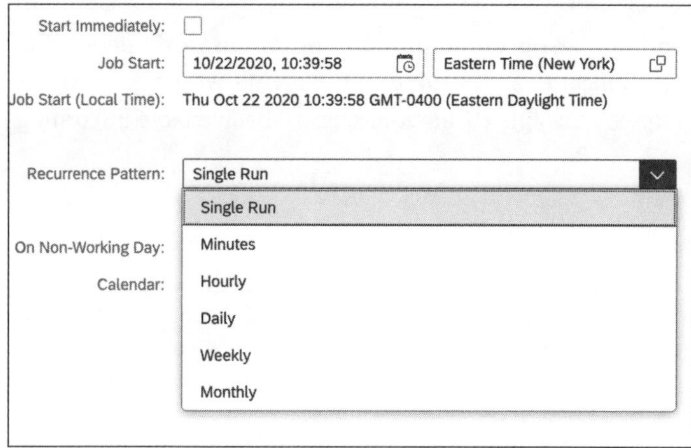

Figure 4.83 Recurrence Pattern

In the final step, you'll set some values for predefined job parameters. For example, for the Lock Unused Business Users job template, you can select the number of days users have not logged on, as shown in Figure 4.84.

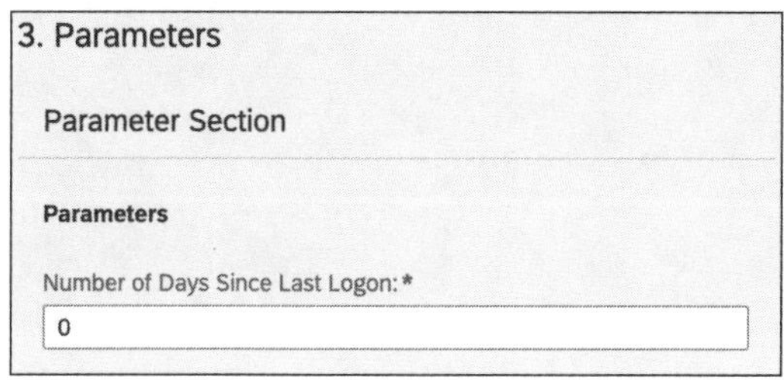

Figure 4.84 Parameter Options for a Job

Once your job is scheduled, you can monitor and track the job's progress on the initial screen of the app, which also features various filter options so that targeted jobs can be viewed.

4.5   Summary

Operating SAP BTP, ABAP environment applications is extremely critical for successful implementation of your cloud project. Operating applications is not solely an administrator's responsibility. As a developer, you should closely work with your administrators to configure and execute transport management, debugging, tracing, user access and authorization, monitoring, security, and administration tasks. The success of your cloud project will significantly depend on the effective use of these features and functionalities.

SAP provides numerous options for operating, monitoring, tracing, and debugging both standard and custom applications. Continuously evolving with new and enhanced features, you can expect many new functionalities to be available in coming days.

Chapter 5
Next Steps

In this chapter, we'll look into the future development and features planned for the SAP Business Technology Platform (SAP BTP), ABAP environment.

In this book, we covered all the different perspectives of the SAP BTP, ABAP environment, starting with an introduction to the platform itself and its tools, and continuing through development and operations. We showed you how to leverage the currently available features and capabilities and also went beyond the platform, to give you an idea of the new world of possibilities.

Since its earliest days, SAP BTP, ABAP environment has constantly evolved and improved, adding new features and capabilities with each new release. Customers and partners can play an important role in this evolution by submitting feature requests. With this feedback, SAP can understand what you need and what gaps exist that hinder your company's journey to innovation.

In this chapter, you'll learn what to expect in future releases.

5.1 Future Roadmap

As we look at the future roadmap, we'll start with recapping the latest features and innovations. Each of the following topics represent different perspectives of the SAP BTP, ABAP environment:

- **Partner development**
  In the area of partner development, the most recent, and probably most important, innovation was the enablement of multitenancy based on `CLIENT` field. With this innovation and other features from the latest release, SAP partners can offer solutions based on the SAP BTP, ABAP environment to their customers following the software as a service (SaaS) model by leveraging a single service instance. The next step in the roadmap is to introduce extensibility for this multitenancy, which will enable you to introduce you own user interface (UI) adaptations, enhance field extensibility, and accommodate custom code.

- **ABAP RESTful application programming model**
  In the most recent release, innovations including OData V4 support, early numbering of managed business objects, and much more. The next feature on the roadmap is to provide Transaction SM30- like generation of SAP Fiori apps. A similar feature is already available with custom business object extensions in SAP S/4HANA and is expected to improve productivity in this area.

- **ABAP programming language and development tools**
  Major planned improvements will focus on improving the ABAP Test Cockpit (ATC), as a key component. Innovations planned in this area include mass ATC runs and the approval of ATC exemptions.

- **Analytics**
  This area has only recently come into focus. You can now expect the integration of the SAP Analytics Cloud client, SAP Smart Business, and SAP NetWeaver Enterprise Search.

- **Integration and security**
  In this area, you can expect additional features, the most important being event consumption. With this new feature, integration will achieve a completely new level. From a security perspective, the main innovation that will be introduced is read access logging.

- **Released reuse services**
  Probably the most important innovation in this area is the extension of capabilities from the powerful XCO library, with support for custom entities. You can also expect support for SAP Fiori apps for factory calendars.

- **Infrastructure**
  In this area, the most recent innovations included the introduction of an Amazon Web Services (AWS) data center in Tokyo and new sizing options for runtime and persistence memory. In the near future, you can expect the introduction of additional AWS data centers as well as support for additional hyperscalers (for example, Microsoft Azure). Future plans include dynamic scaling based on Kubernetes environments, high availability and disaster recovery, and the reduction of planned maintenance windows. All these new features will improve the scalability and availability of SAP BTP.

More details can be found at the SAP Road Map Explorer, shown in Figure 5.1, available at *https://roadmaps.sap.com*.

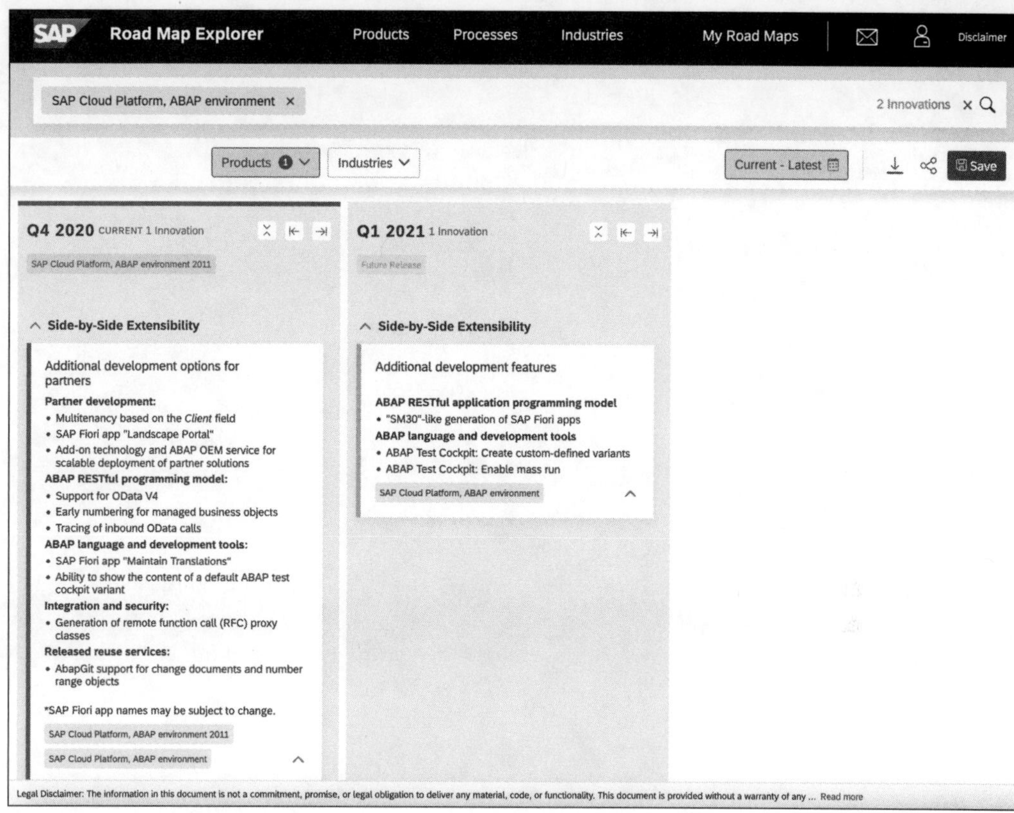

Figure 5.1 SAP Road Map Explorer

5.2 Summary

At this point, we've covered the most important innovations expected in the near future for the SAP BTP, ABAP environment, which will improve several aspects of the development environment and the runtime.

To close, we'd like to remind you of the important role customers and partners play by providing valuable feedback to SAP. One way to provide feedback is through the feature requests submitted via SAP's Customer Influence and Adoption programs, which are available at *https://www.sap.com/about/customer-involvement/influence-adopt.html*.

The Authors

Gairik Acharya is a senior technical architect and associate partner at IBM with more than 20 years of IT experience. He is a recognized expert in ABAP, SAP HANA, OData, SAP Gateway, SAP S/4HANA, SAP Fiori, SAPUI5, SAP Business Technology Platform (SAP BTP), and SAP Mobile Platform. As senior architect at the IBM SAP S/4HANA Center of Excellence group in North America, he has architected solutions for several global implementations for several of IBM's top-tier clients.

A certified and recognized expert in SAP application consulting, Gairik has been responsible for building new capabilities within the group. Gairik is a frequent speaker at SAP TechEd, covering a variety of new technical areas. His sessions are always highly rated and considered to be among the most popular and appreciated. Loving to spend his free time with his wife Lipi and his daughter Sneha, Gairik can be found on Twitter @AcharyaGairik and on LinkedIn at *https://www.linkedin.com/in/ gairik-acharya/*.

Aleksandar Debelic is an SAP technical architect at IBM with more than 17 years of SAP experience, primarily as a technical solution architect and technology team lead. He is a managing consultant and member of the Global SAP Center of Competence and a global development team lead at SAP Innovation by IBM. His specialties include the design and development of solutions that leverage the latest technologies, such as blockchain, the Internet of Things (IoT), and machine learning. Aleks' current focus is building assets for SAP BTP extensions and DevOps for SAP.

Shubhangi (Deshmukh) Joshi is an SAP technical architect at IBM with more than 15 years of SAP experience in the areas of project delivery, asset building, and SAP S/4HANA migration projects. In addition to her ABAP experience, her main areas of expertise are SAP S/4HANA, machine learning with SAP, SAP Conversational AI, SAP CoPilot, SAP BTP, and SAP Analytics Cloud.

Shubhangi has performed implementation projects for IBM clients from the chemical; petroleum, oil, and gas; automotive; and pharmaceuticals industries. She has presented webinars and knowledge-sharing sessions on various topics including ABAP platform-as-a-service (PaaS), machine learning use cases, SAP CoPilot, and SAP Analytics Cloud. Her machine learning applications have been presented and well received at SAPPHIRE and SAP TechEd.

Aayush Dhawan is a managing consultant at IBM with more than 14 years of SAP experience in the areas of technical development, functional consulting, and asset building. As part of the SAP Innovation team at IBM, he has contributed towards many co-innovation projects on emerging technologies including cloud development, digital advisors, blockchain, machine learning, and data science. Additionally, he has delivered many challenging client projects across several industries, including consumer products, pharmaceuticals, chemicals, and petroleum.

Aayush is a certified SAP consultant, and his research papers have been published in both national and international journals. His mobile application ideas have been recognized in the SAP Fiori UX Design and Build Challenge for two consecutive years. He has presented numerous webinars and knowledge-sharing sessions, including SAP TechEd speaker sessions.

Index

- Learn to write readable and maintainable ABAP code

- Improve the way you work with classes, interfaces, methods, variables, tables, and more

- Understand how to choose useful names, write strong comments, and format code correctly

Haeuptle, Hoffman, Jordão, Martin, Ravinarayan, Westerholz

Clean ABAP

A Style Guide for Developers

ABAP developers, are you looking to clean up your code? Then pick up this official companion to the Clean ABAP GitHub repository. This book is brimming with best practices, straight from the experts, to help you write effective ABAP code. Start by learning when to apply each clean ABAP practice. Then, dive into detailed code examples and explanations for using classes, methods, names, variables, internal tables, and more. From writing code to troubleshooting and testing, this is your complete style guide!

351 pages, pub. 11/2020

E-Book: $69.99 | **Print:** $79.95 | **Bundle:** $89.99

www.sap-press.com/5190